METHODS IN CELL BIOLOGY

VOLUME XIII

Contributors to This Volume

MICHAEL BALLS

DENNIS BROWN

PETER J. BRUNS

D. E. BUETOW

HAROLD I. CALVIN

R. FAROOKHI

NORMAN FLEMING

LESTER GOLDSTEIN

PAUL V. C. HOUGH

ALOYS HÜTTERMANN

CHRISTINE KO

HIROSHI MIYAMOTO

EDUARDO ORIAS

WOLF PRENSKY

DAVID M. PRESCOTT

LEIF RASMUSSEN

A. L. ROSNER

BRUCE K. SCHRIER

PER O. SEGLEN

JERRY SHAY

C. SONNENSCHEIN

A. M. SOTO

WALTER E. STUMPF

GEORGE VEOMETT

GERTRUD WENDLBERGER

SAMUEL H. WILSON

ERIK ZEUTHEN

Methods in
Cell Biology

Edited by

DAVID M. PRESCOTT

DEPARTMENT OF MOLECULAR, CELLULAR AND
DEVELOPMENTAL BIOLOGY
UNIVERSITY OF COLORADO
BOULDER, COLORADO

VOLUME XIII

1976

ACADEMIC PRESS • New York   San Francisco   London
A Subsidiary of Harcourt Brace Jovanovich, Publishers

ACADEMIC PRESS, INC.
111 Fifth Avenue, New York, New York 10003

United Kingdom Edition published by
ACADEMIC PRESS, INC. (LONDON) LTD.
24/28 Oval Road, London NW1

LIBRARY OF CONGRESS CATALOG CARD NUMBER: 64-14220

ISBN 0-12-564113-3

PRINTED IN THE UNITED STATES OF AMERICA

CONTENTS

LIST OF CONTRIBUTORS ix

PREFACE xi

1. *Large-Scale Enucleation of Mammalian Cells*
George Veomett, Jerry Shay, Paul V. C. Hough, and David M. Prescott

    I.   Introduction 1
    II. Cells and Materials 2
    III. Techniques 2
        References 6

2. *Reconstruction of Cultured Mammalian Cells from Nuclear and Cytoplasmic Parts*
George Veomett and David M. Prescott

    I.   Introduction 7
    II. Preparation of Components for Cell Reconstruction 8
    III. Criteria for Determination of Reconstructed Cells 10
    IV. Protocol for Cell Reconstruction 11
    V. Conclusions 13
        References 14

3. *Recording of Clonal Growth of Mammalian Cells through Many Generations*
Hiroshi Miyamoto, Leif Rasmussen, and Erik Zeuthen

    I.   Introduction 15
    II. Culture Medium 16
    III. Preparation of the Cells 16
    IV. Microphotographic Recording of Clonal Growth 18
    V. Clonal Growth 19
    VI. Applications 22
        References 26

4. Preparation of Isolated Rat Liver Cells
Per O. Seglen

I.	Introduction	30
II.	Quality Evaluation	30
III.	Nonenzymatic Methods for Liver Cell Preparation	32
IV.	Enzymatic Treatment of Liver Slices	34
V.	Collagenase Perfusion	35
VI.	Purification of Parenchymal Cells	57
VII.	Incubation of Cell Suspensions	63
VIII.	Properties of Isolated Parenchymal Cells	64
IX.	*In Vitro*-Culture of Parenchymal Liver Cells	71
X.	Preparation of Nonparenchymal Liver Cells	73
	References	78

5. Isolation and Subfractionation of Mammalian Sperm Heads and Tails
Harold I. Calvin

I.	Introduction	85
II.	Isolation of Sperm Heads and Tails	89
III.	Subfractionation of the Sperm Head	95
IV.	Subfractionation of the Tail	100
	References	103

6. On the Measurement of Tritium in DNA and Its Applications to the Assay of DNA Polymerase Activity
Bruce K. Schrier and Samuel H. Wilson

I.	Introduction	105
II.	Evaluation of Methods for Precipitation, Collection, and Counting of Tritium-Labeled DNA	106
III.	Determination of Optimal Conditions for Collection and Counting of DNA Polymerase Assay Products	114
IV.	Selection of Methods for a Particular System	117
	References	120

7. The Radioiodination of RNA and DNA to High Specific Activities
Wolf Prensky

I.	Introduction	121
II.	Radioiodine	122
III.	Commerford's Reaction: General Description	136

IV.   Equipment Needs   138
V.   Radioiodination Reaction   141
VI.   Comments and Discussion   147
     References   152

8. Density Labeling of Proteins
Aloys Hüttermann and Gertrud Wendlberger

I.   Introduction   153
II.   Density Labeling   156
III.   Equilibrium Density Gradient Sedimentation   158
IV.   Ultracentrifugation   163
V.   Use of Density Markers   164
VI.   Combination with Other Biochemical Methods   167
VII.   Conclusions   169
     References   170

9. Techniques for the Autoradiography of Diffusible Compounds
Walter E. Stumpf

I.   Introduction   171
II.   Dry-Mount Autoradiography   174
III.   Thaw-Mount Autoradiography   185
IV.   Smear-Mount Autoradiography   186
V.   Touch-Mount Autoradiography   187
     References   192

10. Characterization of Estrogen-Binding Proteins in Sex Steroid Target Cells Growing in Long-Term Culture
A. M. Soto, A. L. Rosner, R. Farookhi, and C. Sonnenschein

I.   Introduction   195
II.   Materials and Methods   196
III.   Discussion   207
     References   210

11. Long-Term Amphibian Organ Culture
Michael Balls, Dennis Brown, and Norman Fleming

I.   Introduction   214
II.   Animals   214

III. Methods 215
IV. Results 224
V. Concluding Remarks 235
    References 236

12. *A Method for the Mass Culturing of Large Free-Living Amebas*
Lester Goldstein and Christine Ko

I. Introduction 239
II. The Basic Ameba Culture Procedure 240
III. The Culturing and Harvesting of the Food Organism
Tetrahymena pyriformis 241
IV. The Cleaning and Harvesting of Ameba Cultures 243
V. Concluding Remarks 245
    References 246

13. *Induction and Isolation of Mutants in Tetrahymena*
Eduardo Orias and Peter J. Bruns

I. Introduction 248
II. Elements of *Tetrahymena* Genetics 249
III. Strains 253
IV. Media 256
V. Routine Methods 257
VI. Calibration and Suggested Doses of Mutagens 269
VII. Strategies and Protocols for Mutant Isolation 271
VIII. Genetic Analysis of the Mutants 275
IX. Additional Information for Nongeneticists 278
    References 281

14. *Isolation of Nuclei from Protozoa and Algae*
D. E. Buetow

I. Introduction 284
II. Monitoring the Preparation 284
III. *Tetrahymena* 285
IV. *Paramecium* 292
V. *Blepharisma* 297
VI. *Didinium* 298
VII. *Spirostomum* 298
VIII. *Amoeba* 299
IX. *Euglena* 301
X. Algae 303
    References 310

SUBJECT INDEX 313
CONTENTS OF PREVIOUS VOLUMES 318

LIST OF CONTRIBUTORS

Numbers in parentheses indicate the pages on which the authors' contributions begin.

MICHAEL BALLS, Department of Human Morphology, The Medical School, University of Nottingham, Nottingham, England (213)

DENNIS BROWN,[1] School of Biological Sciences, University of East Anglia, Norwich, England (213)

PETER J. BRUNS, Section of Genetics, Development and Physiology, Cornell University, Ithaca, New York (247)

D. E. BUETOW, Department of Physiology and Biophysics, University of Illinois, Urbana, Illinois (283)

HAROLD I. CALVIN, International Institute for the Study of Human Reproduction, Columbia University College of Physicians and Surgeons, New York, New York (85)

R. FAROOKHI, Tufts University School of Medicine, Tufts Cancer Research Center, Boston, Massachusetts (195)

NORMAN FLEMING,[2] School of Biological Sciences, University of East Anglia, Norwich, England (213)

LESTER GOLDSTEIN, Department of Molecular, Cellular and Developmental Biology, University of Colorado, Boulder, Colorado (239)

PAUL V. C. HOUGH,[3] Department of Molecular, Cellular and Developmental Biology, University of Colorado, Boulder, Colorado (1)

ALOYS HÜTTERMANN, Forstbotanisches Institut der University Göttingen, Göttingen, West Germany (153)

CHRISTINE KO, Department of Molecular, Cellular and Developmental Biology, University of Colorado, Boulder, Colorado (239)

HIROSHI MIYAMOTO,[4] The Biological Institute of the Carlsberg Foundation, Copenhagen, Denmark (15)

EDUARDO ORIAS, Section of Biochemistry and Molecular Biology, University of California at Santa Barbara, Santa Barbara, California (247)

WOLF PRENSKY, Molecular Cytology Laboratory, Memorial Sloan-Kettering Cancer Center, New York, New York (121)

DAVID M. PRESCOTT, Department of Molecular, Cellular and Developmental Biology, University of Colorado, Boulder, Colorado (1,7)

LEIF RASMUSSEN, The Biological Institute of the Carlsberg Foundation, Copenhagen, Denmark (15)

A. L. ROSNER,[5] Tufts University School of Medicine, Tufts Cancer Research Center, Boston, Massachusetts (195)

BRUCE K. SCHRIER, Behavioral Biology Branch, National Institute of Child Health and Human Development, and Laboratory of Biochemistry, National Cancer Institute, National Institutes of Health, Bethesda, Maryland (105)

[1] *Present address:* Institut d'Histologie et d'Embryologie, École de Médecine, Université de Genève, Geneva, Switzerland.

[2] *Present address:* Anatomisches Institut, Universität Zürich, Zürich, Switzerland.

[3] *Present address:* Biology Department, Brookhaven National Laboratory, Upton, Long Island, New York.

[4] *Present address:* Department of Physiology, Kinki University School of Medicine, Nishiyama, Sayama-cho, Minamikawachi-gun, Osaka, Japan.

[5] *Present address:* Department of Pathology, Beth Israel Hospital, Boston, Massachusetts.

PER O. SEGLEN, Norsk Hydro's Institute for Cancer Research, Department of Tissue Culture, The Norwegian Radium Hospital, Montebello, Oslo, Norway (29)

JERRY SHAY, Department of Molecular, Cellular and Developmental Biology, University of Colorado, Boulder, Colorado (1)

C. SONNENSCHEIN, Tufts University School of Medicine, Tufts Cancer Research Center, Boston, Massachusetts (195)

A. M. SOTO, Tufts University School of Medicine, Tufts Cancer Research Center, Boston, Massachusetts (195)

WALTER E. STUMPF, Departments of Anatomy and Pharmacology, University of North Carolina, Chapel Hill, North Carolina (171)

GEORGE VEOMETT, Department of Molecular, Cellular and Developmental Biology University of Colorado, Boulder, Colorado (1,7)

GERTRUD WENDLBERGER, Forstbotanisches Institut der University Göttingen, Göttingen, West Germany (153)

SAMUEL H. WILSON, Behavioral Biology Branch, National Institute of Child Health and Human Development, and Laboratory of Biochemistry, National Cancer Institute, National Institutes of Health, Bethesda, Maryland (105)

ERIK ZEUTHEN, The Biological Institute of the Carlsberg Foundation, Copenhagen, Denmark (15)

PREFACE

Volume XIII of this series continues to present techniques and methods in cell research that have not been published or have been published in sources that are not readily available. Much of the information on experimental techniques in modern cell biology is scattered in a fragmentary fashion throughout the research literature. In addition, the general practice of condensing to the most abbreviated form materials and methods sections of journal articles has led to descriptions that are frequently inadequate guides to techniques. The aim of this volume is to bring together into one compilation complete and detailed treatment of a number of widely useful techniques which have not been published in full detail elsewhere in the literature.

In the absence of firsthand personal instruction, researchers are often reluctant to adopt new techniques. This hesitancy probably stems chiefly from the fact that descriptions in the literature do not contain sufficient detail concerning methodology; in addition, the information given may not be sufficient to estimate the difficulties or practicality of the technique or to judge whether the method can actually provide a suitable solution to the problem under consideration. The presentations in this volume are designed to overcome these drawbacks. They are comprehensive to the extent that they may serve not only as a practical introduction to experimental procedures but also to provide, to some extent, an evaluation of the limitations, potentialities, and current applications of the methods. Only those theoretical considerations needed for proper use of the method are included.

Finally, special emphasis has been placed on inclusion of much reference material in order to guide readers to early and current pertinent literature.

DAVID M. PRESCOTT

Chapter 1

Large-Scale Enucleation of Mammalian Cells

GEORGE VEOMETT, JERRY SHAY, PAUL V. C.
HOUGH,[1] and DAVID M. PRESCOTT

Department of Molecular, Cellular and Developmental Biology,
University of Colorado,
Boulder, Colorado

I. Introduction 1
II. Cells and Materials 2
III. Techniques 2
    A. Without Centrifuge Inserts 2
    B. With Acrylic Inserts 3
    References 6

I. Introduction

Carter (1967) was the first to realize the potential use of cytochalasin B (CB), a metabolite from *Helminthosporium dematoideum*, for experimental enucleation of cultured mammalian cells. A major step in developing this potential was introduced by Prescott *et al.* (1972) and Wright and Hayflick (1973), who subjected cells to a centrifugal force in the presence of CB. Under these conditions the nucleus is forced into a cytoplasmic stalk which subsequently severs spontaneously.

Prescott *et al.* (1972) and Prescott and Kirkpatrick (1973) used glass and subsequently tissue culture–grade plastic cover slips as a substrate for cell growth. The cover slips are easily handled and can be placed sterilely into centrifuge tubes for the enucleation procedure. The number of enucleated cells obtained is limited by the surface area of the cover slips used for growth. Several other techniques have utilized the same principle of combined

[1] *Present address*: Biology Department, Brookhaven National Laboratory, Upton, Long Island, New York.

centrifugation and CB treatment of tissue culture cells, but have varied the surface area for cellular growth. For example, Wright and Hayflick (1973) grow cells on the inner surface of centrifuge tube inserts, while Croce and Koprowski (1973) have described a technique in which the cells are grown directly on the inner surface of the centrifuge tubes. Follett (1974) has described a technique in which 35-mm tissue culture dishes used for cell growth are inverted in the wells of a centrifuge rotor and used for the enucleation procedure. Large numbers (approximately $0.5-1 \times 10^7$) of enucleated cells can be obtained by these procedures. The purity and viability of the cytoplasts (enucleated cells) using mass-enucleation procedures has been shown (Wright and Hayflick, 1973; Croce and Koprowski, 1973; Follett, 1974), but the karyoplasts (nucleated cellular fragments) have not been well characterized.

We describe here an alternate, relatively simple method for obtaining large numbers of enucleated cells and karyoplasts. This method utilizes 25-cm^2 tissue culture flasks both as substrate for cellular growth and as centrifuge tubes. The method is described for mouse L cells, but should be applicable to any cell that grows in monolayers.

II. Cells and Materials

Mouse L cells were grown in 25-cm^2 tissue culture flasks (Fisher or Falcon brand) in Ham's F12 medium supplemented with 10% fetal bovine serum and antibiotics.

The medium for enucleation consists of growth medium supplemented with 10 μg/ml of CB (Aldrich Chemical Co., Milwaukee, Wisconsin).

III. Techniques

A. Without Centrifuge Inserts

Cells are grown in the tissue culture flasks for at least 24 hours to allow the cells to attach firmly. The tissue culture medium is then removed, and medium containing CB is added to the flask until it is completely filled. The cap of the flask is then replaced and tightened.

The enucleation procedure is performed in a Sorvall RC-2B centrifuge, utilizing the GSA rotor, both of which are prewarmed to 37°C. Approximately 125 ml of water at 37°C is added to each rotor well; a filled flask is then added to each well. The rotor, with its cover off, is then spun at approximately 500 rpm, and water is continuously added to the rotor wells until all the wells contain the maximum volume of water, which is apparent when water is ejected from the rotor. The rotor is then stopped, the cover is replaced, and centrifugation is performed at 8500 rpm for 30 minutes.

The percentage of enucleation after one centrifugation is good, generally greater than 95% (Fig. 1). The karyoplasts form a pellet in the corner of the flask and are easily recovered. The trypan blue exclusion test indicates 80–90% viability. With this technique more than 90% of the karyoplasts can be recovered. Approximately $1-1.5 \times 10^6$ enucleated cells can be obtained from a single flask, or $6-9 \times 10^6$ from a single centrifugation run.

The major drawback of this technique is flask breakage, which is about 10%. Fisher flasks tend to crack and shear at the neck, whereas Falcon flasks shear along the side seams. In both cases, flask breakage is minimized by balancing the rotor as described.

B. With Acrylic Inserts

Acrylic inserts were constructed for the GSA rotor and were designed to accommodate Falcon bent-neck tissue culture flasks. The design of these inserts is shown in Fig. 2. They fit very tightly into the rotor wells and may have to be custom-trimmed for some rotors. Approximately 125 ml of water and a filled flask are added to each assembly used. The assemblies are balanced by weight, the tops are screwed on tightly, and centrifugation is performed in the GSA rotor in a Sorvall RC-2B centrifuge at 37°C for 20 minutes at 10,000–11,000 rpm. The results of the enucleation are similar to those obtained by the technique described in Section III,A, i.e., generally greater than 95% enucleation. However, karyoplast recovery is poorer. The reasons for poorer karyoplast recovery are unknown, but may be related to the alterations in centrifugal force and angle at which the flasks are held during centrifugation.

The inserts greatly reduce the breakage of Falcon flasks; only 1–2% of the flasks fail under these conditions, generally by leaking at the neck. The straight-necked Fisher flasks are unsuitable for use with this technique.

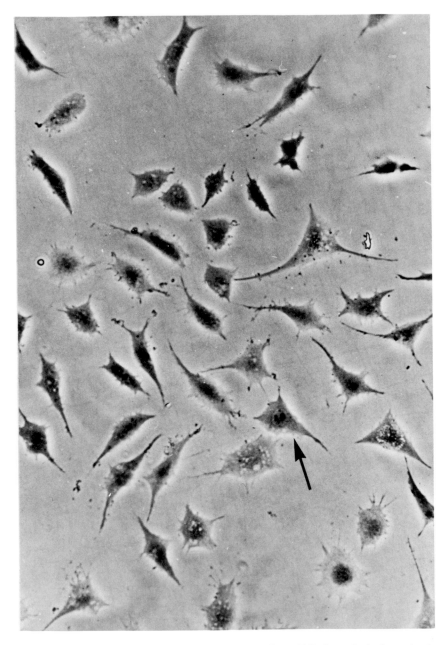

Fig. 1.　Phase-contrast micrograph of enucleates from T-flask method of enucleation. (Arrow indicates nucleated cell for reference.) After fixation and staining, this preparation was shown to be 97% enucleated.

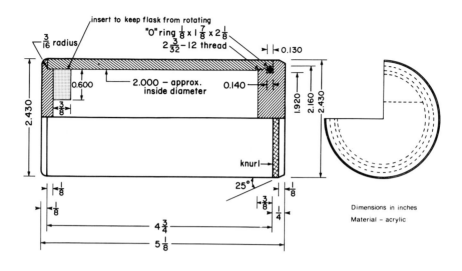

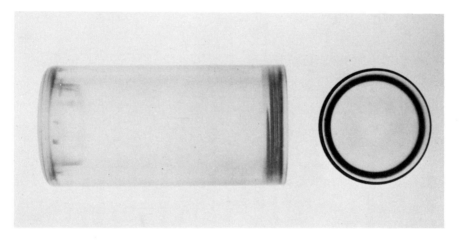

FIG. 2. Diagram and photograph of acrylic inserts for Sorvall model GSA rotor for use with T-flask enucleation procedure.

ACKNOWLEDGMENTS

This work was supported by NIH Postdoctoral Fellowship no. 1 F22 CA00797-01 to G. V., a grant from the Muscular Dystrophy Association of America to J. S., and grant no. 1 P01 CA13419-01 from the National Cancer Institute on which Dr. Prescott is a co-investigator.

We wish to acknowledge the assistance of E. Nettesheim in the design and construction of the acrylic inserts.

REFERENCES

Carter, S. B. (1967). *Nature (London)* **213**, 261.

Croce, C. M., and Koprowski, H. (1973). *Virology* **51**, 227.

Follett, E. A. C. (1974). *Exp. Cell Res.* **84**, 72.

Prescott, D. M., and Kirkpatrick, J. B. (1973). *In* "Methods in Cell Biology" (D. M. Prescott, ed.), Vol. 7, pp. 189–202. Academic Press, New York.

Prescott, D. M., Myerson, D., and Wallace, J. (1972). *Exp. Cell Res.* **71**, 480.

Wright, W. E., and Hayflick, L. (1973). *Proc. Soc. Exp. Biol. Med.* **144**, 587.

Chapter 2

Reconstruction of Cultured Mammalian Cells from Nuclear and Cytoplasmic Parts

GEORGE VEOMETT AND DAVID M. PRESCOTT

Department of Molecular, Cellular and Developmental Biology,
University of Colorado,
Boulder, Colorado

I.	Introduction	7
II.	Preparation of Components for Cell Reconstruction	8
	A. Preparation of Enucleated Cytoplasm	8
	B. Preparation of Nucleated Fragments	9
	C. Source of Sendai Virus	10
III.	Criteria for Determination of Reconstructed Cells	10
IV.	Protocol for Cell Reconstruction	11
V.	Conclusions	13
	References	14

I. Introduction

The techniques of cell fusion using inactivated Sendai virus and cell enucleation using cytochalasin B (CB) can be coupled to reconstruct mammalian cells in tissue culture. The technique is straightforward and should be valuable in the analysis of nuclear–cytoplasmic interactions involved in the determination of cell behavior.

Nuclear exchange or transplantation systems have already been developed for (1) protozoa, (2) fungi, (3) algae, (4) insects, and (5) amphibians (Goldstein, 1965; Jeon and Danielli, 1971; Tartar, 1953; Wilson, 1963; Zetsche, 1962; Zaolocar, 1971; King, 1966). Advantageous structural features and/or the large size of the cells involved in these systems allow the use of micromanipulation and microinjection techniques which are difficult but not impossible to apply to relatively small tissue culture cells (Diacumakos, 1973, 1975). Much of the success of nuclear transplantation with amebas can be attributed to the unique nuclear structure of this protozoan (Jeon and Danielli, 1971). The nucleus is reinforced and protected by the honeycomb layer just inside the nuclear envelope, hence it is not so readily damaged during the nuclear transfer with a microneedle. The advantages of large cell size involved in such transplantation systems are obvious.

An important generalization has emerged from the nuclear exchange systems already described: The nuclei must be protected from the external aqueous environment. Brief exposures of the nuclei to the external aqueous environment damage them irreversibly (Goldstein, 1965; King, 1966). Although the development of an organic milieu (e.g., Kopac, 1955) for nuclear isolation might alleviate these problems, relatively little has been done in this area. Therefore, until such nonaqueous systems are developed, nuclei used for transplantations or cell reconstructions should be protected from the external aqueous environment to minimize nuclear damage.

II.   Preparation of Components for Cell Reconstruction

A.   Preparation of Enucleated Cytoplasm

The development of a technique for nuclear exchange in vertebrate cells is dependent on a source of enucleated cells and of separated nuclei. Enucleated vertebrate cells in tissue culture can be obtained by micrurgy (Goldstein *et al.*, 1960) or by treatment of cells with CB (Prescott and Kirkpatrick, 1973; Wright, 1973; Poste, 1973). In the former method portions of cytoplasm are severed from the main nucleated region of the cell using a microneedle. Since the process is performed on one cell at a time, it is impossible to obtain large numbers of enucleates. However, large numbers of enucleated mammalian cells can be obtained after treatment of the cells with CB. The procedures and most current hypotheses concerning the mode of action of CB have been the subject of several communications (e.g., Prescott and Kirkpatrick, 1973; Wright, 1973; Poste, 1973). Briefly, CB appears to disorganize the cellular cortex by affecting microfilaments. The denser portions of the cell (e.g., the nucleus) can be removed from the bulk of the cytoplasm by isolating them in thin cytoplasmic threads which subsequently sever spontaneously. This technique requires only that the cells adhere to a substrate firmly enough to remain attached during centrifugation in CB.

With the CB technique large numbers ($0.8-9 \times 10^6$) of enucleates (cytoplasts) can be obtained in about 30 minutes with less than 5% whole-cell contamination. (The number of enucleated cells obtained is dependent upon the area of cell growth substrate. See Chapter 1, this volume.) The cytoplasts obtained by the micrurgy and CB techniques survive for over 24 hours and initially support protein synthesis (Goldstein *et al.*, 1960; Prescott *et al.*, 1972; Follett, 1974). The cytoplasts obtained after CB treatment have been more thoroughly characterized. They contain the Golgi apparatus, the centrioles, the microtubules, and the bulk of the other cytoplasmic constituents (Wise and Prescott, 1973; Shay *et al.*, 1973), and they support viral DNA (Prescott *et al.*, 1971; Pennington and Follett, 1974) and virion (Pollack and Goldman, 1973; Follett *et al.*, 1974) synthesis.

B. Preparation of Nucleated Fragments

The second requirement for a general procedure of nuclear exchange in vertebrate cells is a supply of nuclei. Nuclei can be obtained in large numbers using techniques for cell fractionation, or by enucleation of cells with CB. In the former methods, nuclei are generally obtained after cellular breakage in an aqueous hypotonic medium or after cellular breakage in an aqueous medium containing detergent (see, e.g., Penman, 1969; Perry and Kelly, 1968). In either case the nuclei can be obtained relatively free of cytoplasmic contamination and are surrounded by the inner (detergent methods) or double (nondetergent methods) nuclear membrane. Although nuclei thus obtained would theoretically be ideal for nuclear exchange systems because cytoplasmic contamination is minimal, they present several problems for the renucleation of cells. They have been exposed to the external aqueous environment and, by analogy to nuclei used in other exchange systems, probably are damaged irreversibly. However, a second major obstacle involves the actual process of cell reconstruction. If the nuclei were ingested by a cytoplast, a process that does appear to occur (Poste, 1973), the nuclei would be enclosed in a phagocytic vesicle; nuclei ingested by phagocytosis appear to be degraded (Poste, 1973). If the isolated nuclei are capable of being fused via Sendai virus to cytoplasts, the nuclear membrane would presumably be incorporated into the cytoplasmic membrane and the nuclear contents would presumably become mixed with the cytoplasmic contents. Thus success of such a method seems doubtful.

A second source of nuclei is the nucleated fragments of cells obtained after treatment of cells with CB. These nucleated fragments (karyoplasts) have been partially characterized. While they lack the Golgi apparatus, microtubules, and centrioles, they do contain a thin shell of cytoplasm with ribosomes and an occasional mitochondrion. They are initially capable of macromolecular synthesis and survive in culture for several days (Prescott and Kirkpatrick, 1973). A disadvantage of karyoplasts as a source of nuclei is the fact that a small amount of cytoplasm is included in the structure and may mediate any type of "cytoplasmic" inheritance. However, in karyoplasts the nuclei are protected from the external aqueous environment by their shells of cytoplasm. Fusion of karyoplasts to cytoplasts, for example, using inactivated Sendai virus, yields a reconstructed cell with an intact, undamaged nucleus.

A potential disadvantage in the use of karyoplasts, which should be noted, is the presence of a significant fraction of whole cell contaminants in the karyoplast preparation. Generally 5–10% of the nuclei in a karyoplast preparation represents whole cells. This contamination arises from the detachment of whole cells during the centrifugation used to enucleate the cells, and care must be taken to eliminate these whole cells and their fusion products in experiments with cells reconstruction (see Section III for one technique).

Use of the nuclear and cytoplasmic fragments of cells obtained after treatment with CB to reconstruct cells permits reciprocal nuclear exchange.

C. Source of Sendai Virus

The Sendai virus used for fusion of cytoplasts and karyoplasts can be obtained commercially (Connaught Labs, Canada). However, because of the variability in fusing capacity from lot to lot of this virus, it is better to produce, inactivate, and titer it according to standard methods. An excellent review of the techniques has recently appeared (Giles and Ruddle, 1973).

III. Criteria for Determination of Reconstructed Cells

Criteria are required to distinguish reconstructed cells from whole cell contaminants present in the preparation of cytoplasts and karyoplasts. In principle, one must show that the bulk of the cytoplasm is derived from one source and the nuclei from a different source. To date only very general markers have been utilized, namely, latex spheres of uniform size as cytoplasmic markers and incorporated thymidine-^{3}H as a nuclear marker. The rationale for the use of these markers is shown in Fig. 1 (from Veomett et al., 1974). Cells used to obtain cytoplasts and karyoplasts are labeled with latex spheres of different sizes. The ingestion of these particles by phagocytosis produces no detectable deleterious effects on the cells. The karyoplast donor is also labeled with thymidine-^{3}H. The latex spheres, which remain primarily with the cytoplasm during enucleation with CB, are

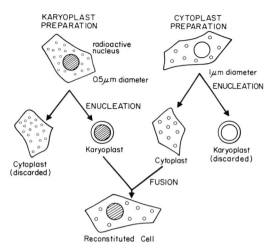

FIG. 1. Schematic representation of the experimental design. Karyoplasts were derived from cells labeled with thymidine-^{3}H and small (0.5-μm-diameter) latex spheres and fused to cytoplasts from cells labeled only with large (1.0-μm-diameter) latex spheres. The latex spheres permitted recognition of the source of the cytoplasm. Mononucleated cells with radioactive nuclei and only large latex spheres in the cytoplasm were considered cells reconstructed from components derived from different sources. Fusions resulting from whole-cell contaminants in the karyoplast preparation were recognized by the numerous small spheres in the cytoplast and were excluded from consideration as reconstructed cells.

readily observed by microscopic examination of the cells, and the radioactive nuclei can be identified after autoradiography. After fusion of karyoplasts and cytoplasts, cells having a radioactive nucleus and cytoplasm containing latex spheres only of the size used to label the cytoplast donors are considered reconstructed cells. In the first work using this system (Veomett *et al.*, 1974), the latex spheres used had diameters of 0.5 μm (cytoplast donors) and 1.0 μm (karyoplast donors). However, other bead sizes can be used. Currently we use 2.0 and 0.8 μm spheres, because both types are readily observed with 100 × phase-contrast microscopy. This is advantageous for identification and isolation of reconstructed cells for clonal analysis.

IV. Protocol for Cell Reconstruction

1. The cells are grown on an appropriate substrate [e.g., discs punched from the bottom of plastic petri dishes (Prescott and Kirkpatrick, 1973)].

Cells used as karyoplast donors are labeled with 0.8-μm latex spheres and thymidine-^{3}H (0.01 μCi/ml, 1 Ci/mmole). Karyoplast donors are labeled with the smaller-sized latex spheres, since the cells tend to take up more of the smaller spheres and only 0.1–0.5% fail to take up any spheres. Since there are more whole-cell contaminants in a karyoplast preparation than in a cytoplast preparation, and since these karyoplast contaminants must be identified and eliminated from consideration in cell reconstruction, cytoplasmic labeling of karyoplast donors should be kept to a maximum to afford easy identification of whole-cell contaminants. Cytoplast donors are labeled with the larger (2.0-μm) latex spheres. The cells are grown in the presence of the marker spheres for 24–48 hours to ensure good uptake.

2. Karyoplast and cytoplast donors are prepared using the established procedures (e.g., see Prescott and Kirkpatrick, 1973). Briefly, cells attached to a substrate are centrifuged in growth medium (cells oriented in the direction of the centrifugal force) at 27,000 g for 15 minutes to remove loosely attached cells. The cells are then centrifuged for 20 minutes at 18,000 g in medium containing 10 μg/ml of CB to induce enucleation. The best concentration of CB for enucleation needs to be determined for each kind of cell. The cytoplasts remain attached to the substrate, and the karyoplasts can be recovered in a pellet at the bottom of the centrifuge tube and from the CB medium used for enucleation. The enucleation process can be repeated to obtain a higher percentage of enucleated cells (greater than 99%). All centrifugations are performed at temperatures as close to 37°C as possible.

3. The karyoplasts are collected and washed twice by centrifugation (1000 g) in Earle's balanced salt solution (EBSS). The cytoplasts are allowed to recover from the effects of CB by incubation for 30 minutes in fresh medium free from CB. The cytoplasts are then removed from the substrate by normal trypsinization techniques and washed twice by centrifugation (1000 g) in EBSS.

4. The concentrations of cytoplasts and karyoplasts are determined (e.g., by hemocytometer counts). Equal numbers of the components are mixed in a centrifuge tube and washed once by centrifugation (1000 g) in EBSS. The two components are resuspended in a volume of EBSS sufficient to yield a concentration of 0.5 to 1 $\times$ 10^7/ml of each component.

5. The mixture is placed in an ice-water bath and treated with inactivated Sendai virus according to procedures for whole-cell fusions (e.g., Veomett et al., 1974; Klebe et al., 1970). The following regimen has been used frequently in our laboratory: (a) The mixture is incubated in the ice-water bath for 5 minutes. (b) Inactivated Sendai virus is added to the mixture (1 hemagglutinating unit per 10^4 cellular components) and the mixture incubated an additional 3–5 minutes with shaking by hand in the ice-water bath. (c) The

mixture is then placed at 37°C and shaken periodically for 20–30 minutes. We have obtained our best results by repeating steps a–c a second time.

6. After incubation of the mixture at 37°C, the fusion reaction mixture is diluted to the desired concentration, and the suspension is seeded into a culture vessel.

V. Conclusions

Reconstructed cells obtained using this procedure are viable. We have observed such cells in mitotic configurations (Veomett *et al.*, 1974), and we have followed such cells through a division cycle (G. Veomett and D. M. Prescott, unpublished observations). The markers described above (latex spheres and thymidine-^{3}H) are suitable for such limited observations, because dilution of the latex spheres and radioactive label by cellular division is not severe. In fact, latex spheres are an excellent marker, since their distribution into the daughter cells allows positive identification of the daughters. For example, one cell with 13 spheres divided into two—one daughter with 8 spheres and one with 5—permitting rapid and positive identification of the daughters among the others present in the same area.

The number of whole-cell contaminants present in cytoplast and karyoplast preparations can possibly be reduced by further purification of the components. For example, cytoplasts can be further purified using Ficoll or serum gradients (Poste, 1973). Other nuclear and cytoplasmic markers can be utilized. For example, in reconstruction experiments using different cell lines, the karyotype may be a sufficient nuclear marker. It may be possible to develop cytoplasmic biochemical markers (such as mitochondrial mutants) to allow the direct selection of reconstructed cells by growth in selective medium.

In summary, the technique described above for the reconstruction of tissue culture cells using the components produced by enucleation of the cells using CB (1) uses well-established techniques, (2) is applicable to most tissue culture cells, (3) permits reciprocal nuclear exchanges, and (4) should be applicable in studying several aspects of nucleocytoplasmic interactions.

ACKNOWLEDGMENTS

This work was supported by NIH Postdoctoral Fellowship no. 1 F22 CA00797-01 to G. V., and by grant no. 1 PO1 CA13419-01 from the National Cancer Institute. Dr. Prescott is a co-investigator on the NCI grant.

References[1]

Diacumakos, E. G. (1973). *In* "Methods in Cell Biology" (D. M. Prescott, ed.), Vol. 7, pp. 287–311. Academic Press, New York.

Diacumakos, E. G. (1975). *In* "Methods in Cell Biology" (D. M. Prescott, ed.), Vol. 10, pp. 147–156. Academic Press, New York.

Follett, E. A. C. (1974). *Exp. Cell Res.* **84**, 72.

Follett, E. A. C., Pringle; C. R., Wunner, W. H., and Skehel, J. J. (1974). *J. Virol.* **13**, 394.

Giles, R. E., and Ruddle, F. H. (1973). *In Vitro* **9**, 103.

Goldstein, L. (1965). *In* "Methods in Cell Physiology" (D. M. Prescott, ed.), Vol. 1, pp. 97–108. Academic Press, New York.

Goldstein, L., Cailleau, R., and Crocker, T. T. (1960). *Exp. Cell Res.* **19**, 332.

Jeon, K. W., and Danielli, J. F. (1971). *In* "International Review of Cytology" (G. H. Bourne and J. F. Danielli, eds.), Vol. 30, pp. 49–89. Academic Press, New York.

King, T. J. (1966). *In* "Methods in Cell Physiology" (D. M. Prescott, ed.), Vol. 2, pp. 1–36. Academic Press, New York.

Klebe, R. J., Chen, T., and Ruddle, F. H. (1970). *J. Cell Biol.* **45**, 74.

Kopac, M. J. (1955). *Trans. N. Y. Acad. Sci.* [2] **17**, 257.

Penman, S. (1969). *In* "Fundamental Techniques in Virology" (K. Habel and N. P. Salzman, eds.), Vol. 1, pp. 35–48. Academic Press, New York.

Pennington, T. H. E., and Follett, E. A. C. (1974). *J. Virol.* **13**, 488.

Perry, R. P., and Kelly, D. E. (1968). *J. Mol. Biol.* **35**, 37.

Pollack, R., and Goldman, R. (1973). *Science* **179**, 915.

Poste, G. (1973). *In* "Methods in Cell Biology" (D. M. Prescott, ed.), Vol. 7, pp. 211–249. Academic Press, New York.

Prescott, D. M., and Kirkpatrick, J. B. (1973). *In* "Methods in Cell Biology" (D. M. Prescott, ed.), Vol. 7, pp. 189–202. Academic Press, New York.

Prescott, D. M., Kates, J., and Kirkpatrick, J. B. (1971). *J. Mol. Biol.* **59**, 505.

Prescott, D. M., Myerson, D., and Wallace, J. (1972). *Exp. Cell Res.* **71**, 480.

Shay, J. W., Porter, K. R., and Prescott, D. M. (1973). *J. Cell Biol.* **59**, 312c.

Tartar, V. (1953). *J. Exp. Zool.* **124**, 63.

Veomett, G., Prescott, D. M., Shay, J., and Porter, K. R. (1974). *Proc. Nat. Acad. Sci. U.S.* **71**, 1999.

Wilson, J. F. (1963). *Amer. J. Bot.* **50**, 780.

Wise, G. E., and Prescott, D. M. (1973). *Exp. Cell Res.* **81**, 65.

Wright, W. E. (1973). *In* "Methods in Cell Biology" (D. M. Prescott, ed.), Vol. 7, pp. 203–210. Academic Press, New York.

Zaolocar, M. (1971). *Proc. Nat. Acad. Sci. U.S.* **68**, 1539.

Zetsche, K. (1962). *Nature (London)* **49**, 404.

[1] Note: No attempt at a complete literature survey has been attempted.

Chapter 3

Recording of Clonal Growth of Mammalian Cells through Many Generations

HIROSHI MIYAMOTO,[1] LEIF RASMUSSEN, AND ERIK ZEUTHEN

*The Biological Institute of the Carlsberg Foundation,
Copenhagen, Denmark*

I.	Introduction	15
II.	Culture Medium	16
III.	Preparation of the Cells	16
	A. Subcultures	16
	B. Experimental Cultures	17
IV.	Microphotographic Recording of Clonal Growth	18
	A. Microscopes	18
	B. Recording Equipment	19
	C. Thermostats Use for Photographic Recording	19
V.	Clonal Growth	19
	A. Assigning of Division Time	19
	B. Tracing of Cell Lines	20
	C. Cell Multiplication	21
VI.	Applications	22
	A. Short-Term Inhibitory Agents	23
	B. Long-Term Inhibitory Agents	24
	C. Other Applications	26
	References	26

I. Introduction

Many investigators have found clones of free-living cells—ciliates, amebas, and so on—useful tools in cell biology (e.g., Cleffmann, 1968; Hamburger, 1974; Løvlie, 1963; Løvlie and Farfaglio, 1965; Prescott, 1955; Thormar, 1959; Zeuthen, 1953). When we wanted to study the multiplication

---

[1] *Present address:* Department of Physiology, Kinki University School of Medicine, Nishiyama, Sayama-cho, Minamikawachi-gun, Osaka, Japan.

of clones of mammalian cells, we met with several problems, some solved, some unsolved. Thus, on the one hand, Puck and Fisher (1956) had already reported on the technique of obtaining mammalian cell clones, and Lockardt and Eagle (1959) and Ham (1972) had pointed out that special precautions had to be taken with respect to the composition of the culture medium. On the other hand, since we recorded cell morphology and clonal growth under the microscope, curtailing of evaporation of culture medium was essential. In the following discussion we describe a method we have used to study the reaction of mouse fibroblast cells cultured in monolayer to temperature shocks applied at various stages of the cell cycle (Miyamoto *et al.*, 1973b). Our method allowed us to follow the progeny of one mouse fibroblast cell through seven to eight generations (Miyamoto *et al.*, 1973a). Many parameters of cell growth can be obtained with this technique which we hope will be of use to cell biologists.

II.   Culture Medium

As already observed by Lockardt and Eagle (1959) Eagle's minimal essential medium (MEM) is insufficient to support growth of single cells. We found that MEM supplemented with yeast extract–lactalbumin hydrolyzate–saline solution (YLH) and calf serum supported growth in experiments with only 250 initial cells/ml, yielding reproducible, high growth rates and high final population densities. Tables I and II show the composition of this supplemented culture medium. To make 1 litre of medium, we used 450 ml of modified MEM, 450 ml of YLH, and 100 ml of calf serum (Flow Laboratories Ltd.). The acidity of the medium was maintained by approximately 22 mM N-2-hydroxethylpiperazine-N^1-2-ethanesulfonic acid (HEPES) purchased from Sigma (Eagle, 1971). This concentration kept the pH at 7.0 throughout the experiment, and this decreased contact inhibition (Ceccarini and Eagle, 1971).

III.   Preparation of the Cells

A.   Subcultures

Cells were propagated in Flow tissue culture flasks, size no. 50, containing 20 ml of medium. The size of the inoculum was chosen to give an initial population density of approximately 50,000 cells/ml. Cells were grown for 2–3 days, at the end of which time they were treated with 1:250 trypsin (Difco). The trypsin was diluted to a final concentration of 0.05% with a phosphate buffer–saline solution at pH 7.4 and added to the culture (Magee *et al.*, 1958). Excess trypsin solution was discarded. After 4-minute incuba-

TABLE I

COMPOSITION OF MODIFIED MEM[a]

Substance	Solution A (mg/liter)	Solution B (mg/liter)	Solution C (mg/liter)
$CaCl_2 \cdot 2H_2O$	264.0	—	—
KCl	400.0	—	—
$MgSO_4 \cdot 7H_2O$	200.0	—	—
NaCl	6800.0	—	—
$NaH_2PO_4 \cdot H_2O$	140.0	—	—
$NaHCO_3$	2000.0	—	—
Phenol red	17.0	—	—
Glucose	1000.0	—	—
L-Arginine	126.4	6320	—
L-Cystine	24.0	1200	—
L-Histidine · HCl	38.3	1915	—
L-Isoleucine	52.5	2625	—
L-Leucine	52.5	2625	—
L-Lysine	73.1	3655	—
L-Methionine	14.9	745	—
L-Phenylalanine	33.0	1650	—
L-Threonine	47.6	2380	—
L-Tryptophan	10.2	510	—
L-Tyrosine	36.2	1810	—
L-Valine	46.9	2345	—
D-Biotin	—	—	100
D-Calcium pantothenate	1.0	—	100
Choline chloride	1.0	—	100
Folic acid	1.0	—	100
Isoinositol	2.0	—	200
Nicotinamide	1.0	—	100
Pyridoxal HCl	1.0	—	100
Riboflavin	0.1	—	10
Thiamine HCl	1.0	—	100

[a] Add the following ingredients to 1 liter of solution A: 20 ml of solution B, 20 ml of solution C, 5 ml of 20% glucose, 20 ml of 0.2 M glutamine, 30 ml of 1 M HEPES, and 15 ml of 0.1 N HCl. Solutions A, B, and C were purchased from Flow Laboratoties Ltd. (Scotland).

tion at 37.2°C, 10 ml of fresh culture medium was added, and the cells were separated from each other by gentle pipetting. Cells obtained in this way were used to inoculate both experimental and stock cultures. The handling of cultures was carried out in a sterile box, "Clean bench," regulated to between 37° and 38°C.

B. Experimental Cultures

Trypsinized cells were transferred to sterile Falcon tissue culture dishes (diameter 35 mm) containing 2 ml of nutrient medium. The initial population

TABLE II

COMPOSITION OF YEAST EXTRACT-LACTALBUMIN HYDROLYZATE-SALINE SOLUTION[a]

Substance	Solution D (mg/liter)	Solution E (mg/liter)
$CaCl_2 \cdot 2H_2O$	264	—
KCl	400	—
$MgSO_4 \cdot 7H_2O$	200	—
NaCl	8,000	—
KH_2PO_4	60	—
$Na_2HPO_4 \cdot 2H_2O$	60	—
$NaHCO_3$	—	14,000
Phenol red	30	—
Glucose	5,500	—
Yeast extract	1,000	—
Lactalbumin hydrolyzate	5,000	—
Streptomycin	100	—
Penicillin	100,000[b]	—

[a]To 1 liter of solution D add 25 ml of solution E and 20 ml of 1 M HEPES.
[b]International units.

density was adjusted to 500 cells per dish, and each dish was sealed with three layers of stretched sterile Parafilm in order to decrease evaporation of water from the culture. Brief gentle horizontal shaking secured even distribution of the cells throughout the dishes. The cultures were then placed at $37.2 \pm 0.5°C$ in an air incubator in which the relative humidity was maintained near 100%. The culture dishes were closed, and the requirement for CO_2 for cells growing in HEPES as reported by Itagaki and Kimura (1974) was fulfilled.

IV. Microphotographic Recording of Clonal Growth

A. Microscopes

We used two types of microscopes: stereomicroscopes (Zeiss and Nikon) and an inverted Reichert microscope. The stereomicroscopes had a large field, excellent contrast, and a long focal distance and were used for selecting the cells whose growth was followed in the experiments. The location of the selected cells was marked with a felt-tip marker. Cell multiplication was photographically recorded with the aid of either a stereomicroscope or the inverted microscope.

B.   Recording Equipment

Population growth was followed by time-lapse photomicroscopy or time-lapse videorecording microscopy. In the case of photomicroscopy we used a Robot Star II fully automatic camera attached to the microscope described above. We used 35-mm Ilford Pan F film and made prints employing Ilfobrom 4 (13 × 18 cm). In the case of videorecording, an Ikegami television camera was attached to the Nikon stereomicroscope. The camera was connected to a videorecorder (International Video Corporation, Model 601), and pictures were examined on a television set (Luxor Industri, A. B., Sweden).

C.   Thermostats Used for Photographic Recording

In order to obtain clear photographs of the cells, we found it necessary to prevent the condensation of vapor on the inside lid of the culture dish. This was accomplished by employing two sets of thermostats, an outer *air* thermostat which regulated at a slightly higher temperature than an inner *water* thermostat. The air thermostat was equipped with a blower and maintained the air temperature between 37.5° and 38.4°C. The water thermostat that controlled the temperature at which the cells were growing was a double-walled brass chamber into which the culture dish fitted (Fig. 1). It was constructed to permit passage of light from a conventional light source to a microscope. The chamber temperature was maintained by circulation of water from an 8-liter reservoir regulated to $37.2 \pm 0.1°C$.

V.   Clonal Growth

A.   Assigning of Division Time

We recorded the growth of a selected clone of from 1 or 4 to 128 cells. The cultures were followed over a period of 6–7 days by exposures made every hour. Since cell division occurred at an unknown time between two photographic recordings spaced an hour apart, we needed a method of assigning the division time to these cells in as accurate a manner as possible. Assigning of the division time made use of several morphological criteria. For instance, when one frame showed one cell in the clone and the next frame showed two cells in its place, it was obvious that cell division had occurred between the time of the two recordings. If the two cells observed in the latter frame had moved apart, and if they had obviously advanced far into inter-

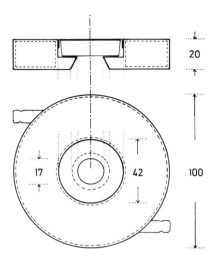

FIG. 1. Cross section and top view of the brass chamber that holds the Falcon tissue culture dish. Water from a thermoregulated reservoir is pumped in through one side tube and returned through the other. Numerals indicate lengths in millimeters. The chamber may be purchased from the Biological Institute of the Carlsberg Foundation. Reproduced from Miyamoto *et al.* (1973a) by courtesy of the Company of Biologists.

phase, the time halfway between the taking of the two frames was considered the division time. If, however, the daughter cells had not moved apart or had not advanced into interphase, then the time at which the latter frame was taken was given as the division time.

B.   Tracing of Cell Lines

Photographic exposures of the clones were made sufficiently often to trace one cell and its progeny to the 128-cell stage. Thus we could calculate the duration of the successive generation times for each cell from the assigned time points of cell division and could construct "family trees" including all the cells of the clone. An example illustrates how the cell lines were followed. A clone (Fig. 2) originated from a cell which divided between the sixteenth and seventeenth recordings. The two daughter cells are designated 17a and 17b. They were found and marked on all frames from the seventeenth to to the thirty-eighth. Cell 17a gave rise to the cells designated 39a and 39b, and cell 17b to 40a and 40b. In turn, cell 39b gave rise to 64a and 64b, and cell 39a to 65c and 65d, and so on.

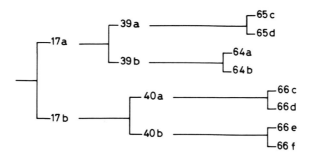

FIG. 2. Part of one of the progeny "trees" which form the basis for calculation of the generation time of successive cell generations. A photograph of the clone was taken every hour. The offsprings of a cell assigned to divide at 17 hours after the beginning of the recording were designated 17a and 17b, and so on. The clone was exposed to a cold shock for 1 hour between the forty-ninth and fiftieth hours. Reproduced from Miyamoto *et al.* (1973b) by courtesy of the Company of Biologists.

C. Cell Multiplication

A clone was followed through six cell generations. Figure 3E shows the cell numbers as a function of time, starting with recordings that began shortly after the clone increased from two to four cells. Cell number is also shown as a function of time in the four subclones that arose from the four cells present when recording was initiated (Fig. 3A–D). The cells divided in relative synchrony in the early part of the experiment and, as time passed, this synchrony deteriorated. In all cases we observed an exact doubling of the cell number of each generation.

We calculated the average duration of the four successive cell generations shown in Fig. 3. The values vary between 20.5 and 26 hours, and statistical analysis shows that there is no significant difference between them (Miyamoto *et al.*, 1973a).

Correlations between generation times were tested. Although the generation times of sister-sister cells and daughter-parent cells were correlated (P less than 1%) in both cases), this correlation appeared to be slightly better between sister and sister cells than between parent and daughter cells (Miyamoto *et al.*, 1973a).

We have studied the frequency distribution of the generation times obtained in the present study. The mean generation time was found to be 21.5 ± 3.3 hours ($n = 578$ cells) and included minimum and maximum values of 14 and 37 hours, respectively. The shape of the curve approximately fits the normal distribution but is slightly skewed toward the high values (Miyamoto *et al.*, 1973a).

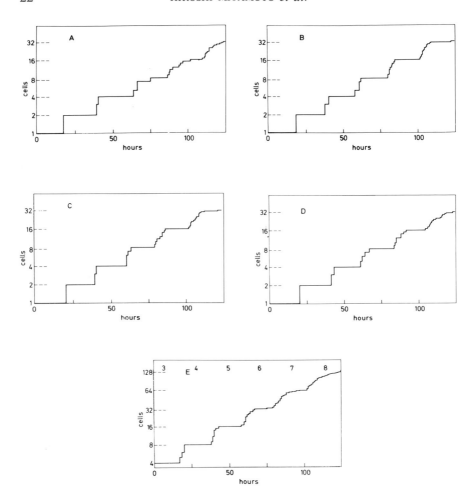

FIG. 3. Cell multiplication in each of the four subclones (A–D) composing a clone (E) of mouse fibroblast cells (line 929). Time zero indicates the beginning of the recording. Inoculation took place approximately 48 hours before time zero. (E) represents multiplication of the total clone, i.e., it is the sum of (A) through (D). Numerals indicate the number of generations since inoculation (the one-cell stage is considered generation 1). Reproduced from Miyamoto *et al.* (1973a) by courtesy of the Company of Biologists.

VI. Applications

The method described here can be used for several different types of experiments. Below we outline some of those we have found useful. Using a single camera we have, on the one hand, the choice of following cell multiplication in a clone with short time intervals (e.g., 1 hour) between each photographic exposure. Such an approach may result in growth curves described by frequent cell counts. On the other hand, since we could mark the position of a selected clone in a petri dish with a felt-tip marker, we could also follow cell multiplication in many dishes containing different inhibitory agents using long time intervals (e.g., 24 hours) between each photographic exposure. In all cases the number of cells per clone reflects the true growth of the clone, because the field of observation was so large that we never saw a cell either leave or enter it.

Working with clones we have the advantage that cell divisions are inherently synchronized during the first few generations and that this synchrony gradually decays. Thus it is possible to use young clones when synchrony of cell division is desired. Older clones simulate the situation prevailing in a mass culture with respect to random cell division.

A. Short-Term Inhibitory Agents

We studied the effect of temperature shocks on preparation for cell division (Miyamoto *et al.*, 1973b). One aspect of this study was to establish the duration of one cell cycle as a function of standard temperature shocks applied at different stages of the cell cycle. We were faced with the difficulty of estimating the exact position between two successive mitoses of the cells at the time they were exposed to the temperature treatment. Since we had found a strong correlation between generation times of parent and daughter cells, we chose to express the age of four treated cells (see Fig. 2) relative to a unit time of 100%, which was the average generation time of the two untreated mother cells. In the case of the progeny "tree" shown in Fig. 2, a 1-hour temperature shock was applied between the forty-ninth and fiftieth hours. The control mother generation time was 22 plus 23 hours divided by 2, which equals 22.5 hours, or 100%. It follows that cells 39a and 39b were 10 hours old (45%), and cells 40a and 40b were 9 hours old (40%), when the temperature shock began. Of course, by adopting this procedure we introduce errors in estimations of both age when treated and duration of the treated generation, apart from minor ones arising from the assigning of division times. However, since the standard deviation of the generation times of cells within *one* clone is relatively low ($\pm 9\%$), we feel justified in using this procedure.

One feature of the method described here is that it yields photographs that give information on cell morphology. In the case of the temperature studies it was observed that some cells rounded up in response to the treatment, whereas others did not show this behavior. So far, this aspect has not been investigated.

B.   Long-Term Inhibitory Agents

We also wanted to see how treatments known to phase cell division in other systems (Xeros, 1962; Galavazi and Bootsma, 1966; Studzinski and Lambert, 1969; Bostock *et al.*, 1971) would affect cell multiplication in our system. Thus the effect of 10^{-2} *M* thymidine for 15 hours is shown in Fig. 4. The clone in question consisted of 13 cells when the photographic recording started, and had increased to 32 cells 25 hours later when the drug was added. At this time the synchrony of cell division in this clone had deteriorated to

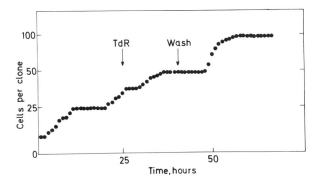

FIG. 4.   Cell multiplication in a clone treated with 10^{-2} *M* thymidine for 15 hours. Microphotographs were taken at hourly intervals.

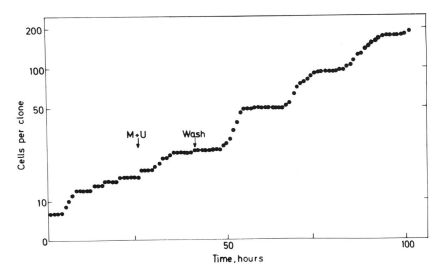

FIG. 5.   Cell multiplication in a clone treated with methotrexate (5×10^{-6} *M*) and uridine (10^{-2} *M*) for 15 hours. Microphotographs were taken at hourly intervals.

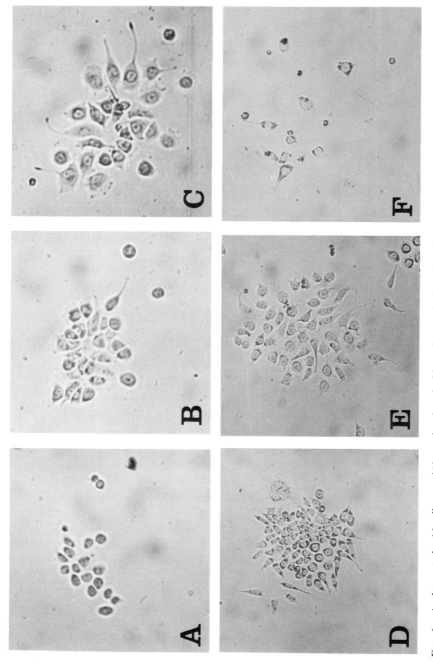

FIG. 6. A clone at the 16-cell stage (A) was incubated with methotrexate ($5 \times 10^{-6}\ M$) plus uridine ($10^{-2}\ M$) for 72 (B) and 120 hours (C). Cell multiplication is blocked early, but the cells enlarge greatly. In the nutrient medium used methotrexate and uridine added singly have no effect. Also excess thymidine blocks cell multiplication while permitting growth (E, $10^{-3}\ M$; F, $10^{-2}\ M$) except in very low concentrations (D, $10^{-4}\ M$). (D) through (F) show 16-cell stage clones after 72 hours of treatment.

some degree. After the addition of thymidine, cell multiplication continued for about 12 hours and was then blocked. However, the drug was removed after an exposure time of 15 hours, and 8 hours later a fairly synchronous burst of cell division occurred, resulting in a doubling of the cell number. About a generation time later a new burst of synchronous cell division was observed (not shown in Fig. 4).

We carried out a similar type of experiment using the inhibitor pair methotrexate and uridine (Fig. 5). In the ciliate *Tetrahymena pyriformis*, these inhibitors are known to block DNA synthesis in complex nutrient media containing thymidine (Zeuthen, 1968). Again, the inherent synchrony of cell division was little pronounced at the time of addition of the inhibitors. However, upon their removal rather good division synchrony ensued. The data presented in Figs. 4 and 5 permit detailed analysis of the duration of *treated* cell cycles, as well as both previous and future *untreated* cell cycles, for each single cell in the clones.

Cell morphology as seen from photographic recordings has been used to characterize the effects of inhibitory agents (Zeuthen and Rasmussen, 1976). Thus it was observed that treatment with methotrexate plus uridine resulted in large cells destined to thymineless death (Fig. 6A–C). Similarly, we observed that treatment with thymidine (10^{-3} and 10^{-2} M) also produced large cells (Fig. 6E and F).

C. Other Applications

Since it is possible to identify all the cells in a clone by microphotometry, this method lends itself exceedingly well to autoradiography (Zetterberg and Killander, 1965), microinterferometry (Killander and Zetterberg, 1965), and microspectrophotometry (Kiefer *et al.*, 1969). Thus this technique can easily be extended to determinations of macromolecular quantities, and so on, at the cellular level.

ACKNOWLEDGMENTS

This work was supported by grants from the Japanese Ministry of Education (to H. Miyamoto) and from the Danish Natural Sciences Research Council (to E. Zeuthen and L. Rasmussen).

REFERENCES

Bostock, C. J., Prescott, D. M., and Kirkpatrick, J. B. (1971). *Exp. Cell Res.* **68**, 163.
Ceccarini, C., and Eagle, H. (1971). *Proc. Nat. Acad. Sci. U.S.* **68**, 229.
Cleffmann, G. (1968). *Exp. Cell Res.* **50**, 193.
Eagle, H. (1971). *Science* **174**, 500.
Galavazi, G., and Bootsma, D. (1966). *Exp. Cell Res.* **41**, 438.
Ham, R. G. (1972). *In* "Methods in Cell Physiology". (D. M. Prescott, ed.), Vol. 5, 37. Academic Press, New York.

Hamburger, K. (1975). C. R. Trav. Lab. Carlsberg 40, 175.
Itagaki, A., and Kimura, G. (1974). Exp. Cell Res. 83, 351.
Kiefer, G., Zeller, W., and Sandritter, W. (1969). Histochemie 20, 1.
Killander, D., and Zetterberg, A. (1965). Exp. Cell Res. 38, 272.
Lockardt, R. Z., Jr., and Eagle, H. (1959). Science 129, 252.
Løvlie, A. (1963). C. R. Trav. Lab. Carlsberg 33, 377.
Løvlie, A., and Farfaglio, G. (1965). Exp. Cell Res. 39, 418.
Magee, W. E., Sheek, M. R., and Sagik, B. P. (1958). Proc. Soc. Exp. Biol. Med. 99, 390.
Miyamoto, H., Zeuthen, E., and Rasmussen, L. (1973a). J. Cell Sci. 13, 879.
Miyamoto, H., Rasmussen, L., and Zeuthen, E. (1973b). J. Cell Sci. 13, 889.
Prescott, D. M. (1955). Exp. Cell Res. 9, 328.
Puck, T. T., and Fisher, H. (1956). J. Exp. Med. 104, 427.
Studzinski, G. P., and Lambert, W. C. (1969). J. Cell. Physiol. 73, 109.
Thormar, H. (1959). C. R. Trav. Lab. Carlsberg 31, 207.
Xeros, N. (1962). Nature (London) 194, 682.
Zetterberg, A., and Killander, D. (1965). Exp. Cell Res. 40, 1.
Zeuthen, E. (1953). J. Embryol. Exp. Morphol. 1, 239.
Zeuthen, E. (1968). Exp. Cell Res. 50, 37.
Zeuthen, E., and Rasmussen, L. (1976). Exp. Cell Res. In press.

Chapter 4

Preparation of Isolated Rat Liver Cells

PER O. SEGLEN

Norsk Hydro's Institute for Cancer Research, Department of Tissue Culture,
The Norwegian Radium Hospital,
Montebello, Oslo, Norway

I.	Introduction	30
II.	Quality Evaluation	30
III.	Nonenzymatic Methods for Liver Cell Preparation	32
	A. Mechanical Dispersion	33
	B. The Use of Chelators	33
IV.	Enzymatic Treatment of Liver Slices	34
V.	Collagenase Perfusion	35
	A. General	35
	B. A Quantitative Assay of Liver Dispersion	35
	C. Collagenase	36
	D. Ca^{2+} and the Two-Step Procedure	39
	E. pH, Oxygenation, and Buffer Compositions	44
	F. Perfusate Flow	45
	G. Liberation of Cells	49
	H. Other Factors	49
	I. Liver Perfusion Apparatus	50
	J. Operative Technique and a Routine Procedure for Liver Cell Preparation	52
VI.	Purification of Parenchymal Cells	57
	A. The Goals of Purification	57
	B. Preincubation	58
	C. Filtration	58
	D. Differential Centrifugation	59
	E. Gradient and Cushion Techniques	59
	F. A Routine Procedure for the Purification of Parenchymal Cells	61
VII.	Incubation of Cell Suspensions	63
VIII.	Properties of Isolated Parenchymal Cells	64
	A. Ultrastructure	65
	B. Carbohydrate Metabolism	65
	C. Lipid Metabolism	68
	D. Protein and Nucleic Acid Metabolism	69
	E. Other Properties	70
IX.	*In Vitro*-Culture of Parenchymal Liver Cells	71
X.	Preparation of Nonparenchymal Liver Cells	73
	References	78

I. Introduction

The liver occupies a central position in body metabolism, and its size, soft-ness, and relative homogeneity have made it a favorite organ for biochemical investigation. For the study of intact liver functions under controlled condi-tions the isolated, perfused rat liver has been extensively used. This experi-mental system is excellent for many purposes, but has several major shortcomings: (1) the liver as an organ is not completely homogeneous, containing up to 40% nonparenchymal cells (Daoust, 1958); (2) it is difficult to obtain many identical samples from one liver, and impossible to test different experimental treatments simultaneously; and (3) the viability of an isolated liver can be maintained only for a limited period of time (8–10 hours). Numerous attempts have been made to overcome these problems by the isolation and purification of intact parenchymal rat liver cells.

The early mechanical and chemical methods for liver cell preparation were relatively successful in converting liver tissue to a suspension of isol-ated cells, but unfortunately nearly all such cells are damaged (the presence of a few intact cells capable of proliferation *in vitro* does not invalidate this general characterization). The introduction of collagenase as a liver-dispers-ing enzyme by Howard *et al.* (1967) greatly facilitated the preparation of intact cells, and when Berry and Friend (1969) introduced the use of physio-logical liver perfusion to make the tissue uniformly accessible to the action of collagenase, it became possible to prepare intact liver cells in high yield. Isolated parenchymal cells are now increasingly being used as an experi-mental tool, and are replacing perfused liver in most laboratories.

The successful preparation of intact liver cells by perfusion with colla-genase is technically quite difficult, and still remains mostly an art. Almost every worker has incorporated his own modifications, hence a variety of methods for liver cell preparation has been published. In this article the approach is taken that "collagenase perfusion" is essentially *one* method. The numerous modifications introduced at various steps in the cell prepara-tion procedure are discussed in a systematic fashion, with no attempt to compare, as such, the different assemblies of modifications that have been published.

The collagenase perfusion technique can also be used for preparation of nonparenchymal cells, a subject that is treated separately in Section X.

II. Quality Evaluation

The objective of any cell preparation method is to produce as many *intact* cells as possible. The quality of a method can therefore be expressed by the total number of intact cells obtained, taking into consideration both the quality of the cell suspension (i.e., the percentage of intact cells) and the total yield of cells (relative to the initial amount of tissue).

The efficiency of a method for tissue dispersion should be evaluated from the *initial* cell suspension obtained. Subsequent purification by filtration and centrifugation alters both the yield and the percentage of intact cells, and the final cell preparation is as much a reflection of the purification procedure as of the method for tissue dispersion. The failure of most published procedures to distinguish between these two aspects makes a comparative methodological evaluation exceedingly difficult. Several methods exist for the separation of intact cells from dead ones, so even the poorest preparation may eventually be turned into a small quantity of "mostly intact" cells.

The most commonly employed criterion of cellular integrity (or viability) is the trypan blue exclusion test. Cells with an intact plasma membrane exclude dyes such as trypan blue, nigrosin, and eosin (Paul, 1972), whereas damaged cells become stained, particularly intensively in the nucleus. Several other viability tests have been proposed, but dye exclusion remains the simplest and most reliable. Metabolic capabilities may give an overall impression of preparation quality, but they show great biological variability (cf., e.g., the disappearance of glycogen-synthesizing ability during starvation, Seglen, 1973c, 1974a) and cannot be used as a quantitative index. Respiratory rates have been used for quality evaluation (LaBrecque *et al.*, 1973), but these are of limited general value because both isolated mitochondria and dead cells can respire quite well (Murthy and Petering, 1969; Howard *et al.*, 1973). Many other metabolic processes can also take place in subcellular structures, hence in damaged cells under appropriate conditions, in particular isotope incorporation (e.g., of amino acids, LaBrecque *et al.*, 1973).

Cellular K^+ content and membrane potential have been suggested as viability criteria (Baur *et al.*, 1975), but the loss and uptake of K^+ are reversible processes (Quistorff *et al.*, 1973; Barnabei *et al.*, 1974; Tolbert and Fain, 1974), in contrast to the structural damage measured by the trypan blue test. The leakage of soluble enzymes such as lactate dehydrogenase (LDH) is also an expression of structural membrane damage, and can be regarded as equivalent to the trypan blue test (Berg *et al.*, 1972).

However, a cell may have internal metabolic lesions or small surface alterations not revealed by the trypan blue test, but from which death may result at a later time. The trypan blue test therefore does not measure *viability* in the strictest sense, but rather gross structural integrity at the moment of testing. This may still be the most relevant parameter to measure, since the further survival capacity of a cell depends on many factors (functional limitations of the cell, incubation conditions, nutrients and hormones, and so on) and is not a very good index for evaluating the method of cell isolation.

The trypan blue test has often been misinterpreted or erroneously employed, and a word of caution seems appropriate. The test must be performed under conditions in which the damaged cells are really stainable, which is not the case, e.g., in the presence of certain polymers such as polyethylene glycol or polyvinylpyrrolidone. Under ordinary conditions a dye concentration of 0.2% is sufficient, but it should be noted that the staining intensity may show considerable variation depending on components in the cell suspension or in the particular batch of trypan blue (Schreiber and Schreiber, 1973). We routinely use an isotonic 0.6% trypan blue solution (150 mg trypan blue plus 120 mg NaCl in 25 ml H_2O, filtered through a 0.22-μm Millipore filter and stored as frozen 2-ml portions), of which 300 μl is mixed with a 100-μl cell suspension for counting (i.e., 0.45% trypan blue finally). Even the slightest nuclear staining is indicative of damage, contrary to the erroneous notions of some workers. In the case of parenchymal liver cells, the intact cells with their yellow color, well-defined outline, and refractile appearance are readily distinguished from the flattened, ground-glass-looking damaged cells even without trypan blue (cf. e.g., Figs. 6 and 7 in Howard et al., 1967).

When the cell suspension is counted in a Bürker chamber (hemocytometer), the cover glass must first be mounted, and the cell suspension applied at the edge of the cover glass to be drawn into the chamber by capillary force. Sample application must be rapid, in order to prevent selective sedimentation of intact cells in the pipette. If, on the contrary, a drop of cell suspension is first placed on the chamber and the cover glass mounted subsequently, the preparation will usually look excellent because the damaged cells are selectively squeezed out to the periphery of the chamber.

III. Nonenzymatic Methods for Liver Cell Preparation

Before the development of enzymatic methods for liver cell preparation, nonenzymatic methods were widely used. Although the cells produced by such methods are almost uniformly damaged, they may be useful, e.g., in the analysis of nuclear numbers and ploidies. The damaged cells retain some gross cellular structure, and certain particle-associated processes may therefore be profitably investigated with such a preparation. However, the metabolic properties of these dead macromolecular aggregates should not be taken to represent the properties of living cells.

A. Mechanical Dispersion

Regardless of how the liver has been pretreated, some kind of mechanical procedure has to be employed to disrupt the organ and liberate the individual cells from the connective and vascular tissue. If the pretreatment has caused the cells to separate nearly completely *in situ* (such as after collagenase perfusion), a very mild mechanical treatment may suffice, e.g., gentle shaking aided by careful raking with a comb, spatula, or forceps. With insufficient pretreatment, however, very drastic mechanical procedures are needed to liberate the cells, resulting in cellular damage.

A variety of mechanical methods has been used for the dispersion of liver tissue without any particular pretreatment. These methods include homogenization (Palade and Claude, 1949; Harrison, 1953), forcing of the tissue through screens of stainless steel, silk, or cheesecloth (Schneider and Potter, 1943; Kaltenbach, 1954), pipetting (Longmuir and ap Rees, 1956), shaking with glass beads (St. Aubin and Bucher, 1952), and other procedures (Branster and Morton, 1957; for more references, see Schreiber and Schreiber, 1973).

Although many claims to the contrary have been made, the cells obtained by such techniques are uniformly damaged and dead. The metabolic activities observed, e.g., mitochondrial respiration, apparently do not require complete cellular integrity.

B. The Use of Chelators

Ca^{2+} is known to play a role in cellular adhesion (Moscona *et al.*, 1965; Gingell *et al.*, 1970), and divalent metal chelators have consequently been extensively employed in attempts to separate liver cells prior to mechanical treatment. The most commonly used chelators are citrate (Anderson, 1953; Jacob and Bhargava, 1962) and EDTA (Coman, 1954; Leeson and Kalant, 1961). A variety of procedures has been described, combining Ca^{2+} chelation and mechanical treatment (for references, see Schreiber and Schreiber, 1973).

In a quantitative study of liver dispersion, it was found that Ca^{2+} removal facilitated the subsequent separation of cells by collagenase perfusion (Seglen, 1972b). The effect of Ca^{2+} removal was irreversible, and probably due to the detachment and washout of a Ca^{2+}-dependent adhesion factor (Modjanova and Malenkov, 1973; Seglen, 1973a). Removal of Ca^{2+} was accomplished either with chelators (EGTA) or by efficient washout (perfusion) with a Ca^{2+}-free buffer (Seglen, 1973a). Recent studies by Amsterdam and Jamieson (1974) suggest that the Ca^{2+}-dependent adhesion factor is identical with the central plaque material of the desmosomes, and that the removal of this material causes the hemidesmosomes to move apart.

However, although the removal of Ca^{2+} (and the Ca^{2+}-dependent adhesion factor) may improve the separability of the liver cells to some extent, the mechanical force required for *complete* separation of the cells still results in virtually uniform damage (Seglen, 1973a). Preliminary Ca^{2+} removal may thus be useful in the preparation of homogeneous suspensions of damaged cells, but for the preparation of *intact* cells such chemical–mechanical methods are useless. The not infrequent claims to the contrary are based on faulty viability criteria.

The chelation of K^+ by tetraphenylboron has been reported to favor liver dispersion (Rappaport and Howze, 1966; Casanello and Gerschenson, 1970), but a number of workers have been unable to confirm this (Jezyk and Liberti, 1969; Mills and Zucker-Franklin, 1969; Murthy and Petering, 1969; Gallai-Hatchard and Gray, 1971; Lipson *et al.*, 1972; Müller *et al.*, 1972; Seglen, 1973a).

The minute quantities of *intact* cells obtainable by nonenzymatic methods ($< 1\%$) may be sufficient, e.g., for the initiation of cell lines (Garvey, 1961; Casanello and Gerschenson, 1970).

IV.   Enzymatic Treatment of Liver Slices

Several enzymes, alone or in combination, have been tried in attempts to disperse liver slices into single cells. *Trypsin* and other proteolytic enzymes appear to promote dispersion to some extent, but they usually also destroy the parenchymal cells (St. Aubin and Bucher, 1952; Easty and Mutolo, 1960; Laws and Stickland, 1961; Pisano *et al.*, 1968). [A contrasting report by Gallai-Hatchard and Gray (1971) has stated that the inclusion of trypsin improved the viability of collagenase-prepared cells; however, the cellular yield in this case was substantially reduced.] *Pronase* (a mixture of *Streptomyces* proteases) destroys parenchymal cells with particular efficiency and, since the nonparenchymal cells are left structurally intact (Roser, 1968; Mills and Zucker-Franklin, 1969), this enzyme is useful for the preparation of pure nonparenchymal cells (Section X).

Lysozyme has been reported to aid in the dispersion of both adult and fetal liver slices (Hommes *et al.*, 1970, 1971), but no dye-exclusion data were given. In the liver dispersion assay of Seglen (1973b), lysozyme was found to be completely ineffective, and the effect of trypsin was also negligible (Fig. 2).

Howard *et al.* (1967) introduced the use of collagenase and hyaluronidase for the dispersion of liver slices, and succeeded in obtaining suspensions of largely intact cells in yields amounting to 5% of the tissue (Howard and Pesch, 1968). The utility of hyaluronidase was not well documented, and in a liver dispersion assay this enzyme was found to be ineffective, either alone or in combination with collagenase (Seglen, 1973b). However, the excellent ability of even purified collagenase alone to disperse liver tissue has been amply confirmed (Seglen, 1973b). The method of Howard *et al.* was the first method capable of producing significant numbers of intact cells, and it has been employed with considerable success by other workers (Lentz and DiLuzio, 1971; Wright and Green, 1971; Holtzman *et al.*, 1972; Hook *et al.*, 1973). A thorough and readable discussion of the method has been presented (Howard *et al.*, 1973).

V.   Collagenase Perfusion

A.   General

In the method of Howard *et al.*, the liver is first perfused with collagenase solution in the cold, and then sliced and incubated in enzyme solution at 37°C. The total yield of intact cells by this procedure is relatively small, and consistently good results may be difficult to obtain (Murthy and Petering, 1969; Gallai-Hatchard and Gray, 1971; Crisp and Pogson, 1972). It therefore could be considered a major breakthrough when Berry and Friend (1969) perfused the liver with collagenase under physiological conditions (at 37°C), with resulting high yields of intact cells. By this method of collagenase perfusion it has become possible to convert the whole liver to a suspension of intact cells, i.e., both the initial yield and the cellular viability approach 100% under ideal circumstances (Seglen, 1972b). The collagenase perfusion method is vastly superior to all previous techniques, and is likely to remain the ultimate method for the preparation of rat liver cells.

B.   A Quantitative Assay of Liver Dispersion

Since the introduction of collagenase perfusion by Berry and Friend (1969), numerous modifications of the technique have appeared. However, a systematic, quantitative study of the various parameters of collagenase perfusion has been undertaken only in our laboratory (Seglen, 1972b, 1973a,b). Advantage has been taken of the fact that extensive liver swelling occurs during perfusion with collagenase, and this increase in liver volume

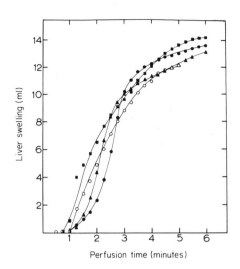

FIG. 1.   Swelling of the liver during perfusion with collagenase. The liver was first preperfused for 5–6 minutes with Ca^{2+}-free buffer or EGTA, and then perfused with collagenase
(and 5 mM Ca^{2+}) from time zero. The figure shows four different perfusions and gives an
indication of the reproducibility of this type of measurement. From Seglen (1973a).

can be quantitatively measured by recording the decrease in extrahepatic
perfusate volume in a closed perfusion circuit. Figure 1 indicates the reproducibility of these swelling measurements. The swelling of the liver is not
due to circulatory failure or swelling of the individual cells, since the portal
pressure remains low and constant, and the isolated cells have a normal
ultrastructure and water content (Seglen, 1973b). Apparently, liver swelling
is due to expansion of the extracellular space when the collagenaceous intercellular cement is dissolved and the cells move apart; the rate of swelling
therefore provides a quantitative assay of liver dispersion.

The lack of precise quantitative information makes it virtually impossible
to evaluate and compare the methodologies of other laboratories (cf. the discussion of quality criteria in Section II). Nevertheless an attempt is made
in the following to discuss the various aspects of collagenase perfusion
which, on the basis of published statements and results, appear to be significant for successful dispersion of the liver. The purification of parenchymal
cells (including the enrichment of *intact* cells in the preparation) is discussed
separately in Section VI.

C.   Collagenase

The collagenase preparations commonly employed for tissue dispersion
are rather crude extracts of *Clostridium histolyticum*, which contain many

proteolytic activities. It has been suggested that enzymes other than collagenase might be responsible for tissue dispersion (Hilfer and Brown, 1971), but experiments have shown that efficient dispersion is obtained even with highly purified collagenase (Fig. 2). The Ca^{2+} requirement for effective dispersion (Seglen, 1973a) probably reflects the Ca^{2+} requirement of collagenase (Seifter and Harper, 1970). Thus, although the commercially available highly purified collagenase used in the cited experiment may still have contained some proteolytic contaminants (Peterkofsky and Diegelmann, 1971), the overall evidence is in favor of collagenase as the active dispersing agent.

It has been the experience of many workers that different batches of purchased crude collagenase may vary in their dispersion efficiency (Garrison and Haynes, 1973; Howard *et al.*, 1973). Some manufacturers market several categories of crude collagenase with different contents of proteolytic impurities, claimed to have selective advantages in the dispersion of different tissues. In the absence of quantitative documentation such claims are hard to evaluate, but it seems surprising, e.g., that nonspecific proteases should improve the dispersion of liver tissue. In the Ca^{2+}-supplemented collagenase perfusion we employ it has not been possible to detect any difference in the dispersion efficiencies of different enzyme blends or batches, and the possibility remains that, e.g., a variable Ca^{2+} content in commercial collagenase may play a role.

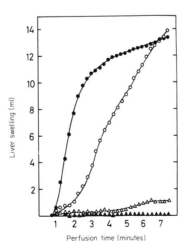

FIG. 2.   Effect of various enzymes on liver dispersion. Livers were preperfused with Ca^{2+}-free buffer for 5–6 minutes, and then perfused with the enzymes indicated added at time zero. ▲—▲, Lysozyme (Sigma grade I, 0.5 mg/ml); △—△, trypsin (Sigma type III, 0.2 mg/ml); ●—●, crude collagenase (Sigma type I, 1.0 mg/ml) plus 5 mM Ca^{2+}; ○—○, purified collagenase (Sigma type III, 0.4 mg/ml) plus 5 mM Ca^{2+}. From Seglen (1973b).

Crude collagenase has been used in concentrations ranging from 0.01 to 0.08%. By the quantitative dispersion assay, 0.05% was found to be optimal with 50 ml of perfusate (Fig. 3). The optimal concentration may possibly depend upon the total volume of recirculating perfusate in this dynamic system.

The commercially available crude collagenase dissolves incompletely in buffer, whereas purified collagenase dissolves easily, i.e., the insoluble material probably represents various contaminants. High-speed centrifugation (18,000 rpm for 30 minutes) of the crude collagenase suspension (10 × concentrated) results in the sedimentation of a dark material, and filtration of the supernatant through a Millipore filter (0.22 μm) yields a clear, yellow liquid. Such a particle-free enzyme solution was necessary for the quantitative studies performed in our laboratory (particles would have increased vascular resistance and induced nonenzymatic swelling), however, the supernatant obtained after high-speed centrifugation seems to disperse the liver just as efficiently, i.e., the cumbersome Millipore filtration (which with this material requires frequent changes of filter) can be omitted.

Hyaluronidase has traditionally been used along with collagenase for liver dispersion, but several workers have found it to be unnecessary (Ingebretsen and Wagle, 1972; Clark *et al.*, 1973; Veneziale and Lohmar, 1973; Seglen, 1973b). At high concentrations, hyaluronidase is in fact inhibitory to dispersion (Fig. 4), and it has been reported to enhance the degradation of hepatic glycogen (Wagle, 1974).

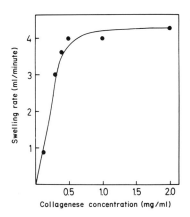

FIG. 3. Effect of collagenase concentration on liver dispersion. Livers were preperfused with Ca^{2+}-free buffer for 5–6 minutes, and then perfused with 5 mM Ca^{2+} and collagenase (Sigma type I) at the concentration indicated. The average swelling rate of the period 45 seconds to 3 minutes after enzyme addition was recorded at each concentration of collagenase. Hyaluronidase (Sigma type V, 0.075 mg/ml) was also present in this experiment. From Seglen (1973b).

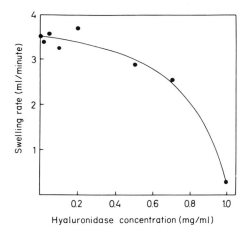

Hyaluronidase concentration (mg/ml)

FIG. 4.   Effect of hyaluronidase on liver dispersion. Livers were preperfused with Ca^{2+}-free buffer for 5–6 minutes, and then perfused with Ca^{2+} (5 mM), collagenase (1 mg/ml), and hyaluronidase (Sigma type V) at the concentrations indicated. The average swelling rate of the period 45 seconds to 3 minutes after enzyme addition was recorded at each concentration of hyaluronidase. From Seglen (1973b).

D. Ca^{2+} and the Two-Step Procedure

Ca^{2+} has been found to play a dual role in liver dispersion. Since collagenase is a Ca^{2+}-requiring enzyme, the inclusion of Ca^{2+} during perfusion with collagenase enhances enzymatic activity and accelerates dispersion. The importance of Ca^{2+} is most clearly revealed by addition of the Ca^{2+}-chelator EGTA; in this case collagenase is completely inactive in liver dispersion (Fig. 5). In a quantitative study, the optimal Ca^{2+} concentration was found to be about 5 mM (Fig. 6). Mg^{2+} inhibits dispersion (Fig. 7), probably by competition with Ca^{2+}, and should be omitted from the enzyme medium. Other ions, such as K^+, SO_4^{2-}, or PO_4^{3-}, did not affect the enzymatic reaction (Seglen, 1973b).

It would seem to be a straightforward matter to include Ca^{2+} throughout the cell preparation procedure. However, this does not work. For the enzymatic dispersion to be effective, Ca^{2+} must first be removed from the tissue by a preperfusion, and then added anew upon perfusion with collagenase (Seglen, 1972b). Figure 8 shows the different dispersion efficiencies obtained after preperfusion with a relatively small volume (50 ml) of *recirculating* Ca^{2+}-free buffer. The addition of EGTA enhances dispersion, while the addition of Ca^{2+} suppresses it. The removal of Ca^{2+} can be accomplished by preperfusion with EGTA or EDTA (Seglen, 1972b) or, provided the perfusion is

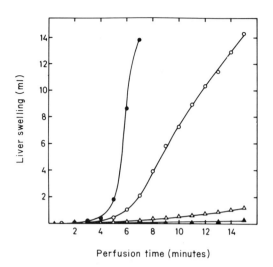

FIG. 5.   Activation of collagenase by Ca^{2+}. Livers were preperfused 6–7 minutes with Ca^{2+}-free buffer, and then perfused with collagenase in the presence or absence of Ca^{2+} or EGTA. ○—○, No Ca^{2+}; ●—●, 1 mM Ca^{2+}; △—△, 0.5 mM EGTA; ▲—▲, no enzyme and no Ca^{2+} (control). From Seglen (1972b).

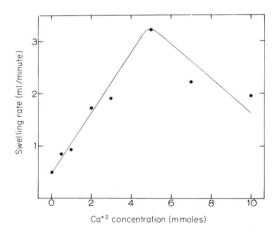

FIG. 6.   Dose-dependent activation of collagenase by Ca^{2+}. Livers were preperfused with Ca^{2+}-free buffer for 5–6 minutes, and then perfused with collagenase (0.25 mg/ml) and Ca^{2+} at the concentrations indicated. The average swelling rate of the period 45 seconds to 3 minutes after enzyme addition was recorded at each concentration of Ca^{2+}. From Seglen (1973a).

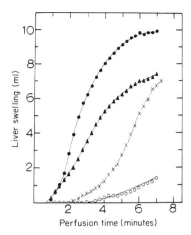

FIG. 7. Inhibition of collagenase activity by Mg^{2+}. Livers were preperfused with Ca^{2+}-free buffer for 5–6 minutes, and then perfused with collagenase from time zero together with the ions indicated. x—x, No divalent cation; ○—○, 5 mM Mg^{2+}; ●—●, 5 mM Ca^{2+}; ▲—▲, 5 mM Ca^{2+} plus 5 mM Mg^{2+}. From Seglen (1973a).

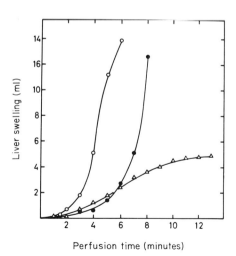

FIG. 8. Effect of preliminary Ca^{2+} removal on enzymatic dispersion. Livers were preperfused 7–8 minutes with Ca^{2+} or EGTA as indicated, and then enzyme-dispersed in the presence of Ca^{2+} (1 mM). ●—●, preperfused without Ca^{2+}; ○—○, preperfused with 0.5 mM EGTA; △—△, preperfused with 1 mM Ca^{2+}. From Seglen (1972b).

effective, simply by preperfusion with a Ca^{2+}-free buffer, thus combining the washout of blood and Ca^{2+} (Seglen, 1973a). Since the effect of Ca^{2+} removal is irreversible, it must be due to some conformational change in the intercellular matrix which favors dispersion. Modjanova and Malenkov (1972) have described a Ca^{2+}-dependent hepatic adhesion factor which is washed out of the mouse liver upon perfusion with a Ca^{2+}-free medium. It would therefore seem likely that the dispersion-promoting effect of preliminary Ca^{2+} removal is due to the detachment and washout of this adhesion factor, in particular since the divalent cation specificity is the same in both cases (see also p. 33).

The introduction of this *two-step procedure* of Ca^{2+} removal followed by addition of Ca^{2+} and collagenase has eliminated much of the variability previously encountered in collagenase perfusion. Although preperfusion with a Ca^{2+}-free buffer is a routine step in most preparative procedures, it has been regarded primarily as a washout of blood, and it was not realized that a successful liver cell preparation might be critically dependent upon the exact manner in which this preperfusion was carried out. Open, one-way flow instead of recirculation, high flow rate, large volume, and long duration of the preperfusion step (Ingebretsen and Wagle, 1972; Zahlten and Stratman, 1974) are all factors favoring dispersion.

The minimal concentration of Ca^{2+} in preperfusion capable of suppressing dispersion has not been exactly determined, but it appears to be very low, since Ca^{2+} released from blanched liver tissue into a recirculating preperfusate may suffice to suppress dispersion markedly (Seglen, 1973a). Any *addition* of Ca^{2+} at the preperfusion step inhibits dispersion; although some swelling of the liver may occur during subsequent perfusion with collagenase (Fig. 8), only damaged cells and cell clumps can be released from the tissue. Recent reports stating that "Ca^{2+} addition is detrimental" (Zahlten *et al.*, 1973; Zahlten and Stratman, 1974) are probably due to a failure to distinguish between the two steps of collagenase perfusion, since *too early* addition of Ca^{2+} is indeed detrimental.

The effect of Ca^{2+} is ion-specific; the presence of, e.g., Mg^{2+}, during preperfusion does not suppress subsequent dispersion (Fig. 9). Among other ions tested, the presence of K^+ during preperfusion was found to be advantageous (Seglen, 1973a).

Since the introduction of the two-step procedure for collagenase perfusion, several workers have adopted the practice of including Ca^{2+} along with collagenase (Berg and Boman, 1973; Barnabei *et al.*, 1974; Baur *et al.*, 1975; Christoffersen and Berg, 1974; Edwards and Elliott, 1974; Nilsson *et al.*, 1974; von Bahr *et al.*, 1974; Williams and Gunn, 1974). Cells isolated by the two-step technique have been reported to retain K^+ better, and to have a higher K^+ uptake capacity, than cells prepared without Ca^{2+}

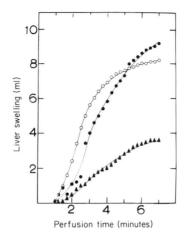

FIG. 9. Selective suppression by Ca^{2+} (and not by Mg^{2+}) of the dispersion-promoting effect of Ca^{2+}-free preperfusion. Livers were preperfused for 5–6 minutes with a buffer containing the divalent cations indicated, and then perfused with collagenase and 5 mM Ca^{2+} from time zero. ○—○, No divalent cation in the preperfusate; ●—●, 1 mM Mg^{2+} in the preperfusate; ▲—▲, 1 mM Ca^{2+} in the preperfusate. From Seglen (1973a).

addition (Barnabei et al., 1974); this has also been shown to be the case with cells prepared from liver slices incubated with collagenase in the presence of Ca^{2+} (Howard et al., 1973). Cells prepared with Ca^{2+}-activated collagenase furthermore have, on the average, a higher membrane potential than cells prepared without Ca^{2+} (Baur et al., 1975). It is not clear whether these effects on the ionic balance are due to general improvement of the cell preparation procedure or to specific effects of Ca^{2+} on the cell membrane. In liver cell suspensions, Ca^{2+} is required for gluconeogenesis from lactate (Tolbert and Fain, 1974), and may stimulate basal respiration (Howard et al., 1973; Dubinsky and Cockrell, 1974); however, its effects on respiration in the presence of substrates or uncouplers are complex.

The presence of Ca^{2+} during perfusion with collagenase has also been reported to be advantageous in the preparation of isolated cells from pancreas (Amsterdam and Jamieson, 1972) and kidney (Balzer et al., 1974), and it is very likely that the two-step collagenase perfusion technique has general applicability.

Two-step perfusion in the reverse order (first collagenase, and then EDTA) has been described (Berry and Friend, 1969; Quistorff et al., 1973), but its utility has not been convincingly documented. Since the irreversible effect of Ca^{2+} removal appears to be a prerequisite for optimal collagenase action, reversal of the order of addition does not seem to be very logical.

E. pH, Oxygenation, and Buffer Compositions

The activity of collagenase in liver dispersion exhibits a sharp optimum at pH 7.5 (Fig. 10), and the pH during collagenase perfusion should not be allowed to drop below 7.3. Since the perfused liver acidifies the perfusion medium quite rapidly (Fig. 11), a strong buffering system is required for pH maintenance even during perfusion periods of 10–15 minutes. pH maintenance can be accomplished by means of a pH-stat which infuses NaOH continuously (Seglen, 1972a,b); by means of the CO_2/bicarbonate system (continuous gassing with 5% CO_2) (Berg et al., 1972); or by means of organic buffers like HEPES at a high concentration (Seglen, 1973b). The problem of pH control during perfusion has been extensively treated previously (Seglen, 1972a). All the methods mentioned are fully acceptable; HEPES buffer being technically the simplest. The pH-stat method requires special equipment, but NaOH is a considerably cheaper chemical than HEPES. CO_2/bicarbonate has the advantage that it combines perfusate oxygenation (95% O_2, 5% CO_2) with a very high buffering capacity (better than 0.1 M HEPES, cf. Fig. 12); among its disadvantages are a tendency toward the formation of microbubbles which may clog the hepatic venules. Continuous oxygenation is of vital importance during long-term perfusion of the liver, but short periods of anoxia (up to 30 minutes) are tolerable. The respiratory activity of the perfused liver is completely restored after a period of anoxia (Fig. 13), and cells prepared under partially hypoxic conditions are fully

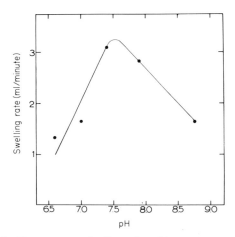

FIG. 10. Effect of pH on enzymatic dispersion. Livers were preperfused with Ca^{2+}-free buffer (pH 7.4) for 5–6 minutes, and then perfused with collagenase and Ca^{2+} (5 mM) at the pH indicated. The average swelling rate of the period 45 seconds to 3 minutes after enzyme addition was recorded at each pH value. From Seglen (1973a).

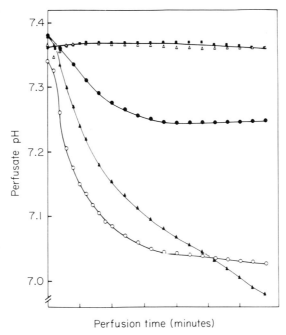

FIG. 11. Fall in pH during perfusion of the isolated liver with various buffers. Livers were perfused (without enzymatic dispersion) with a balanced salt solution buffered with either 10 mM HEPES (○), 40 mM HEPES (●), 10 mM Tricine (▲), 25 mM HCO$_3^-$–5% CO$_2$ (△), or 25 mM HCO$_3^-$–5% CO$_2$ plus 10 mM HEPES (■). From Seglen (1972a) with permission.

viable (Table I). The short time period needed for perfusion with collagenase in the two-step procedure makes it possible to omit oxygenation altogether, hence the equipment needed for gassing of the medium (Seglen, 1973b). In this case a pH-stat or an organic buffer like HEPES at high concentration must be used, since bicarbonate is effective only in equilibrium with 5% CO$_2$. Such a simplified procedure has been used successfully in the preparation of sterile liver cells (Seglen, 1973b, 1974c).

It should be pointed out that hepatic anoxia has important metabolic consequences (Seglen, 1973d, 1974a), but the effects are—as far as we know—not immediately harmful to the cells. To what extent anoxic tolerance depends upon metabolic conditions, e.g., the presence in liver cells of a glycolysable substrate such as glycogen, has not been systematically investigated.

F.   Perfusate Flow

The ability of the perfused liver to tolerate anoxia should make it possible to alter another methodological parameter: the perfusate flow. A flow rate

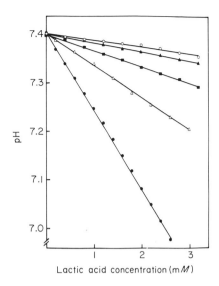

FIG. 12.   Buffering capacities of HEPES and the $HCO_3^- -CO_2$ system. Balanced salt solutions buffered as indicated were continuously gassed with oxygen (HEPES buffers) or 5% CO_2 (bicarbonate buffer) and titrated stepwise with lactic acid (0.2 M solution). ○—○, 25 mM $HCO_3^- -5\%$ CO_2; ●—●, 10 mM HEPES; △—△, 25 mM HEPES; ■—■, 50 mM HEPES; ▲—▲, 100 mM HEPES. From Seglen (1972a) with permission.

of 40–50 ml per minute with oxygen saturated perfusate is necessary for maximal oxygenation of an 8- to 10-gm liver (300-gm rat) when an erythrocyte-free medium is used (Fig. 14). This high flow rate necessitates—for economic reasons—recirculation of the collagenase medium. However,

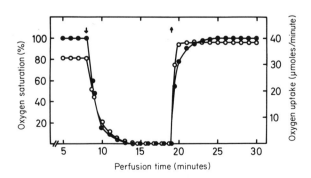

FIG. 13.   Effect of anoxia on hepatic respiration. An isolated rat liver was perfused without enzymatic dispersion, and a cycle of anoxia and reoxygenation was initiated by switching the gassing from oxygen to nitrogen (first arrow), then back to oxygen after 11 minutes (second arrow). ●—●, oxygen saturation; ○—○, oxygen uptake. From Seglen (1973b).

TABLE I

EFFECT OF PERFUSION HYPOXIA AND CENTRIFUGATION ON THE PREPARATIVE
CHARACTERISTICS OF RAT LIVER CELL SUSPENSIONS[a,b]

	Yield (% of liver weight recovered as suspended cells)	Viability (% of viable parenchymal cells)	Nonparenchymal cells (% of total cell number)
Initial suspension			
High centrifugation force			
Normal perfusion	88.4 ± 2.0 (5)	92.3 ± 1.2 (5)	15.4 ± 1.5 (5)
Hypoxic perfusion	85.8 ± 1.2 (4)	92.9 ± 0.9 (4)	16.3 ± 3.7 (4)
Low centrifugation force	83.8 ± 1.3 (5)	94.9 ± 0.5 (5)	10.7 ± 0.4 (5)
Final suspension (purified parenchymal cells)			
High centrifugation force			
Normal perfusion	55.3 ± 4.7 (5)	94.4 ± 0.6 (5)	5.0 ± 1.0 (5)
Hypoxic perfusion	55.1 ± 2.3 (3)	94.7 ± 0.6 (3)	5.1 ± 0.5 (3)
Low centrifugation force	36.2 ± 1.6 (5)	93.2 ± 0.6 (5)	1.4 ± 0.2 (5)

[a] From Seglen (1973b).

[b] Parenchymal cells were purified by differential centrifugation at high (sedimented four times at 500 rpm for 3 minutes) or low (sedimented four times at 200 rpm for 5 minutes) centrifugation force. "Hypoxic perfusion" signifies that continuous oxygenation was omitted during perfusion with collagenase.

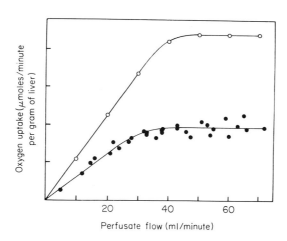

FIG. 14. Dependence of respiration on perfusate flow. Isolated rat livers were perfused without enzymatic dispersion at pH 7.4 (○) or pH 7.0 (●) at various flow rates. From Seglen (1972a) with permission.

since the liver can tolerate hypoxia, it is possible to lower the flow rate to, e.g., 5 ml per minute and use a technically simple open-flow (nonrecirculating) perfusion technique. This has been done by Capuzzi et al. (1971, 1974) and Quistorff et al. (1973), and the results were reported to be satisfactory. In our experience, it is difficult to achieve uniform perfusion of the tissue at such a low flow rate, and intermittent short pulses of rapid flow work better in this respect. The beneficial effects of erythrocytes at low flow rates (Quistorff et al., 1973) may be related to the increase in perfusate viscosity, which results in a higher perfusion pressure (higher vascular resistance), hence a more uniform penetration of the collagenase medium into the tissue (Schimassek, 1968). Albumin, which is routinely included in the perfusion buffers by many workers, may have a similar effect. The inclusion of albumin at suboptimal flow rates has been reported to induce a more uniformly distributed oxygen uptake in the perfused liver (Höller and Breuer, 1974). However, at least when extracorporeal liver perfusion (i.e., perfusion performed after removal of the liver from the animal) is used, it is a simple matter to establish recirculation, and the high flow rate that can thereby be realized makes it relatively easy to obtain consistently good results.

At a flow rate of 50 ml per minute the perfusion pressure is normally 10–15 cm H_2O; this low pressure is maintained even during perfusion with collagenase. The flow rate, hence the portal pressure, must not be excessively high; any distension of the liver caused by high pressure may result in the disruption of cells. The danger of this may be particularly acute following Ca^{2+} removal; indeed some of the old "chelation" techniques produce high yields of isolated, damaged cells in this way.

It must be stressed that a uniform perfusate flow, which insures that the collagenase reaches all cells in the tissue, is extremely important for a successful liver cell preparation. All factors that may impede the flow must therefore be avoided. These include, e.g., the *clogging* of hepatic venules by solid particles or microbubbles (Kessler and Schubotz, 1968). The latter may easily form during bubbling with 5% CO_2, upon heating of gas-saturated buffers, or at widenings or irregularities in the perfusion circuit. A good bubble-trap and a filter (e.g., a cottonwool plug) may reduce these problems.

Vascular collapse may occur if rats are subjected to too deep ether anesthesia, or if the portal blood flow is arrested for too long (as many occur during the operation if the cannulation of the vena porta is not immediately successful). Hepatic *vasoconstriction* may be induced by catecholamines in anxious, stressed animals, and care should be taken to assure that the rats are calm and relaxed when anesthetized. This is most easily accomplished by transferring the rat gently to, e.g., a 5-liter beaker, and then pouring ether onto a filter paper placed as cover above the opening of the beaker.

This gradual introduction of ether from above is much less frightening to the rat than to be put directly into an ether-filled vessel, or to be given an injection of anesthetic agent.

G. Liberation of Cells

After 5–10 minutes of perfusion with collagenase, the liver has swelled to more than twice its original size and consists essentially of single cells embedded in a connective and vascular tissue matrix. Upon disruption of the surrounding liver capsule, the superficial cells can be released by gentle shaking. In order to liberate the inner cells, the connective and vascular tissue must be broken up with a minimum of mechanical force. This can be accomplished with a spatula or forceps (Berry and Friend, 1969) or, more efficiently, by means of a comb with widely spaced teeth (Seglen, 1973b). Gentle homogenization or mincing with scissors, vigorous shaking of the tissue, or agitation by gas bubbling has also been used for the liberation of cells (Ingebretsen and Wagle, 1972; Nilsson *et al.*, 1973a; Quistorff *et al.*, 1973), but no critical comparison of the different methods has been made.

H. Other Factors

We have made liver cell preparations from well-fed, lightly fasted, and 3-days' starved rats with equally good results in terms of cellular viability; hence the nutritional condition of the liver donor seems to be of little importance from the methodological point of view. Animal age also seems to make little difference, provided the perfusate flow rate is adjusted to the size of the liver, although collagen content increases with age (Harkness, 1958). Preparation of isolated cells from regenerating or regenerated liver has been reported to be as easy as, or even easier than, from normal livers (Bissell *et al.*, 1973; Bonney *et al.*, 1973, 1974). However, strain differences may be important. With Wistar rats, the results are always good, but in a series of cell preparations made from the livers of a local strain of hooded rats, the cells although well dispersed, showed extensive blebbing and were easily destroyed during preincubation. It is possible that liver cells from this strain are particularly sensitive to collagenase or to a protease contaminant of the crude enzyme preparation.

The optimal method now available for the preparation of isolated liver cells (the two-step collagenase perfusion technique) works so satisfactorily that little methodological progress can be expected in terms of overall cellular yield and viability. However, a still almost unexplored area is the metabolic priming of cells during preparation. During most of the preparation procedure the cells are exposed to a temperature of 37°C, hence are metabolically active, and the content of metabolites and hormones in the perfusion and preincubation buffers may greatly influence the functional state of the final cell suspension. For example, we used perfusion hypoxia or the addition of glycogenic substrates during the 30-minute preincubation period to control the final cellular glycogen level, and Crisp and Pogson (1972) used insulin during perfusion to preserve cellular glucokinase activity. The presence of amino acids, K^+, and Ca^{2+} during perfusion and preincubation may similarly prevent the depletion of these substances, hence preserve, e.g., the maximal capacity for gluconeogenesis or glycogenesis (Barnabei et al., 1974; Cornell et al., 1974; Zahlten et al., 1974).

I.   Liver Perfusion Apparatus

The technically most difficult part of the collagenase perfusion method is successful perfusion of the liver. Since its modern introduction by Miller et al. (1951), the method of liver perfusion has been widely employed and several technically different procedures have been developed (for reviews, see Staib and Scholz, 1968; Bartosek et al., 1973). In most procedures the liver is perfused with a recirculating, constant volume of perfusate, which is maintained at 37°C and continuously oxygenated. The liver may be perfused in situ, in which case both the portal and the caval vein have to be cannulated in order to establish recirculation (Mortimore, 1961), or it can be perfused isolated from the animal, in which case cannulation of the vena porta suffices (Seglen and Jervell, 1969). The composition of the perfusate may range from whole blood to saline but, at least during the short time periods needed for collagenase perfusion, a buffered salt solution serves the purpose adequately (Seglen, 1972a). The short perfusion time also permits the omission of bile duct cannulation (Seglen, 1972a) and even of oxygenation (Seglen, 1973b).

In general, any good liver perfusion method can be used for preparation of isolated liver cells by collagenase perfusion. We use the very simple perfusion apparatus shown in Fig. 15. Its most important unit is the water-jacketed coiled tube (1), which maintains the perfusate temperature at 37°C (2×4 mm glass tube in 22 to 24 coils; coil diameter 30 mm; coil length 100 mm). A combined filter unit and bubble trap (2) is made of a short piece of silicone rubber tubing with a rubber stopper in each end; stainless-steel cannulas are used to pierce the stoppers. A cotton plug is used as a filter; the cotton is carefully packed at the "portal" end of the fluid-filled filter unit, taking care to avoid the trapping of bubbles within the cotton. The filter unit carries at one end the portal cannula (a piece of autoclavable nylon tubing connected to the stainless-steel cannula with a short piece of 1×3 mm silicone rubber tubing); at the other end the filter

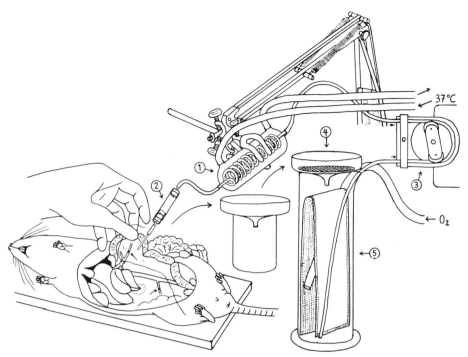

FIG. 15. Experimental arrangement for isolation and perfusion of the rat liver. 1, Water-jacketed (37°C) coiled tube fixed onto flexible lamp holder; 2, filter and bubble trap; 3, pump; 4, liver dish; 5, oxygenator cylinder and buffer reservoir. Arrows point to the sites for heparin injection (lower) and for insertion of the portal cannula (upper).

unit is connected to the coiled tube with a 7-cm piece of 2 × 4 mm silicone rubber tubing. From the other end of the coiled tube a long stretch (90 cm) of 2 × 4 mm silicone rubber tubing leads through the pump (3) to the buffer reservoir. The coiled tube is fixed with a clamp onto a flexible lamp holder. The whole coiled tube–filter–cannula assembly can thus be moved freely in all directions, which facilitates exact positioning of the portal cannula and easy transfer of the cannulated liver from the body to the liver dish. The latter is a flat glass dish (diameter 80 mm) with a conical outlet (4); the isolated liver is placed on a stainless-steel net in the bottom of the dish.

When the perfusate is recirculated, the liver dish is placed on top of the oxygenator (5), a 500-ml measuring cylinder (low form) containing a stainless-steel net shaped like an inverted V (total dimensions 6 × 40 cm), covered with a nylon net (Nytal, 61-μm mesh openings, from Schweizer Seidengazefabrik AG, Thal, Switzerland). This double oxygenator net is easily wettable, and provides sufficient surface area for complete oxygenation of the perfusate (Seglen, 1972a, 1973b). The measuring cylinder is used as a buffer reservoir, and the end of the silicone rubber tubing leading through the pump to the coiled tube can be inserted into the cylinder at the spout and fixed to the edge with autoclave tape. Oxygen (from the pressure flask through a washing bottle) is likewise led into the measuring cylinder at the spout, through silicone rubber tubing.

The pump (3) is capable of producing flow rates ranging from 10 to 70 ml per minute using a 2 × 4 mm silicone rubber tubing.

The perfusion apparatus described here can be used for several types of liver perfusion. When open, one-way (nonrecirculating) perfusion is used, the oxygenator (5) can be omitted and, if the liver is to be perfused *in situ*, the liver dish (4) can also be omitted. In the routine procedure of collagenase perfusion used in our laboratory, the liver is perfused with recirculation only for 7–10 minutes, and oxygenation during this short period can be omitted with no harmful effects, as is done during sterile preparation. However, the oxygenator net has the additional advantage of carrying the effluent perfusate back to the reservoir without any splashing or bubble formation, so we have retained it for convenience even if no oxygenation is used.

J.   Operative Technique and a Routine Procedure for Liver Cell Preparation

The operative technique we presently use is a somewhat simplified version of the technique described previously (Seglen and Jervell, 1969), a major difference being that the bile duct is not cannulated (Seglen, 1972a). Many workers cannulate the vena cava as well as the vena porta; however, cutting the vena cava at both ends permits higher perfusate efflux rates with maintenance of low pressure. For the purpose of liver cell preparation, the simple operative routine procedure described in the following is therefore at least as good as the more complicated operation procedures (e.g., Miller, 1973).

1. The rat is placed in a 5-liter beaker with filter paper in the bottom; 8 ml of ether is poured onto a four-layered filter paper covering the opening of the beaker, and a lid (flat metal plate) is placed on top. After $2-2\frac{1}{2}$ minutes (with 270-gm male Wistar rats), when the rat has just fallen asleep, it is rapidly transferred to the operating table (a stainless-steel tray), where a *very light* anesthesia is maintained by placing the opening of a 100-ml beaker, containing a filter paper with $2\frac{1}{2}$ ml of ether, in front of the rat's nose (the rat's head must not be pushed too deep into the beaker, as this will cause too deep anesthesia).

2. The rat should lie on its back on the operating table with its head to

the left (for right-handed people). The abdomen is opened by a U-shaped transverse incision, and the intestines are displaced to the left side of the abdominal cavity. The loose tissue covering the vena iliolumbalis dextra is torn apart with filter paper, and 0.25 ml of heparin (5000 IU/ml) is injected into the vein while the latter is held distally with forceps. The injection site (Fig. 15, lower arrow) is closed with a small arterial clamp to prevent bleeding.

3. A loose ligature is placed around the vena porta at the site indicated in Fig. 15. The portal cannula is then placed in its correct position adjacent to the portal vein, and the perfusate flow (Ca^{2+}-free perfusion buffer, Table II) is started at a rate of 20–30 ml per minute. The oxygenator cylinder is used as a reservoir for the perfusion buffer (500 ml), which has been preheated to 37°C and saturated with oxygen.

4. The intestinal segment containing the vena porta is held firmly in the left hand, with the thumb covering the distal end of the vena porta and also fixing one of the thread ends of the ligature. With fine scissors in the right hand a deep cut is made in the lower vena cava (for perfusate efflux), and another cut is then rapidly made in the vena porta at the site indicated in Fig. 15 (upper arrow), cutting halfway through the vein. The scissors are dropped, and with the right hand the portal cannula is rapidly inserted into the vein and pushed until the orifice of the cannula is 2 mm inside the

TABLE II

COMPOSITION OF BUFFERS USED FOR COLLAGENASE PERFUSION[a]

	Ca^{2+}-free perfusion buffer	Collagenase buffer	Washing buffer	Suspension buffer
NaCl	8,300	3,900	8,300	4,000
KCl	500	500	500	400
$CaCl_2 \cdot 2H_2O$	—	700	180	180
$MgCl_2 \cdot 6H_2O$	—	—	—	130
KH_2PO_4	—	—	—	150
Na_2SO_4	—	—	—	100
HEPES	2,400	24,000	2,400	7,200
TES	—	—	—	6,900
Tricine	—	—	—	6,500
1 M NaOH	5.5	66.0	5.5	52.5
Collagenase	—	500	—	—
pH	7.4	7.6	7.4	7.6

[a]Salt concentrations are given in milligrams per 1000 ml of final solution, and the concentration of NaOH (1 M) as milliliters per 1000 ml of final solution. The two strongly buffered solutions (collagenase buffer and suspension buffer) are designed to withstand continuous acidification by the liver cells, and therefore have a high initial pH.

ligature. If the cannula is placed in this position (on the outside) before cannulation, it will not slip out of its position when the right hand is removed to tie the ligature. The ligature must be tied rather tightly (the nylon cannula will not be compressed) and secured with an extra knot.

The pointed orifice of the nylon cannula should be rather blunt (cut at an angle of 65°), since a very pointed cannula will have to be pushed into the region of portal branching (hence obstructing perfusate flow to the branches) in order to make possible the tying of the ligature. The cut edges of the orifice should furthermore be smoothed with a file in order to prevent perforation of the portal wall.

5. When the ligature has been secured, the upper vena cava (between the liver and the diaphragm) is cut, and the perfusate flow is increased to 50 ml per minute. The first 2–3 minutes of preperfusion (with Ca^{2+}-free perfusion buffer) should be performed while the liver remains in situ, since in this nearly natural position there is less danger of flow interruption due to compression, bending, or twisting of the blood vessels than when the liver is being cut loose from the carcass. However, since the rat is lying on its back, and the natural intraabdominal topology (intestinal arrangement) has been disturbed, some compression may occur. The washout of blood (and Ca^{2+}) can therefore be somewhat improved by gently moving the liver lobes with the fingers (without the use of forceps, and with no pressure exerted).

6. While preperfusion with Ca^{2+}-free buffer is continued, the liver is removed from the carcass by cutting the vascular and biliary supply distally to the ligature (one cut), both ends of the vena cava as close to the liver as possible, and the thin, clear ligaments to the intestines and the abdominal wall. Care must be exercised to avoid twisting of the hepatic lobes, cutting into the liver tissue, or rupture of the liver capsule due to stretching of the ligaments. This surgical removal of the liver is not easy without training, and it is advisable to have a large volume of preperfusate. If the lobes have been twisted, "untwisting" is most easily accomplished while the liver is momentarily hanging freely from the portal cannula after its excision.

7. The liver is transferred to the liver dish (temporarily placed on top of a 250-ml beaker) and placed in a position similar to its in situ position. In the correct position there should be a rapid perfusate efflux with no swelling, and the whole liver should have a uniform, light-tan color. When most of the Ca^{2+}-free perfusion buffer in the oxygenator-reservoir has been used, the flow (pump) is momentarily stopped, the remaining buffer is rapidly poured out of the reservoir, and 50 ml of collagenase buffer (Table II) is rapidly poured onto the oxygenator net. The perfusate flow is then switched on again to its maximal rate (50 ml per minute), and the liver dish is placed

on top of the oxygenator cylinder. During the seconds it takes to move the liver dish to this new position, most of the Ca^{2+}-free perfusion buffer in the tubings and the filter unit are flushed out.

8. The liver is perfused with recirculating collagenase buffer (with or without oxygenation) at 50 ml per minute for 7–10 minutes; during this time it swells to more than twice its original size. Usually rupture of the vena porta occurs by the end of this time period, hence naturally terminating the perfusion.

9. The liver is usually flushed with suspension buffer (Table II) in order to remove the collagenase, although this does not seem to be necessary (no difference in cell quality is detectable if this step is omitted). The liver is then transferred to a square (10 × 10 cm) plastic petri dish (Sterilin) containing 75 ml of suspension buffer (cold or warm depending on the further use of the cells). While the liver is being held in the portal region with forceps and gently shaken, the cells are liberated from the connective-vascular tissue by careful raking with a stainless-steel comb (a dog comb, 3 mm between teeth). Lobes that have been poorly perfused are removed before the liberation of cells. If the perfusion has been perfect, the liver remnant will be just a white, gelatinous mass of connective and vascular tissue. Areas of poorly perfused tissue (infarcts) will remain as small brown clumps, and disruption of these, which liberate damaged cells, should be avoided.

10. The connective-vascular tissue remnant is lifted out, and the cell suspension is filtered through a coarse (250-μm mesh openings) nylon filter to remove connective tissue debris and clumps of infarctious tissue. The resulting *initial cell suspension* is a mixture of parenchymal and nonparenchymal cells, and can be used for further purification of any cell type.

A differential count of the initial cell suspension is routinely made. For this purpose 100 μl of cell suspension is diluted with 300 μl of isotonic 0.6% trypan blue and counted in a Bürker chamber (hemocytometer). Intact parenchymal cells, damaged cells, and nonparenchymal cells are counted.

The composition of the initial cell suspension is indicated in Table I. Approximately 85–90% of the liver mass is brought into suspension; the remainder is the connective-vascular tissue remnant. Virtually 100% of the parenchymal cells are in suspension, and their viability is usually 90–95%. The number of nonparenchymal cells is quite variable (10–30% by cell number), and may depend both on animal age and on the vigor of combing.

The shapes of the parenchymal cells in the initial suspension are rather irregular, and may to some extent reflect the shape of the cell *in situ* (Fig. 16A). However, many of the cells have deep constrictions which resemble

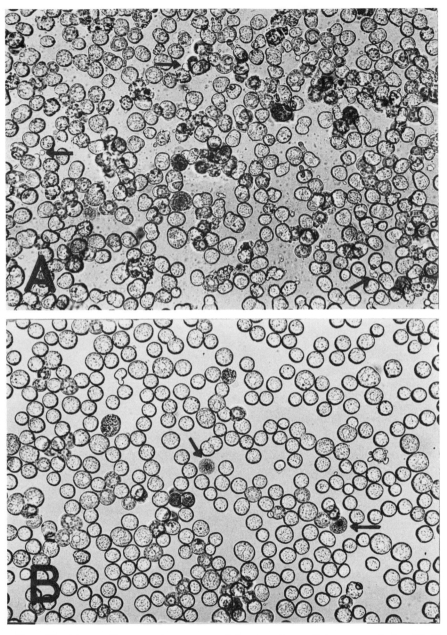

Fig. 16. Liver cell suspensions prepared by two-step collagenase perfusion (Ca^{2+} removal and Ca^{2+} readdition). (A) Initial cell suspension containing nonparenchymal cells (arrows) and parenchymal cells of somewhat irregular shape. Most of the nonparenchymal cells are out of focus and not readily seen. (B) Final cell suspension containing practically only parenchymal cells of rounded shape. A few damaged cells staining darkly with trypan blue can be seen (arrows). Cells that are out of focus may appear dark in the picture, but they have unstained nuclei and are structurally intact.

cleavage furrows, and it must be considered that these mostly polyploid cells possess a latent cleavage potential which is partially triggered when the cells are released from cell-to-cell and cell-to-matrix contacts (Seglen, 1973b). It may be difficult during differential cell counting to distinguish between single cells with deep furrows and two or three tightly apposed cells, and the individual judgment exercised here may affect the counting values quite significantly.

VI. Purification of Parenchymal Cells

A. The Goals of Purification

The initial cell suspension obtained after collagenase perfusion contains, in addition to intact parenchymal cells, variable numbers of nonparenchymal and damaged cells, some cell clumps, pieces of connective and vascular tissue, and subcellular debris. The purpose of purification is to remove all these contaminants and end up with only intact parenchymal cells. This goal can be achieved (or at least approximated) in many ways, the most commonly used procedures being combinations of filtration and differential centrifugation.

It is recommended that all purification work be done at low temperatures (0°–4°C) to minimize aggregation and to make the cells metabolically dormant, hence less sensitive to various types of stress. Low temperatures do not seem to be harmful to the cells (Table III).

TABLE III

EFFECT OF COLD PRETREATMENT ON THE GLUCONEOGENIC
CAPACITY OF ISOLATED RAT LIVER CELLS[a,b]

	Gluconeogenic rate (μmoles glucose/gm per hour)	
	Without lactate	With lactate
Preincubated at 0°C	6.1 ± 0.1 (3)	26.1 ± 1.2 (5)
Preincubated at 37°C	6.3 ± 0.1 (3)	24.2 ± 1.4 (5)

[a] Adapted from Seglen (1973b).

[b] Parenchymal rat liver cells were incubated for 60 minutes with or without lactate (23 mM) *after* a preincubation period of 20 minutes at 0° or 37°C.

B. Preincubation

Preincubation with gentle shaking of the initial cell suspension at 37°C for 30 minutes was used in the original procedure of Berry and Friend (1969) and serves several purposes:

1. In cells that have been structurally damaged, the damage is brought to completion, i.e., the cells release their soluble contents and become lighter (lower absolute weight); many of them are also fragmented. This makes it easier to separate the damaged cells from the intact ones by differential centrifugation.

2. Intact parenchymal cells, which may have a very irregular outline when freshly isolated, assume a more rounded shape during incubation (Fig. 16B). This may reduce cellular aggregation and thus make filtration easier. To some extent small aggregates may break up into single cells.

3. During preincubation, large aggregates of cells and debris may form, probably cemented together by DNA and other sticky materials released from damaged cells. Kupffer cells have a particular affinity for such aggregates, and a considerable proportion of these can therefore be selectively removed by filtration. However, many intact cells are also lost in such aggregates.

Sometimes collagenase has been included in the preincubation fluid, but we have not been able to detect any significant effect of the enzyme at this stage.

It should be remembered that liver cells are metabolically active during preincubation, and the metabolic state of the finally purified cells may be influenced by the contents of the preincubation medium. If, e.g., preservation of glycogen is desirable, glucose must be present at a high concentration during preincubation.

C. Filtration

Filtration serves to remove aggregates and tissue fragments. Both gauze (cheesecloth) and nylon stocking are suitable filter materials for the removal of large aggregates, but special nylon mesh fabrics such as Nytal (Schweizer Seidengazefabrik AG, Thal, Switzerland) offer a wider choice of exact mesh width. By using filters of successively diminishing mesh width, smaller and smaller aggregates can be removed. Such use of a series of filters prevents overloading of a single filter, which might result in excessive filtration pressure and damage to cells squeezed through the filter. To avoid any pressure above the filter, an internal cylinder of glass or plastic may be used to keep the filter pressed against the bottom and sides of a beaker. The cell suspension is slowly poured into the cylinder, making sure that the fluid does not rise higher on the inside than on the outside of the filter. The cylinder and the filter are then slowly lifted through the suspension with continuous gentle shaking. In our laboratory, a filter with 250-μm mesh openings is first used to remove connective tissue and cell clumps from the initial cell suspension. Two filters with 250- and 100-μm mesh openings are used after preincubation, and a filter with 61-μm mesh openings is used for filtration of the final cell preparation. The resulting suspension contains single cells and some aggregates of two to three cells.

D. Differential Centrifugation

Since nonparenchymal cells are much smaller, and damaged parenchymal cells much lighter (lower absolute cell weight) than intact parenchymal cells, both of the former can be removed by differential centrifugation. The cells are sedimented at a low centrifugation force (sufficient to sediment *some* of the cells, but not *all* of them) three or four times, and the supernatant is discarded each time. Subcellular debris, damaged cells, and nonparenchymal cells are thereby removed along with some intact parenchymal cells. There is inevitably an inverse relationship between the yield and the purity of the parenchymal cells. A yield of 40% of the intact parenchymal cells can be obtained by a differential centrifugation procedure which removes 95% of the nonparenchymal cells (Table I). Most of the initially damaged cells are also removed by differential centrifugation, but new damage is very easily inflicted on the cells unless great caution is exercised, and it is difficult to achieve better than 95% intact cells in the final preparation. A major problem is that the sedimented cells may form too hard-packed aggregates, particularly at high temperatures (e.g., room temperature). The mechanical force required to resuspend the cell pellets will then damage some of the cells. The use of a very low centrifugation speed and a low temperature (0°–4°C) is advantageous; the use of flat-bottomed beakers (50 ml) instead of centrifuge tubes furthermore gives a thin sediment which is very easily resuspended. Albumin or serum may help to prevent aggregation and possibly offer protection against mechanical damage through a lubricating action. The use of albumin or serum is of course practical only if these substances are to be included in the final experimental incubation, otherwise an additional series of centrifugations will be needed for their removal.

E. Gradient and Cushion Techniques

The separation of cell types according to *size* by means of rate zonal (velocity) sedimentation in a stabilizing gradient represents an analytical amplification of the differential centrifugation principle, but it is not very suitable for the preparation of large quantities of cells. Cells can also be separated according to *density* (specific weight) by isopycnic gradient centrifugation; a partial separation of parenchymal and nonparenchymal rat liver cells is in fact observed in gradients of Metrizamide (Fig. 17). There is considerable overlapping of distributions in the region of the parenchymal cells, so it is unlikely that isopycnic separation offers any advantage over differential centrifugation for the purification of parenchymal cells on a preparative scale.

However, isopycnic sedimentation methods may be useful for the separation of intact and damaged cells. During the experiments with Metrizamide gradients, it was observed that damaged cells always sedimented to the bottom of the tube (Munthe-Kaas and Seglen, 1974). When the cell membrane of a damaged cell becomes freely permeable to water, the cell—as a particle—approaches the density of its macromolecular constituents, which is much higher (e.g., protein = 1.30 gm/cm³) than the density of a water-filled intact cell (1.12 for parenchymal liver cells). By centrifuging a cell suspension above a buffered, isoosmotic 30% Metrizamide cushion (density = 1.16 gm/cm³), the damaged cells will sediment through the

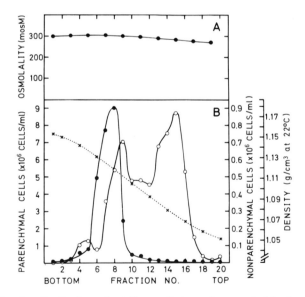

FIG. 17. Density distribution of rat liver cells in an isoosmotic Metrizamide gradient. An initial cell suspension was incorporated into an 8-ml Metrizamide gradient (density 1.05–1.16 gm/cm³) and centrifuged for 20 minutes at 5000 rpm. (A) Osmolality (B) Density (x--x), parenchymal cells (●—●), and nonparenchymal cells (○—○). Notice 10-fold scale difference. From Munthe-Kaas and Seglen (1974).

cushion, while the intact cells will remain at the interface (Fig. 18). By removing the supernatant above the intact cell layer and continuing the centrifugation, it may even be possible to sediment the subcellular debris into the Metrizamide cushion and thus retain a clean suspension of 100% intact parenchymal cells (note that nonparenchymal cells are not removed by this procedure). The same principle has been applied to purification of liver cells by flotation in dense albumin solutions (Roser, 1968; Ohuchi and Tsurufuji, 1972). Metrizamide has the unique advantage of forming solutions of high density at a low viscosity and osmolality (Munthe-Kaas and Seglen, 1974), and the single-step method outlined should be generally useful for the purification of intact cells from any cell suspension containing many damaged cells.

F. A Routine Procedure for the Purification of Parenchymal Cells

The following routine procedure is used in our laboratory for the purification of parenchymal liver cells (Seglen, 1973b).

1. The initial cell suspension (in 75 ml of *suspension buffer* (Table II), after filtration through a coarse nylon filter (250-μm mesh width), is pre-

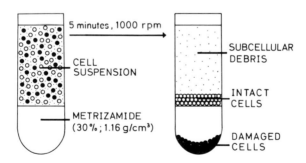

FIG. 18.   One-step separation of intact cells from damaged cells. A cell suspension con-containing a mixture of intact and damaged cells is layered above a cushion of buffered 30% Metrizamide (Nyegaard and Co. A/S, Oslo, Norway) (300 gm of Metrizamide, 2400 mg of HEPES, 500 mg of KCl, 180 mg of $CaCl_2 \cdot 2H_2O$, and 5.5 ml of 1 M NaOH per 1000 ml of solution; pH 7.6 at 37°C). The interface is smoothed with a mixing device (a thin metal thread coiled at one end into a flat spiral, bent at an angle of 90° so that the length of the thread serves as a shaft) before centrifugation to minimize tight packing of cells. After 5 minutes of centrifugation at 1000 rpm (a higher centrifugation force may be used if required), the damaged cells sediment to the bottom of the tube, while the intact cells band at the interface. By removing the supernatant and centrifuging a second time at higher speed, the subcellular debris in the intact cell layer also moves into the Metrizamide cushion, and a very clean preparation of intact cells can be obtained.

incubated at 37°C for 30 minutes in a large (20-cm diameter) petri dish with gentle, reciprocating shaking.

2. The petri dish is placed on ice and shaken for another 5 minutes to lower the temperature of the cell suspension to 0°C.

3. The suspension is filtered through a double layer consisting of two nylon filters with mesh widths of 250 and 100 μm. The suspension is then distributed equally in two 50-ml glass beakers maintained at 0°C (on ice).

4. The beakers are centrifuged in a bench-top centrifuge (Heraeus-Christ Universal Junior 1) at 200 rpm for 5 minutes. The supernatants are sucked off and discarded.

A refrigerated centrifuge would be preferable, but most commercially available centrifuges do not run at such low speeds. Maintenance of a low temperature can be improved by keeping the centrifuge buckets on ice, or by putting the whole centrifuge into a refrigerator.

5. The cells are *very carefully* resuspended by swirling (on ice) in 2 × 40 ml of ice-cold *washing buffer* (Table II). The cells are most easily resuspended in a small initial volume of buffer and then diluted.

6. The cells are sedimented again by centrifugation at 200 rpm for 5 minutes.

7. This washing procedure is repeated two times more (a total of four sedimentations).

8. The last pellet is resuspended in 25–50 ml of *suspension buffer* and filtered through a single nylon filter (61-μm mesh width).

The filtered suspension is the *final cell suspension* and is used for experiments (Fig. 16B). It usually contains about 3 gm of 90–95% intact parenchymal cells and only 1–2% nonparenchymal cells (Table I). The final cell suspension is routinely counted with trypan blue in a Bürker chamber in order to assess the quality, and samples are taken for the determination of wet weight. All experimental results are expressed on a wet-weight basis, in accordance with a recommended standard (Knox, 1972). The wet weight is determined by centrifuging four replicate samples (300 μl in tared centrifuge tubes) at 3000 rpm for 5 minutes, and then inverting the tubes; after 15–20 minutes any fluid on the tube walls is wiped off and the tubes are weighed. Such cell samples have, on a wet-weight basis largely the same composition as intact liver tissue (Table IV). The dry weight is 27% of the wet weight. Much lower dry-weight values (11%) have been reported (Crisp and Pogson, 1972; Schreiber and Schreiber, 1973); these are probably due to a high percentage of damaged cells (cf. the viability data of Müller *et al.*, 1972). The glycogen content is negligible, because the cells were prepared from lightly fasted rats further depleted of glycogen during the preparation procedure. The cellularity is lower than the value given for intact tissue (Knox, 1972), because the latter is based on nuclear counts

TABLE IV

BIOCHEMICAL CHARACTERISTICS OF PURIFIED PARENCHYMAL RAT LIVER CELLS AS
COMPARED WITH INTACT LIVER TISSUE[a]

	Parenchymal liver cells	Whole liver (Knox, 1972)	Whole liver (Shibko et al., 1967)
Dry weight (mg/gm)	272 ± 7 (8)	295 ± 15 (27)	—
DNA (mg/gm)	2.47 ± 0.16 (6)	2.51 ± 0.29 (24)	2.06 ± 0.2 (6)
RNA (mg/gm)	11.2 ± 0.4 (6)	—	11.1 ± 0.2 (6)
Protein (mg/gm)	228 ± 16 (6)	192 ± 19 (36)	195 ± 3 (6)
Glycogen (mg/gm)	0.55 ± 0.28 (11)	39.2 ± 11.3 (12)	41.2 ± 0 (6)
Lipid (mg/gm)[b]	44.6 ± 1.4 (6)	32.1 ± 3.2 (12)	34.4 ± 1.4 (6)
Cellularity (parenchymal cells/gm × 10^6)[c]	128 ± 7 (5)	194 ± 43 (16)	—
Binucleated parenchymal cells (%)	20.2 ± 1.5 (3)	—	—

[a] From Seglen (1973b).

[b] Knox (1972) gives the amount of phospholipids.

[c] Knox (1972) gives the nuclearity of the tissue, which includes both binucleated parenchymal cells and nonparenchymal cells.

which include all the small nonparenchymal cells and score each binucleated cell as two cells. Correcting for this, the two values become similar (Seglen, 1973b).

VII.   Incubation of Cell Suspensions

The final preparation of purified parenchymal cells can be incubated in several types of vessel. For short-term experiments (1–2 hours) we prefer disposable glass centrifuge tubes. These are very convenient for further biochemical analysis, because acid precipitation and metabolite extraction can be performed directly, without transferring the contents of the incubation vessel to a second tube. Adequate oxygenation of the cells in such tubes requires a small volume (400 μl) and rather vigorous gyrorotatory shaking (215 rpm in a Metabolyte water bath shaker from New Brunswick). Approximately 20–30 mg of cells are incubated in each tube (300 μl of cell suspension plus 100 μl of various additives). Unfortunately the cells have a tendency to adhere to the tube wall at the air–liquid interface, and the formation of increasingly large aggregates, in which slow deterioration of the cells occurs, eventually limits the practically useful incubation time to approximately 2 hours.

For long-term incubations of several hours' duration, flat-bottomed vessels (Erlenmeyer flasks or petri dishes) and gentle reciprocating shaking should be used. Under such conditions hepatocytes have been shown to be structurally intact for at least 7 hours (Berg et al., 1972). During such long incubations, cells tend to form small biological aggregates, as an attempt toward organization into a tissue.

Albumin or serum is often included in incubation mixtures and may, among other things, serve to protect the cells against mechanical stress. However, with gentle shaking in plastic petri dishes as previously described, the cellular viability is maintained equally well in the presence or absence of albumin.

A variety of buffering systems (bicarbonate and CO_2, phosphate buffers, or organic buffers) has been used for the incubation of hepatocytes. Since phosphate buffers are relatively weak, and the maintenance of a constant atmosphere of 5% CO_2 may be inconvenient (particularly if samples are frequently taken), organic buffers seem advantageous. We have used a mixture of organic buffers (HEPES, TES, and Tricine) with a high buffering capacity; the use of several buffers is a precaution taken to avoid the possible toxic effects of prolonged exposure to a very high concentration of a single buffering substance. An initial pH of 7.6 is chosen, since hepatic metabolism seems to be more active at this slightly superphysiological pH (Seglen, 1972a), and because it then takes a long time before pH falls to intolerably low values.

When organic buffers are used, it must of course be considered that a bicarbonate-free medium may support a metabolic pattern different from that of a bicarbonate-buffered medium, to the extent that bicarbonate is involved in certain metabolic pathways (Berry et al., 1974; Nilsson et al., 1974; Rognstad and Clark, 1974b). However, this is part of the more general problem of cellular nutrition and metabolism, and the choice of medium composition depends on the type of experiment one plans to do.

VIII. Properties of Isolated Parenchymal Cells

Much published work on the properties of isolated parenchymal cells exists, but most of the early work was done with dead and damaged cells. In the following brief review, only work with intact parenchymal cells prepared by collagenase incubation (Howard et al., 1967) or collagenase perfusion (Berry and Friend, 1969) is considered. For references on isolated liver cells up to 1973, a recent review by Schreiber and Schreiber (1973) is recommended.

A. Ultrastructure

Isolated parenchymal liver cells have a completely normal internal ultrastructure (Fig. 19A), i.e., the same structure as in organized tissue (Howard *et al.*, 1967; Berry and Friend, 1969; Capuzzi *et al.*, 1971; Gallai-Hatchard and Gray, 1971; Weigand *et al.*, 1971; Lipson *et al.*, 1972; Müller *et al.*, 1972; Moldéus *et al.*, 1974; Zahlten and Stratman, 1974). Their only remarkable feature is the presence of microvilli all over the cell surface; this is particularly evident in scanning electron micrographs (Fig. 19B). Specialized membrane regions (e.g., bile canaliculi) can no longer be found on the surface of the isolated cell.

B. Carbohydrate Metabolism

Isolated liver cells have a characteristically low ability to utilize glucose, and concentrations of about 20 mM glucose may be needed to observe net glycolysis or glycogen synthesis (Clark *et al.*, 1973, 1974; Seglen, 1973c,d, 1974a; Walli *et al.*, 1974; Walter *et al.*, 1974). The cells have been shown to contain a high-K_M glucokinase, whereas the presence of low-K_M hexokinases is disputed (Crisp and Pogson, 1972; Werner *et al.*, 1972; Bonney *et al.*, 1973; Quistorff *et al.*, 1973). It can be shown that isotopically labeled glucose is rapidly phosphorylated by the cells and incorporated, e.g., into lactate, CO_2, and glycogen (Baquer *et al.*, 1973; Clark *et al.*, 1973), but glucose phosphate is equally rapidly dephosphorylated as a result of the high activity of glucose-6-phosphatase. Thus there is considerable recycling of glucose, hence rapid isotopic equilibration with the glucose-phosphate pool, but no net glucose utilization (Clark *et al.*, 1973, 1974; Rognstad *et al.*, 1973, 1974). This is related to the functional role of the liver as a gluconeogenic tissue, and indeed the isolated liver cells show rapid rates of gluconeogenesis from lactate, pyruvate, fructose, and amino acids (Berry and Friend, 1969; Berry and Kun, 1972; Ingebretsen and Wagle, 1972; Ingebretsen *et al.*, 1972; Cornell *et al.*, 1973, 1974; Garrison and Haynes, 1973; Quistorff *et al.*, 1973; Tolbert *et al.*, 1973; Veneziale and Lohmar, 1973; Zahlten *et al.*, 1973, 1974; Clark *et al.*, 1974; Cornell and Filkins, 1974; Haynes *et al.*, 1974; Rognstad and Clark, 1974a,b; Seglen, 1974a; Tolbert and Fain, 1974; Zahlten and Stratman, 1974).

Isolated liver cells are relatively freely permeable to amino acids (Schreiber and Schreiber, 1972; Heldt *et al.*, 1974) and lose much of their amino acid content during preparation. This results in a low activity of the aspartate

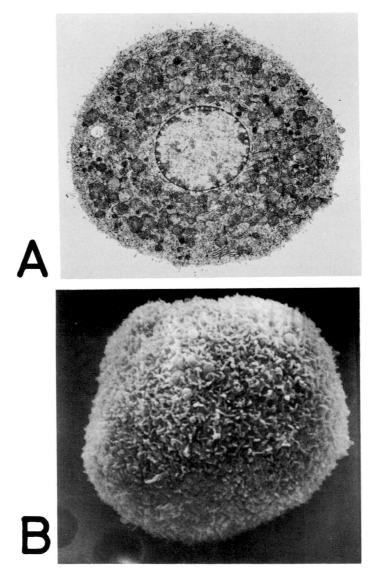

Fig. 19.   Ultrastructure of isolated parenchymal rat liver cells. (A) Transmission electron micrograph, ×4000 (courtesy of A. Reith). (B) Scanning electron micrograph. ×4000 (courtesy of D. Boman). Note uniform distribution of microvilli.

shuttle which is necessary for hydrogen transport across the mitochondrial membrane when lactate is the substrate for gluconeogenesis. Supplementation of the liver cells with certain amino acids therefore stimulates gluconeogenesis from lactate (Cornell et al., 1973, 1974; Zahlten et al., 1974). Hydrogen transport mechanisms have been extensively studied in isolated liver cells (Berry, 1971a,b,c; Berry and Kun, 1972; Berry and Werner, 1973; Berry et al., 1973a, 1974; Grunnet et al., 1973; Cornell et al., 1974; Rognstad and Clark, 1974a,b; Werner and Berry, 1974; Zahlten et al., 1974). In addition to regulation by amino acids, hydrogen shuttles are affected by bicarbonate (Berry et al., 1974), NH_4^+, and Ca^{2+} (Williamson et al., 1974; Zahlten et al., 1974).

Gluconeogenesis is not subject to end-product inhibition by glucose. However, glucose stimulates glycogen synthesis by a regulatory action and can thereby direct the gluconeogenic flow into glycogen (Seglen, 1974a). Whereas very high concentrations of glucose are needed for glycogen synthesis from glucose alone (Seglen, 1973c,d, 1974a), moderate amounts strongly stimulate glycogen synthesis from fructose (Seglen, 1974a). By this regulatory effect glucose can also prevent the degradation of glycogen (Seglen, 1973d, 1974a; Wagle et al., 1973). These studies with isolated liver cells suggest that the liver regulates blood glucose not by consuming or producing glucose, but rather by directing a more-or-less constant gluconeogenic flux *either* into glucose (under hypoglycemic conditions) *or* into glycogen (under hyperglycemic conditions).

Hepatic *glycolysis* can be observed with fructose as a glycolytic substrate. Fructose enters the Embden-Meyerhof pathway at the triose phosphate step, and the lower segment of this pathway (to lactate) is not as unidirectionally gluconeogenic as the upper segment (to glucose). Glycolysis is inhibited by the end-product lactate even under conditions in which the gluconeogenic and oxidative pathways are saturated (Seglen, 1974a), and it is possible that the metabolic flux under such conditions is directed into lipid synthesis.

The glycolysis from fructose shows a strong Pasteur effect (up to a 10-fold increase under anaerobic conditions). Anaerobiosis furthermore inhibits gluconeogenesis and glycogen synthesis and abolishes the regulatory functions of glucose and lactate. The hepatic response to anoxia therefore represents a general coordination of carbohydrate metabolism in the direction of glycolysis, termed the "generalized Pasteur effect" (Seglen, 1974a).

The carbohydrate metabolism of isolated liver cells is sensitive to *glucagon*, which promotes glycogenolysis (Johnson et al., 1972; Garrison and Haynes, 1973; Seglen, 1973d; Wagle et al., 1973; Wagle and Ingebretsen, 1973), inhibits glycogen synthesis and glycolysis (Seglen, 1973d), and stimulates gluconeogenesis (Johnson *et al.*, 1972; Garrison and Haynes, 1973; Tolbert *et al.*, 1973; Veneziale and Lohmar, 1973; Zahlten *et al.*, 1973, 1974; Tolbert and Fain, 1974; Wagle, 1974; Zahlten and Stratman, 1974). Glucagon activates adenylate cyclase in plasma membranes prepared from isolated parenchymal liver cells (Christoffersen *et al.*, 1972), and effects a rapid increase in the cellular levels of 3',5'-cyclic AMP (cAMP) (Ingebretsen *et al.*, 1972; Johnson *et al.*, 1972; Garrison and Haynes, 1973; Barnabei *et al.*, 1974; Christoffersen and Berg, 1974). This accumulation of cAMP is transient, and rather high concentrations of glucagon are necessary for sustained action of the hormone (Garrison and Haynes, 1973; Seglen, 1973d).

Insulin has been reported to affect the glycogen metabolism of isolated liver cells under certain conditions (Johnson *et al.*, 1972; Seglen, 1973c, 1974e; Wagle and Ingebretsen, 1973; Wagle *et al.*, 1973; Walter *et al.*, 1974). However, some of the reported effects are nonsignificant (Johnson *et al.*, 1972; Wagle and Ingebretsen, 1973; Wagle *et al.*, 1973). We have observed an insulin effect only in cells from starved rats, in which the glycogen-synthesizing capacity was strongly depressed (Seglen, 1973c), and never in cells from lightly fasted rats tested under a variety of conditions (Seglen, 1974e). Since insulin has been shown to stimulate significantly the synthesis of glycogen in parenchymal cells after 24 hours in monolayer culture (Bissell *et al.*, 1973; Bernaert *et al.*, 1974), it is possible that the hormone action becomes apparent only under conditions in which some component in the glycogen-synthesizing system has become rate-limiting, e.g., insulin-inducible glycogen synthetase phosphatase (Seglen, 1973c).

C. Lipid Metabolism

Isolated parenchymal rat liver cells can transform free fatty acids (e.g., palmitate administered as an albumin-bound complex) or acetate to fatty acid esters (di- and triglycerides, phospholipids, and so on) or to ketones; a certain amount is also oxidized (Wright and Green, 1971; Ontko, 1972; Berry *et al.*, 1973b; Nilsson *et al.*, 1973a; Sauer and Mahadevan, 1973; Capuzzi *et al.*, 1974; Sundler *et al.*, 1974a). Cells from fed rats synthesize mainly fatty acid esters, and cells from fasted rats mainly ketones (Ontko, 1972). Some triglycerides are secreted as lipoproteins (Sundler *et al.*, 1973). Liver cells apparently cannot utilize triglycerides directly (Felts and Berry, 1971), and there is no evidence that they ever give off free fatty acids to the extracellular medium.

Isolated liver cells can also synthesize fatty acid esters (Nilsson *et al.*, 1973a, 1974; Clark *et al.*, 1974; Sundler *et al.*, 1974b) or ketones (Seglen, 1973c) from endogenous sources. When the cells are incubated with exogenous fatty acids, the utilization of endogenous substrates is diminished

(Berry *et al.*, 1973b; Nilsson *et al.*, 1973a). Certain carbohydrates and ethanol can direct the flow of exogenous fatty acids into esterification instead of into oxidation and ketogenesis in cells from fasted rats (Ontko, 1972, 1973). In the absence of exogenous fatty acids, some carbohydrates inhibit the synthesis of fatty acid esters, whereas others are stimulatory (Clark *et al.*, 1974; Nilsson *et al.*, 1974). In particular, lactate is an excellent lipogenic substrate (Clark *et al.*, 1974; Geelen, 1974), whereas glucose appears to stimulate lipogenesis by a regulatory action (Clark *et al.*, 1974), possibly analogous to its regulatory stimulation of glycogenesis (Seglen, 1974a).

The synthesis of fatty acid esters from exogenous fatty acids is inhibited by dibutyryl-cAMP (Capuzzi *et al.*, 1974). Lipolysis has not been extensively investigated, but lysosomes have been shown to play a role (Nilsson *et al.*, 1973b). On the cell surface, lipid peroxidation has been shown to take place (Högberg *et al.*, 1975).

D. Protein and Nucleic Acid Metabolism

Isolated liver cells can synthesize protein and RNA (Jezyk and Liberti, 1969; Bojar *et al.*, 1974; Seglen, 1974b) and, when incubated in a substrate-free medium, they give off amino acids and urea, both indicative of protein degradation (Rognstad and Clark, 1974b; P. O. Seglen, unpublished observations). DNA synthesis has been observed in cells isolated from regenerating liver (Hirsiger *et al.*, 1972).

The kinetics of general protein synthesis (incorporation of radioactive amino acids) in isolated cells has been studied by Schreiber and Schreiber (1972, 1973). A striking observation is that the liver cells appear to be freely permeable to amino acids (as is also the case with, e.g., glucose). Protein synthesis in isolated cells is inhibited by cycloheximide and puromycin, but not by streptomycin and chloramphenicol, as is to be expected for eukaryotic protein synthesis.

The polysomes of isolated liver cells showed increasing disassembly during incubation with amino acids in the experiments of Grant and Black (1974), but this was apparently due to the progressive deterioration of the cells (East *et al.*, 1973). In contrast, other investigators have reported linear rates of protein synthesis for 3–10 hours (Capuzzi *et al.*, 1974; May *et al.*, 1974; Wagle, 1974). In cells from diabetic rats, the rate of protein synthesis was strongly depressed (Ingebretsen *et al.*, 1972).

The synthesis of several specific proteins has been demonstrated in isolated liver cells. *Albumin* is both synthesized (Weigand *et al.*, 1971; East *et al.*, 1973) and secreted (Weigand and Otto, 1974) by the cells, as shown by immunological methods. Isolated liver cells also synthesize *fatty acid synthetase* (Burton *et al.*, 1969) at a rate that varies with the nutritional and hormonal status of the liver donor rat (Craig and Porter, 1973). It has furthermore been shown that glucocorticoid hormone can induce increased synthesis of *tryptophan oxygenase* in isolated parenchymal rat liver cells (Berg *et al.*, 1972). This induction is blocked both by cycloheximide and by actinomycin and probably represents *de novo-* synthesis of mRNA for the enzyme. Polycyclic hydrocarbons can induce *benzpyrene hydroxylase* both in parenchymal and nonparenchymal isolated liver cells (Cantrell and Bresnick, 1972). Allylisopropylacetamide (AIA) can similarly induce δ-*aminolevulinate synthetase* in isolated parenchymal cells, provided dibuty-ryl-cAMP is also added to the cells (Edwards and Elliott, 1974). Induction of tyrosine aminotransferase by glucocorticoid hormone has also been reported (Haung and Ebner, 1969), but this induction took place even in tetraphenylboron-prepared cells, which are known to be almost uniformly damaged.

In monolayer cultures of parenchymal cells, albumin synthesis, drug induction of *drug-metabolizing enzymes* (Bissell *et al.*, 1973), and glucocorticoid induction of *tyrosine aminotransferase* (Bonney *et al.*, 1974) have been demonstrated.

It was recently shown that anucleate cytoplasts can be prepared from isolated rat liver cells by a method that combines cytochalasin treatment, homogenization, and gradient centrifugation (Seglen, 1974b). These nucleus-free "minicells" are structurally intact and synthesize protein at a normal rate, but they have virtually no RNA synthesis and offer a unique material for the study of nucleo-cytoplasmic interactions in the control of protein synthesis.

E. Other Properties

The respiration (oxygen uptake) of isolated liver cells has been measured by several investigators (Howard *et al.*, 1967, 1973; Howard and Pesch, 1968; Berry and Friend, 1969; Murthy and Petering, 1969; LaBrecque *et al.*, 1973; Qusitorff *et al.*, 1973; Dubinsky and Cockrell, 1974; Seglen, 1974a; Wilson *et al.*, 1974); respiratory rates were found to be in the range 1–3 μmoles oxygen/gm per minute. However, since even completely damaged cells may respire efficiently (Murthy and Petering, 1969; Howard *et al.*, 1973), the absolute respiratory rates are of limited information value. In general, isolated liver cells behave like isolated mitochondria with respect to respiratory control (Dubinsky and Cockrell, 1974; Wilson *et al.*, 1974).

Drug-metabolizing enzymes are present and drug-inducible in isolated rat liver cells (Cantrell and Bresnick, 1972), and such cells can metabolize a variety of drugs (Holtzman *et al.*, 1972; Moldéus *et al.*, 1973, 1974; von Bahr *et al.*, 1974). Isolated liver cells have also been used as a test system for drug hepatotoxicity (Zimmerman *et al.*, 1974).

Isolated liver cells have been used for the isolation of plasma membranes (Solyom *et al.*, 1972), for the study of surface proteins (Evans, 1974), and for the investigation of cortisol uptake (Rao *et al.*, 1974). They have also been used for enzymatic analysis of single cells (Wudl and Paigen, 1974), for isoenzyme analyses (Crisp and Pogson, 1972; Kaneko *et al.*, 1972; Berg and Blix, 1973; Bonney *et al.*, 1973), and for measurements of free-radical concentrations in *carcinogenesis* and regeneration by electron spin resonance spectroscopy (Abe and Kurata, 1974)). Parenchymal cells isolated from livers at an early stage in hepatocarcinogenesis were found to have increased adenylate cyclase responsivity toward epinephrine (Christoffersen and Berg, 1974). A comparison has been made of the carbohydrate metabolism in normal parenchymal liver cells and cultured hepatoma cells (Seglen, 1974d), and an attempt has been made to liberate intact cells from hepatoma tissue (Müller *et al.*, 1972).

IX.   *In Vitro*-Culture of Parenchymal Liver Cells

The cultivation of adult parenchymal liver cells *in vitro* has traditionally been regarded as a difficult problem, for two major reasons: (1) the lack of methods for the preparation of *intact* cells in significant quantities, and (2) the inability of liver cells to proliferate. The first problem is a technical one which has now been solved. The second problem is a biological one: The adult liver is normally a nonproliferating tissue with cellular life-spans of several hundred days (Grisham, 1973), so one would not a priori expect rapid growth of liver cells *in vitro*.

Nevertheless, rapidly growing liver cell lines have been established from fetal (Leffert and Paul, 1972), young (Williams *et al.*, 1971; Diamond *et al.*, 1973; Bauscher and Schaeffer, 1974), and adult (Coon, 1968; Casanello and Gerschenson, 1970; Schwartz, 1974) rat liver. These cells may have some functions in common with normal, adult rat liver parenchyma (e.g., glucocorticoid-inducible tyrosine aminotransferase, Gerschenson *et al.*, 1970), but their histiotypic origin must still be regarded as uncertain. In several cases such cells have been shown to undergo spontaneous malignant transformation in culture (Katsuta *et al.*, 1965; Oshiro *et al.*, 1972; Borek, 1972). They may possibly represent a growth-capable minority fraction of the liver cells able to survive the destructive procedures previously used for liver cell preparation.

Since the introduction of novel collagenase methods for the preparation of intact cells (Howard *et al.*, 1967; Berry and Friend, 1969), several laboratories have succeeded in establishing mass cultures of nongrowing adult parenchymal rat liver cells (Iype, 1971; Alwen and Gallai-Hatchard, 1972; Bissell et al., 1973; Chapman et al., 1973; Pickart and Thaler, 1973; Alwen and Lawn, 1974; Bonney et al., 1974; Rubenstein et al., 1974; Seglen, 1974c; Wanson et al., 1974; Williams and Gunn, 1974).

With the simple collagenase perfusion technique described in Section V,D, the preparation of sterile cell suspensions is usually no problem. All parts of the perfusion apparatus can be autoclaved, as well as the filters and beakers. The buffers can likewise be autoclaved (except the collagenase buffer) or filter-sterilized. If oxygenated buffers are desired, they can be pregassed and filter-sterilized under oxygen pressure. However, it must be cautioned that, when pregassed, sterile perfusion buffers are heated to 37°C, the stoppers must be loosened and the bottles shaken in order to prevent the formation of microbubbles which will clog the capillaries. Oxygenation of the buffers immediately before and during perfusion with sterile gassing equipment is of course also possible, but we generally prefer to omit oxygenation during sterile preparation.

During the operation, care must be taken that the stomach or intestines are not disrupted, as this will cause heavy bacterial contamination. Contamination by rat hair must also be avoided, e.g., by shaving the belly of the animal before the operation. Antiobiotics may be included during the 30-minute preincubation and should also be present in the primary culture dishes. Beakers must of course be covered (e.g., with aluminium foil) during centrifugation. Sterile tubes with stoppers are more convenient to use, but resuspension of cells becomes more difficult, and generally more cell damage occurs.

The behavior of the cells in culture appears to be the same in all cases; they attach readily to the tissue culture dish and flatten out. In 24 hours they assemble to form trabecular aggregates (Fig. 20) and monolayer sheets, with development of bile canaliculi between cells (Chapman et al., 1973; Alwen and Lawn, 1974; Wanson et al., 1974). During the next few days, cells in sparse cultures show progressive spreading of the cytoplasm into a thin sheet, and the cell outlines become less polygonal, sometimes fibroblast-like (Williams and Gunn, 1974; P. O. Seglen, unpublished observations). In dense monolayers the initial "epithelial" morphology is better maintained (Bissell et al., 1973; Bonney et al., 1974).

Optimal conditions (hormones, nutrients, and so on) for long-term maintenance of liver functions have not yet been defined, and in 1–2 weeks most of the cells show progressive structural deterioration and eventually die. However, several islands of small, proliferating cells with an epithelial morphology appear (Bonney et al., 1974; Seglen, 1974c; Williams and Gunn, 1974), and these cells can be subcultured and maintained as permanent lines (Iype, 1974; Williams and Gunn, 1974). They look like the rapidly growing

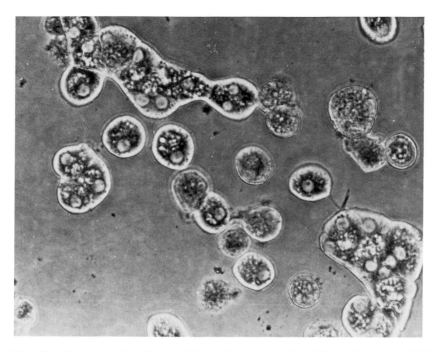

FIG. 20. Parenchymal rat liver cells in early primary culture. The cells were seeded in plastic tissue culture dishes (Falcon) with Dulbecco's modified Eagle's medium and 2.5% fetal calf serum and allowed to attach overnight. After 20 hours most of the cells were attached and flattened, and some of them aggregated into trabecular structures.

cell lines previously established from rat livers. It would be of great interest to elucidate the properties and histiotypic origin of these rapidly growing epithelial liver cells, in particular their relationship to those cultured *in vitro* from rapidly growing, preneoplastic hepatic nodules (Slifkin *et al.*, 1970).

X. Preparation of Nonparenchymal Liver Cells

The major method for preparation of nonparenchymal liver cells is based on the selective sensitivity of parenchymal cells toward proteases. By incubation of rat liver minces with Pronase (a mixture of *Streptomyces* proteases), most of the liver parenchyma is digested, while nonparenchymal cells remain intact and can be recovered from the incubate (Roser, 1968;

Mills and Zucker-Franklin, 1969). Similar results have been reported with trypsin digestion of collagenase-dispersed liver minces (Pisano *et al.*, 1968; Lentz and DiLuzio, 1971), but Pronase appears to be more effective. The most common procedure is to perfuse the liver briefly with Pronase (to achieve good enzyme penetration) before it is minced and incubated with the enzyme. Such direct Pronase methods have been used by several investigators with yields of nonparenchymal liver cells reported to be in the range 2–15 × 10⁶ cells/gm liver (Pisano *et al.*, 1968; Roser, 1968; Mills and Zucker-Franklin, 1969; Lentz and DiLuzio, 1971; Linder *et al.*, 1971; Van Wyk *et al.*, 1971; Bissell *et al.*, 1972; Ohuchi and Tsurufuji, 1972; Van Berkel *et al.*, 1972; Fujiwara *et al.*, 1973; Knook *et al.*, 1974).

Liver cell suspensions prepared by collagenase perfusion are an ideal starting material for Pronase digestion (Cantrell and Bresnick, 1972), and it was shown by Berg and Boman (1973) that yields of up to 30 × 10⁶ cells/gm liver could be obtained by incubating cell suspensions with Pronase for 60 minutes at 37°C (Fig. 21). Optimal results are obtained with 0.1% Pronase (Calbiochem) (pH 7.5) and approximately 2.5 × 10⁶ parenchymal cells/ml, incubated with gentle shaking (Munthe-Kaas *et al.*, 1975). All parenchymal cells are destroyed within 1 hour under these conditions, and nonparenchymal cells can be separated from the debris by repeated washing and centrifugation. Several investigations on such cells have been published by Berg

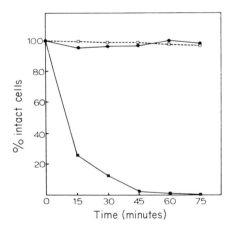

FIG. 21. Selective destruction of parenchymal cells by Pronase. Twenty-five milliliter portions of cell suspension (4 × 10⁶ cells/ml) were incubated with shaking at 37°C in the presence or absence of 0.25% Pronase (Calbiochem). Open squares, intact parenchymal cells (percent initial value) in the absence of Pronase; Solid squares, percent intact parenchymal cells in the presence of Pronase; Solid circles, percent intact nonparenchymal cells in the presence or absence of Pronase. From Berg and Boman (1973) with permission.

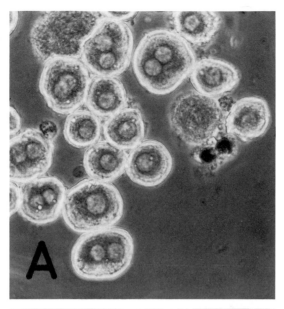

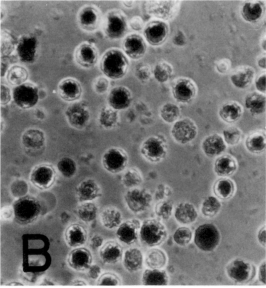

FIG. 22. Nonparenchymal cells purified by Pronase digestion. Isolated liver cells were prepared from a rat preinjected with India ink in order to fill the Kupffer cells with carbon particles. (A) Initial cell suspension with mostly parenchymal cells. Two adhering Kupffer cells filled with black carbon particles can be seen. (B) Nonparenchymal cells after Pronase digestion and washing. No parenchymal cells are present. About 40% of the cells contain carbon particles and are presumably Kupffer cells. ×400. From Munthe-Kaas *et al.* (1975) with permission.

and his colleagues (Berg and Boman, 1973; Berg and Blix, 1973; Arborgh *et al.*, 1973, 1974; Berg and Christoffersen, 1974; Munthe-Kaas *et al.*, 1975).

The cells can also be purified by taking advantage of their low density as compared to damaged cells and debris, e.g., by flotation in a dense albumin solution (Roser, 1968) or sedimentation above a dense cushion of Metrizamide (Section VI,E).

After starting with an initial cell suspension prepared as described in Section V,D, the suspension can simply be diluted with the appropriate amount of washing buffer (Table II) to which is added powdered Pronase. Alternatively, all the supernatants and used filters obtained during parenchymal cell purification can be pooled and incubated with Pronase.

The suspension of pure nonparenchymal cells obtained after washing of the Pronase digest contains about 40% Kupffer cells, identified by their content of phagocytized carbon particles (Fig. 22 and Table V). The remaining 60%, which look like small lymphocytes or plasma cells, are probably mostly of endothelial origin. Further purification of Kupffer cells can be accomplished by means of cell culture techniques. Kupffer cells attach selectively to plastic tissue culture dishes (Linbro or Falcon), and 100% pure cultures of these cells can thus be obtained, with a recovery of about one-third of the Kupffer cells present in the initial cell suspension (Munthe-Kaas *et al.*, 1975).

TABLE V

CELLULAR COMPOSITION AND LYSOSOMAL ENZYME ACTIVITIES OF PARENCHYMAL AND NONPARENCHYMAL CELL PREPARATIONS[a]

Type of cell preparation	Initial cell suspension	Purified parenchymal cells	Purified nonparenchymal cells	Kupffer cell cultures
Cell type (% of total cell number)				
Parenchymal cells	70	98	0	0
Kupffer cells	12	1	40	100
Other cell types	18	1	60	0
Protein content (pg/cell)	—	1918	—	154
Lysosomal enzyme activities (μ moles/minute per mg protein)				
β-Glucuronidase	—	4.1	—	22.8
Acid deoxyribonuclease	—	4.8	—	25.1
Cathepsin D	—	0.9	—	9.0

[a] Table adapted from Munthe-Kaas *et al.* (1975) with permission.

Pure Kupffer cells have 5- to 10-fold higher specific activity of lysosomal enzymes than parenchymal cells (Table V). They can furthermore be shown to exhibit several macrophage-specific properties in culture, such as phagocytosis and surface receptors (Munthe-Kaas *et al.*, 1975).

Nonparenchymal cells can also be purified from liver cell suspensions without the use of Pronase. As shown by the density distribution patterns in Metrizamide gradients (Fig. 17), most nonparenchymal cells are lighter than most parenchymal cells. Purified nonparenchymal cells can therefore be prepared by centrifuging the cell suspension above a Metrizamide cushion with a density that retains the nonparenchymal cells at the interface

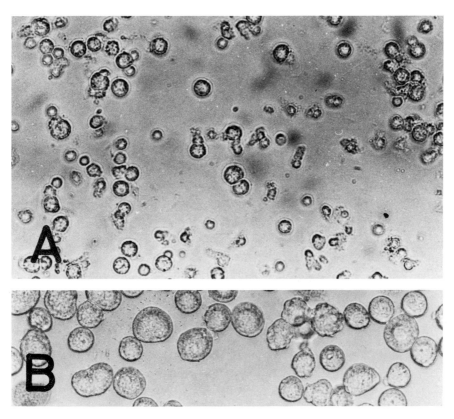

Fig. 23. Nonparenchymal cells purified by discontinuous Metrizamide gradient centrifugation. An initial cell suspension was centrifuged above a 15% buffered Metrizamide cushion (density = 1.08 gm/cm³) for 60 minutes at 3500 rpm. (A) Purified nonparenchymal cells recovered from the Metrizamide interface. (B) Purified parenchymal cells shown at the same magnification for comparison.

while the parenchymal cells sediment through (Fig. 23). However, since there is considerable overlapping in the densities of the two cell categories, the discriminating density has to be as low as 1.08, which gives rather low recoveries of nonparenchymal cells. The density cushion method is therefore inferior to the Pronase method in terms of cellular yield, but if there is reason to suspect that Pronase treatment may disturb some cellular functions under investigation (e. g., surface receptors), density cushion purification may be the method of choice.

Addendum

Several authors have recently published review articles about their methodologies for liver cell preparation or cultivation (Berry, 1974; Gerschenson et al., 1974; Krebs et al., 1974; Lentz and Di Luzio, 1974; Wagle and Ingebretsen, 1975).

ACKNOWLEDGMENTS

I wish to express my sincere thanks to Reidar Oftebro for his continued interest in my work, and to my colleagues Amy Munthe-Kaas and Trond Berg for helpful discussions and experimental data. The electron micrographs were kindly provided by Dag Boman and Albrecht Reith. The work was supported by grants from The Norwegian Cancer Society.

REFERENCES

Abe, H., and Kurata, Y. (1974). *Gann* **65**, 75–78.
Alwen, J., and Gallai-Hatchard, J. J. (1972). *J. Cell Sci.* **11**, 249–260.
Alwen, J., and Lawn, A. M. (1974). *Exp. Cell Res.* **89**, 197–205.
Amsterdam, A., and Jamieson, J. D. (1972). *Proc. Nat. Acad. Sci. U.S.* **69**, 3028–3032.
Amsterdam, A., and Jamieson, J. D. (1974). *J. Cell Biol.* **63**, 1037–1056.
Anderson, N. G. (1953). *Science* **117**, 627–628.
Arborgh, B., Berg, T., and Ericsson, J. L. E. (1973). *FEBS (Fed. Eur. Biochem. Soc.) Lett.* **35**, 51–53.
Arborgh, B., Glaumann, H., Berg, T., and Ericsson, J. L. E. (1974). *Exp. Cell Res.* **88**, 279–288.
Balzer, K., Bojar, H., Boeminghaus, H., and Staib, W. (1974). *Hoppe-Seyler's Z. Physiol. Chem.* **355**, 1172.
Baquer, N. Z., Cascales, M., Teo. B. C., and McLean, P. (1973). *Biochem. Biophys. Res. Commun.* **52**, 263–269.
Barnabei, O., Leghissa, G., and Tomasi, V. (1974). *Biochim. Biophys. Acta* **362**, 316–325.
Bartošek, I., Guaitani, A., and Miller, L. L., eds. (1973). "Isolated Liver Perfusion and its Applications." Raven Press, New York.
Baur, H., Kasperek, S., and Pfaff, E. (1975). *Hoppe-Seyler's Z. Physiol. Chem.* **356**, 827–838.
Bauscher, J., and Schaeffer, W. I. (1974). *In Vitro* **9**, 286–293.
Berg, T., and Blix, A. S. (1973). *Nature (London) New Biol.* **245**, 239–240.
Berg, T., and Boman, D. (1973). *Biochim. Biophys. Acta* **321**, 585–596.
Berg, T., and Christoffersen, T. (1974). *Biochem. Pharmacol.* **23**, 3323–3329.
Berg, T., Boman, D., and Seglen, P. O. (1972). *Exp. Cell Res.* **72**, 571–574.
Bernaert, D., Drochmans, P., Penasse, W., and Ooms, H. (1974). *Hoppe-Seyler's Z. Physiol. Chem.* **355**, 1175–1176.

Berry, M. N. (1971a). *Biochem. Biophys. Res. Commun.* **44**, 1449–1456.
Berry, M. N. (1971b). *Biochem. J.* **123**, 40P.
Berry, M. N. (1971c). *Biochem. J.* **123**, 41P.
Berry, M. N. (1974). *In* "Methods in Enzymology" (S. Fleischer and L. Packer, eds.), Vol. 32, pp. 625–632. Academic Press, New York.
Berry, M. N., and Friend, D. S. (1969). *J. Cell Biol.* **43**, 506–520.
Berry, M. N., and Kun, E. (1972). *Eur. J. Biochem.* **27**, 395–400.
Berry, M. N., and Werner, H. V. (1973). *Biochem. Soc. Trans.* **123**, 190–193.
Berry, M. N., Kun, E., and Werner, H. V. (1973a). *Eur. J. Biochem.* **33**, 407–417.
Berry, M. N., Abraham, S., and Werner, H. V. (1973b). *Biochem. Soc. Trans.* **1**, 963–965.
Berry, M. N., Werner, H. V., and Kun, E. (1974). *Biochem. J.* **140**, 355–361.
Bissell, D. M., Hammaker, L., and Schmid, R. (1972). *J. Cell Biol.* **54**, 107–119.
Bissell, D. M., Hammaker, L. E., and Meyer, U. A. (1973). *J. Cell Biol.* **59**, 722–734.
Bojar, H., Balzer, K., Basler, M., Wittliff, J. L., Staib, W., van der Meulen, N., and Sekeris, C. E. (1974). *Hoppe-Seyler's Z. Physiol. Chem.* **355**, 1180.
Bonney, R. J., Walker, P. R., and Potter, V. R. (1973). *Biochem. J.* **136**, 947–954.
Bonney, R. J., Becker, J. E., Walker, P. R., and Potter, V. R. (1974). *In Vitro* **9**, 399–413.
Borek, C. (1972). *Proc. Nat. Acad. Sci. U.S.* **69**, 956–959.
Branster, M. V., and Morton, R. K. (1957). *Nature (London)* **180**, 1283–1284.
Burton, D. N., Collins, J. M., and Porter, J. W. (1969). *J. Biol. Chem.* **244**, 1076–1083.
Cantrell, E., and Bresnick, E. (1972). *J. Cell Biol.* **52**, 316–321.
Capuzzi, D. M., Rothman, V., and Margolis, S. (1971). *Biochem. Biophys. Res. Commun.* **45**, 421–429.
Capuzzi, D. M., Rothman, V., and Margolis, S. (1974). *J. Biol. Chem.* **249**, 1286–1294.
Casanello, D. E., and Gerschenson, L. E. (1970). *Exp. Cell Res.* **59**, 283–290.
Chapman, G. S., Jones, A. L., Meyer, U. A., and Bissell, D. M. (1973). *J. Cell Biol.* **59**, 735–747.
Christoffersen, T., and Berg, T. (1974). *Biochim. Biophys. Acta* **338**, 408–417.
Christoffersen, T., Mørland, J., Osnes, J. B., Berg, T., Boman, D., and Seglen, P. O. (1972). *Arch. Biochem. Biophys.* **150**, 807–809.
Clark, D. G., Rognstad, R., and Katz, J. (1973). *Biochem. Biophys. Res. Commun.* **54**, 1141–1148.
Clark, D. G., Rognstad, R., and Katz, J. (1974). *J. Biol. Chem.* **249**, 2028–2036.
Coman, D. R. (1954). *Cancer Res.* **14**, 519–521.
Coon, H. G. (1968). *J. Cell Biol.* **39**, 29a.
Cornell, N. W., Lund, P., Hems, R., and Krebs, H. A. (1973). *Biochem. J.* **134**, 671–672.
Cornell, N. W., Lund, P., and Krebs, H. A. (1974). *Biochem. J.* **142**, 327–337.
Cornell, R. P., and Filkins, J. P. (1974). *Proc. Soc. Exp. Biol. Med.* **145**, 203–209.
Craig, M. C., and Porter, J. W. (1973). *Arch. Biochem. Biophys.* **159**, 606–614.
Crisp, D. M., and Pogson, C. I. (1972). *Biochem. J.* **126**, 1009–1023.
Daoust, R. (1958). *In* "Liver Function" (R. W. Brauer ed.), pp. 3–10. Amer. Inst. Biol. Sci., Washington, D.C.
Diamond, L., McFall, R., Tashiro, Y., and Sabatini, D. (1973). *Cancer Res.* **33**, 2627–2636.
Dubinsky, W. P., and Cockrell, R. S. (1974). *Biochem. Biophys. Res. Commun.* **56**, 415–422.
East, A. G., Louis, L. N., and Hoffenberg, R. (1973). *Exp. Cell Res.* **76**, 41–46.
Easty, G. C., and Mutolo, V. (1960). *Exp. Cell Res.* **21**, 374–385.
Edwards, A. M., and Elliott, W. H. (1974). *J. Biol. Chem.* **249**, 851–855.
Evans, W. H. (1974). *Nature (London)* **250**, 391–394.
Felts, J. M., and Berry, M. N. (1971). *Biochim. Biophys. Acta* **231**, 1–7.
Fujiwara, K., Sakai, T., Oda, T., and Igarashi, S. (1973). *Biochem. Biophys. Res. Commun.* **54**, 531–537.

Gallai-Hatchard, J. J., and Gray, G. M. (1971). *J. Cell. Sci.* **8**, 73–86.
Garrison, J. C., and Haynes, R. C. (1973). *J. Biol. Chem.* **248**, 5333–5343.
Garvey, J. S. (1961). *Nature (London)* **191**, 972–974.
Geelen, M. J. H. (1974). *Hoppe-Seyler's Z. Physiol. Chem.* **355**, 1195.
Gerschenson, L. E., Andersson, M., Molson, J., and Okigaki, T. (1970). *Science* **170**, 859–861.
Gerschenson, L. E., Berliner, J., and Davidson, M. B. (1974). *In* "Methods in Enzymology" (S. Fleischer and L. Packer, eds.), Vol. **32**, pp. 733–740. Academic Press, New York.
Gingell, D., Garrod, D. R., and Palmer, J. F. (1970). *In* "Calcium and Cellular Function" (A. W. Cuthbert, ed.), pp. 59–64. Macmillan, London.
Grant, A. G., and Black, E. G. (1974). *Eur. J. Biochem.* **47**, 397–401.
Grisham, J. W. (1973). *In* "Drugs and the Cell Cycle" (A. M. Zimmerman, G. M. Padilla, and I. L. Cameron eds.), pp. 95–136. Academic Press, New York.
Grunnet, N., Quistorff, B., and Thieden, H. I. D. (1973). *Eur. J. Biochem.* **40**, 275–282.
Harkness, R. D. (1958). *In* "Liver Function" (R. W. Brauer, ed.), pp. 59–71. Amer. Inst. Biol. Sci., Washington, D.C.
Harrison, M. F. (1953). *Nature (London)* **171**, 611.
Haung, Y. L., and Ebner, K. E. (1969). *Biochim. Biophys. Acta* **191**, 161–163.
Haynes, R. C., Garrison, J. C., and Yamazaki, R. K. (1974). *Mol. Pharmacol.* **10**, 381–388.
Heldt, H. W., Werdan, K., Walli, A. K., Birkmann, K., and Schimassek, H. (1974). *Hoppe-Seyler's Z. Physiol. Chem.* **355**, 1203.
Hilfer, S. R., and Brown, J. M. (1971). *Exp. Cell Res.* **65**, 246–249.
Hirsiger, H., Gremlich, H. P., and Schindler, R. (1972). *Experientia* **28**, 747–748.
Högberg, J., Orrenius, S., and Larson, R. E. (1975). *Eur. J. Biochem.* **50**, 595–602.
Höller, M., and Breuer, H. (1974). *Hoppe-Seyler's Z. Physiol. Chem.* **355**, 1208–1209.
Holtzman, J. L., Rothman, V., and Margolis, S. (1972). *Biochem. Pharmacol.* **21**, 581–584.
Hommes, F. A., Draisma, M. I., and Molenaar, I. (1970). *Biochim. Biophys. Acta* **222**, 361–371.
Hommes, F. A., Oudman-Richters, A. R., and Molenaar, I. (1971). *Biochim. Biophys. Acta* **244**, 191–199.
Hook, G. E. R., Dodgson, K. S., Rose, F. A., and Worwood, M. (1973). *Biochem. J.* **134**, 191–195.
Howard, R. B., and Pesch, L. A. (1968). *J. Biol. Chem.* **243**, 3105–3109.
Howard, R. B., Christensen, A. K., Gibbs, F. A., and Pesch, L. A. (1967). *J. Cell Biol.* **35**, 675–684.
Howard, R. B., Lee, J. C., and Pesch, L. A. (1973). *J. Cell Biol.* **57**, 642–658.
Ingebretsen, W. R., and Wagle, S. R. (1972). *Biochem. Biophys. Res. Commun.* **47**, 403–410.
Ingebretsen, W. R., Moxley, M. A., Allen, D. O., and Wagle, S. R. (1972). *Biochem. Biophys. Res. Commun.* **49**, 601–607.
Iype, P. T. (1971). *J. Cell. Physiol.* **78**, 281–288.
Iype, P. T. (1974). *In* "Chemical Carcinogenesis Essays" (R. Montesano and L. Tomatis eds.), pp. 119–132. Int. Ag. Res. Cancer, Lyon.
Jacob, S. T., and Bhargava, P. M. (1962). *Exp. Cell Res.* **27**, 453–467.
Jezyk, P. F., and Liberti, J. P. (1969). *Arch. Biochem. Biophys.* **134**, 442–449.
Johnson, M. E. M., Das, N. M., Butcher, F. R., and Fain, J. N. (1972). *J. Biol. Chem.* **247**, 3229–3235.
Kaltenbach, J. P. (1954). *Exp. Cell Res.* **7**, 568–571.
Kaneko, A., Dempo, K., and Onoé, T. (1972). *Biochim. Biophys. Acta* **284**, 128–135.
Katsuta, H., Takaoka, T., Doida, Y., and Kuroki, T. (1965). *Jap. J. Exp. Med.* **35**, 513–544.
Kessler, M., and Schubotz, R. (1968). *In* "Stoffwechsel der isoliert perfundierten Leber" (W. Staib and R. Scholz, eds.), pp. 12–24. Springer-Verlag, Berlin and New York.

Knook, D. L., Sleyster, E. C., and van Noord, M. J. (1975). *In* "Cell Impairment in Aging and Development" (V. J. Christofalo and E. Holečková, eds.), pp. 155–169. Plenum, New York.

Knox, W. E. (1972). "Enzyme Patterns in Fetal, Adult and Neoplastic Rat Tissues." Karger, Basel.

Krebs, H. A., Cornell, N. W., Lund, P., and Hems, R. (1974). *In* "Regulation of Hepatic Metabolism" (F. Lundquist and N. Tygstrup, eds.), pp. 726–750. Munksgaard, Copenhagen.

LaBrecque, D. R., Bachur, N. R., Peterson, J. A., and Howard, R. B. (1973). *J. Cell Physiol.* **82**, 397–400.

Laws, J. O., and Stickland, L. H. (1961). *Exp. Cell Res.* **24**, 240–254.

Leeson, T. S., and Kalant, H. (1961). *J. Biophys. Biochem. Cytol.* **10**, 95–104.

Leffert, H. L., and Paul, D. (1972). *J. Cell Biol.* **52**, 559–568.

Lentz, P. E., and Di Luzio, N. R. (1971). *Exp. Cell Res.* **67**, 17–26.

Lentz, P. E., and Di Luzio, N. R. (1974). *In* "Methods in Enzymology" (S. Fleischer and L. Packer, eds.), Vol. **32**, pp. 647–653. Academic Press, New York.

Linder, M. C., Anderson, G. H., and Ascarelli, I. (1971). *J. Biol. Chem.* **246**, 5538–5540.

Lipson, L. G., Capuzzi, D. M., and Margolis, S. (1972). *J. Cell Sci.* **10**, 167–179.

Longmuir, L. S., and ap Rees, W. (1956). *Nature* **177**, 997.

May, C., Drochmans, P., Wanson, J. C., and Penasse, W. (1974). *Hoppe-Seyler's Z. Physiol. Chem.* **355**, 1229.

Miller, L. L. (1973). *In* "Isolated Liver Perfusion and its Applications" (I. Bartošek, A. Guaitani, and L. L. Miller, eds.), pp. 11–52. Raven Press, New York.

Miller, L. L., Bly, C. G., Watson, M. L., and Bale, W. F. (1951). *J. Exp. Med.* **94**, 431–453.

Mills, D. M., and Zucker-Franklin, D. (1969). *Amer. J. Pathol.* **54**, 147–166.

Modjanova, E. A., and Malenkov, A. G. (1973). *Exp. Cell Res.* **76**, 305–314.

Moldéus, P., Grundin, R., von Bahr, C., and Orrenius, S. (1973). *Biochem. Biophys. Res. Commun.* **55**, 937–944.

Moldéus, P., Grundin, R., Vadi, H., and Orrenius, S. (1974). *Eur. J. Biochem.* **46**, 351–360.

Mortimore, G. E. (1961). *Amer. J. Physiol.* **200**, 1315–1319.

Moscona, A., Trowell, O. A., and Willmer, E. N. (1965). *In* "Cells and Tissues in Culture" (E. N. Willmer, ed.), Vol. 1, pp. 19–98. Academic Press, New York.

Müller, M., Schreiber, M., Kartenbeck, J., and Schreiber, G. (1972). *Cancer Res.* **32**, 2568–2576.

Munthe-Kaas, A. C., and Seglen, P. O. (1974). *FEBS (Fed. Eur. Biochem. Soc.) Lett.* **43**, 252–256.

Munthe-Kaas, A. C., Berg, T., Seglen, P. O., and Seljelid, R. (1975). *J. Exp. Med.* **141**, 1–10.

Murthy, L., and Petering, H. G. (1969). *Proc. Soc. Exp. Biol. Med.* **132**, 931–935.

Nilsson, Å., Sundler, R., and Åkesson, B. (1973a). *Eur. J. Biochem.* **39**, 613–620.

Nilsson, Å., Nordén, H., and Wilhelmsson, L. (1973b). *Biochim. Biophys. Acta* **296**, 593–603.

Nilsson, Å., Sundler, R., and Åkesson, B. (1974). *FEBS (Fed. Eur. Biochem. Soc.) Lett.* **45**, 282–285.

Ohuchi, K., and Tsurufuji, S. (1972). *Biochim. Biophys. Acta* **258**, 731–740.

Ontko, J. A. (1972). *J. Biol. Chem.* **247**, 1788–1800.

Ontko, J. A. (1973). *J. Lipid Res.* **14**, 78–86.

Oshiro, Y., Gerschenson, L. E., and DiPaolo, J. A. (1972). *Cancer Res.* **32**, 877–879.

Palade, G. E., and Claude, A. (1949). *J. Morphol.* **85**, 35–69.

Paul, J. (1970). "Cell and Tissue Culture," 4th ed. Livingstone, Edinburgh.

Peterkofsky, B., and Diegelmann, R. (1971). *Biochemistry* **10**, 988–993.

Pickart, L., and Thaler, M. M. (1973). *Nature (London), New Biol.* **243**, 85–87.

Pisano, J. C., Filkins, J. P., and DiLuzio, N. R. (1968). *Proc. Soc. Exp. Biol. Med.* **128**, 917–922.

Quistorff, B., Bondesen, S., and Grunnet, N. (1973). *Biochim. Biophys. Acta* **320**, 503–516.

Rao, M. L., Rao, G. S., Höller, M., and Breuer, H. (1974). *Hoppe-Seyler's Z. Physiol. Chem.* **355**, 1239–1240.

Rappaport, C., and Howze, G. B. (1966). *Proc. Soc. Exp. Biol. Med.* **121**, 1010–1016.

Rognstad, R., and Clark, D. G. (1974a). *Eur. J. Biochem.* **42**, 51–60.

Rognstad, R., and Clark, D. G. (1974b). *Arch. Biochem. Biophys.* **161**, 638–646.

Rognstad, R., Clark, D. G., and Katz, J. (1973). *Biochem. Biophys. Res. Commun.* **54**, 1149–1156.

Rognstad, R., Clark, D. G., and Katz, J. (1974). *Eur. J. Biochem.* **47**, 383–388.

Roser, B. (1968). *J. Reticuloendothel. Soc.* **5**, 455–471.

Rubenstein, D., Baker, M. R., Stott, E. J., and Tavill, A. S. (1974). *Brit. J. Exp. Pathol.* **55**, 20–25.

St. Aubin. P. M. G., and Bucher, N. L. R. (1952). *Anat. Rec.* **112**, 797–809.

Sauer, F., and Mahadevan, S. (1973). *Can. J. Biochem.* **52**, 1567–1580.

Schimassek, H. (1968). *In* "Stoffwechsel der isoliert perfundierten Leber" (W. Staib and R. Scholz eds.), pp. 1–10. Springer-Verlag, Berlin and New York.

Schneider, W. C., and Potter, V. R. (1943). *J. Biol. Chem.* **149**, 217–227.

Schreiber, G., and Schreiber, M. (1972). *J. Biol. Chem.* **247**, 6340–6346.

Schreiber, G., and Schreiber, M. (1973). *Sub-Cell. Biochem.* **2**, 321–383.

Schwartz, A. G. (1974). *Cancer Res.* **34**, 10–15.

Seglen, P. O. (1972a). *Biochim. Biophys. Acta* **264**, 398–410.

Seglen, P. O. (1972b). *Exp. Cell Res.* **74**, 450–454.

Seglen, P. O. (1973a). *Exp. Cell Res.* **76**, 25–30.

Seglen, P. O. (1973b). *Exp. Cell Res.* **82**, 391–398.

Seglen, P. O. (1973c). *FEBS (Fed. Eur. Biochem. Soc.) Lett.* **30**, 25–28.

Seglen, P. O. (1973d). *FEBS (Fed. Eur. Biochem. Soc.) Lett.* **36**, 309–312.

Seglen, P. O. (1974a). *Biochim. Biophys. Acta* **338**, 317–336.

Seglen, P. O. (1974b). *Abstr. Commun. 9th Meet. Fed. Eur. Biochem. Soc.*, *1974* p. 430.

Seglen, P. O. (1974c). *Hoppe-Seyler's Z. Physiol. Chem.* **355**, 1255.

Seglen, P. O. (1974d). *Abstr. Commun., Int. Cancer Congr., 11th, 1974* Vol. 4, p. 628.

Seglen, P. O. (1974e). *Acta Endocrinol. (Copenhagen)* **77**, Suppl. 191, 153–158.

Seglen, P. O., and Jervell, K. F. (1969). *Hoppe-Seyler's Z. Physiol. Chem.* **350**, 308–316.

Seifter, S., and Harper, E. (1970). *In* "Methods in Enzymology (G. Perlman and L. Lorand, eds.), Vol. 19, pp. 613–635. Academic Press, New York.

Shibko, S., Koivistoinen, P., Tratnyek, C. A., Newhall, A. R., and Friedman, L. (1967). *Anal. Biochem.* **19**, 514–528.

Slifkin, M., Merkow, L. P., Pardo, M., Epstein, S. M., Leighton, J., and Farber, E. (1970). *Science* **167**, 285–287.

Solyom, A., Lauter, C. J., and Trams, E. G. (1972). *Biochim. Biophys. Acta* **274**, 631–637.

Staib, W., and Scholz, R., eds. (1968). "Stoffwechsel der isoliert perfundierten Leber." Springer-Verlag, Berlin and New York.

Sundler, R., Åkesson, B., and Nilsson, Å. (1973). *Biochem. Biophys. Res. Commun.* **55**, 961–968.

Sundler, R., Åkesson, B., and Nilsson, Å. (1974a). *Biochim. Biophys. Acta* **337**, 248–254.

Sundler, R., Åkesson, B., and Nilsson, Å. (1974b). *FEBS (Fed. Eur. Biochem. Soc.) Lett.* **43**, 303–307.

Tolbert, M. E. M., and Fain, J. N. (1974). *J. Biol. Chem.* **249**, 1162–1166.

Tolbert, M. E. M., Butcher, F. R., and Fain, J. N. (1973). *J. Biol. Chem.* **248**, 5686–5692.

Van Berkel, T. J. C., Koster, J. F., and Hülsmann, W. C. (1972). *Biochim. Biophys. Acta* **276**, 425–429.

Van Wyk, C. P., Linder-Horowitz, M., and Munro, H. N. (1971). *J. Biol. Chem.* **246**, 1025–1031.

Veneziale, C. M., and Lohmar, P. H. (1973). *J. Biol. Chem.* **248**, 7786–7791.

von Bahr, C., Vadi, H., Grundin, R., Moldéus, P., and Orrenius, S. (1974). *Biochem. Biophys. Res. Commun.* **59**, 334–339.

Wagle, S. R. (1974). *Biochem. Biophys. Res. Commun.* **59**, 1366–1372.

Wagle, S. R., and Ingebretsen, W. R. (1973). *Biochem. Biophys. Res. Commun.* **52**, 125–129.

Wagle, S. R., and Ingebretsen, W. R. (1975). *In* "Methods in Enzymology" (J. M. Lowenstein ed.), Vol. 35, pp. 579–594. Academic Press, New York.

Wagle, S. R., Ingebretsen, W. R., and Sampson, L. (1973). *Biochem. Biophys. Res. Commun.* **53**, 937–943.

Walli, A. K., Birkmann, K., and Schimassek, H. (1974). *Hoppe-Seyler's Z. Physiol. Chem.* **355**, 1264–1265.

Walter, P., Fasel, P., and de Sagarra, M. R. (1974). *Hoppe-Seyler's Z. Physiol. Chem.* **355**, 1265.

Wanson, J. C., Drochmans, P., Popowski, A., and Mosselmans, R. (1974). *Hoppe-Seyler's Z. Physiol. Chem.* **355**, 1265–1266.

Weigand, K., and Otto, I. (1974). *FEBS (Fed. Eur. Biochem. Soc.) Lett.* **46**, 127–129.

Weigand, K., Müller, M., Urban, J., and Schreiber, G. (1971). *Exp. Cell Res.* **67**, 27–32.

Werner, H. V., and Berry, M. N. (1974). *Eur. J. Biochem.* **42**, 315–324.

Werner, H. V., Bartley, J. C., and Berry, M. N. (1972). *Biochem. J.* **130**, 1153–1155.

Williams, G. M., and Gunn, J. M. (1974). *Exp. Cell Res.* **89**, 139–142.

Williams, G. M., Weisburger, E. K., and Weisburger, J. H. (1971). *Exp. Cell Res.* **69**, 106–112.

Williamson, J. R., Gimpel, J. A., DeLeeuw, G. J., Hensgens, B., Meijer, A. J., and Tager, J. M. (1974). *Hoppe-Seyler's Z. Physiol. Chem.* **355**, 1270.

Wilson, D. F., Stubbs, M., Veech, R. L., Erecińska, M., and Krebs, H. A. (1974). *Biochem. J.* **140**, 57–64.

Wright, J. D., and Green, C. (1971). *Biochem. J.* **123**, 837–844.

Wudl, L., and Paigen, K. (1974). *Science* **184**, 992–994.

Zahlten, R. N., and Stratman, F. W. (1974). *Arch. Biochem. Biophys.* **163**, 600–608.

Zahlten, R. N., Stratman, F. W., and Lardy, H. A. (1973). *Proc. Nat. Acad. Sci. U.S.* **70**, 3213–3218.

Zahlten, R. N., Kneer, N. M., Stratman, F. W., and Lardy, H. A. (1974). *Arch. Biochem. Biophys.* **161**, 528–535.

Zimmerman, H. J., Kendler, J., Libber, S., and Lukacs, L. (1974). *Biochem. Pharmacol.* **23**, 2187–2189.

Chapter 5

Isolation and Subfractionation of Mammalian Sperm Heads and Tails

HAROLD I. CALVIN

*International Institute for the Study of Human Reproduction,
Columbia University College of Physicians and Surgeons,
New York, New York*

I. Introduction	85
A. The Mammalian Spermatozoon: Structural Components	85
B. Methods of Cell Lysis	88
II. Isolation of Sperm Heads and Tails	89
A. Separation of Rat Sperm Heads and Tails by Sonication and Equilibrium Density Gradient Centrifugation	89
B. Separation of Rat Sperm Heads and Tails by Enzymatic Lysis and Equilibrium Centrifugation	93
C. Procedures with Other Species	95
III. Subfractionation of the Sperm Head	95
A. Acrosome and Plasma Membrane	95
B. Nuclear Chromatin	97
C. Perforatorium	98
IV. Subfractionation of the Tail	100
A. Midpieces and Principal Pieces	100
B. Mitochondria	100
C. Keratinlike Structures	101
References	103

I. Introduction

A. The Mammalian Spermatozoon: Structural Components

The mammalian spermatozoon, the end product of spermiogenesis, is a highly differentiated cell of well-delineated architecture, whose structural adaptations clearly reflect its unusual role as a motile cell designed for the protection, transport, and delivery into the ooplasm of the male genome. The unusual variety of stable organelles in this cell suggests that it is readily amenable to subcellular fractionation. However, despite the recent growth of research in mammalian reproductive biology, biochemical analysis of the sperm has progressed relatively slowly and awaits the refinement and further establishment of techniques for the fractionation of its various components. It is the purpose of this article to review briefly the present availability of such procedures, with special attention to the problem of separating sperm heads and tails.

Most of our own methods have been developed during studies with the rat spermatozoon, whose structure is characterized by a sickle-shaped (falciform) head, widely encountered in murine rodents (Friend, 1936), and an unusually long and thick tail (Fig. 1A). A diagrammatic longitudinal cross-section through the median plane of the sperm head, adapted from Piko (1969), appears in Fig. 1B. Of special interest to those concerned with the biochemistry of fertilization is the acrosome, which surrounds the anterior and central portions of the nucleus and is known to contain a variety of hydrolytic enzymes, some of which are believed to play a role in penetrating the outer vestments of the ovum (see Allison and Hartree, 1970; McRorie and Williams, 1974). The subacrosomal structure known as the perforatorium, either absent or of modest dimensions in most mammalian species (see Fawcett, 1975), is especially prominent in rat and other rodent sperm (Austin and Bishop, 1958). Distal to these two elements, bordering the posterior and caudal regions of the nucleus, respectively, are the postacrosomal sheath and the basal plate. Further information about the topography of the rat sperm head is available in the article of Pikó (1969).

Aside from its dimensions, the rat sperm tail follows a morphological pattern that is essentially universal among both metatherian and eutherian mammals (Fawcett, 1970; Phillips, 1970; Fawcett, 1975). The nine doublets of the axonemal complex are complemented by nine dense outer fibers, which in turn are surrounded either by a sheath of mitochondria in the midpiece region or by the fibrous sheath of the principal piece (Fig. 1A). The tapering of the tail reflects the loss and decrease in diameter of these accessory elements, until only the axonemal complex and plasma membrane remain. The medial cross section shown (Fig. 1B) includes a diagram of the neck region, adapted from Woolley and Fawcett (1973), in which the axonemal and dense fibers end, the latter abutting the longitudinal columns of the connecting piece, a ribbed, fibrous structure which adheres to the basal plate of the sperm head.

A striking feature of mammalian spermatozoa is the presence of various

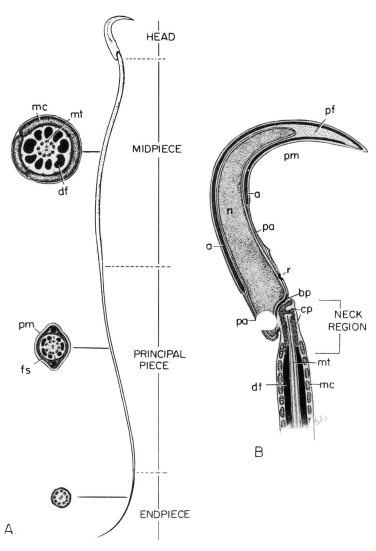

FIG. 1. The rat spermatozoon. (A) Transverse sections through the three principal divisions of the tail; redrawn from electron micrographs provided by Dr. D. M. Phillips. (B) Medial cross section of the head (redrawn from Pikó, 1969) and upper portion of the tail, including the neck region (redrawn from Woolley and Fawcett, 1973). a, acrosome; bp, basal plate; cp, connecting piece; df, dense fibers; fs, fibrous sheath; mc, mitochondrion; mt, microtubule; n, nucleus; pa, postacrosomal sheath; pf, perforatorium; pm, plasma membrane; r, posterior ring. Drawing by R. J. Demarest.

components cross-linked by $-S-S-$ bridges. Such bonding is found in both the heads and tails of eutherian spermatozoa, but is essentially confined to the tail in all marsupials so far examined (Calvin and Bedford, 1971; Bedford and Calvin, 1974a,b). The sperm head structures stabilized by disulfide linkages include the nucleus, perforatorium, and postacrosomal sheath (Calvin and Bedford, 1971; Koehler, 1973). Within the tail, the connecting piece, outer mitochondrial membranes, dense fibers, and fibrous sheath are similarly stabilized (Bedford and Calvin, 1974a). These numerous keratinoid (i.e., $-S-S-$ cross-linked) elements present a barrier to lysis of the mammalian spermatozoon and resolution of its subcellular components. This is partially compensated for, however, by the chemically specific nature of the stability the –S–S– bonding confers upon both the keratinoid cell structures and their association with other cellular components.

B.   Methods of Cell Lysis

The first step in most sperm fractionation procedures has traditionally been the separation of heads and tails. Most commonly, physical methods have been employed to sever the connection between these two parts of the cell, the long, thin tail cleaving readily in response to such treatments as sonication (Henle et al., 1938; Zittle and O'Dell, 1941; Mohri et al., 1965; Stambaugh and Buckley, 1969; Calvin et al., 1973), grinding in the presence of glass particles (Zittle and Zitin, 1942; Morton, 1968) or ice (Nelson, 1954), and shearing in a blender (Coelingh et al., 1969). A more recent variation of this approach is the use of a nitrogen pressure bomb to sever and fragment the tail (Pihlaja et al., 1973). By increasing the intensity of the disrupting agent, the tail can be further disintegrated to varying degrees and its various ultrastructural components released (Morton, 1968; Pihlaja et al., 1973; Pihlaja and Roth, 1973; Price, 1973; Baccetti et al., 1973).

The chemical approach to sperm lysis is exemplified by the studies of Edelman and Millette (1971) and Millette et al. (1973), who have reported that the head–tail linkage in mouse and rat sperm can be labilized by endopeptidases or extremes of pH and recommend brief treatment with trypsin as the least damaging method of cleaving heads from tails, terming this a "chemical dissection" of the cell. In most species, however, there is no satisfactorily mild procedure for head–tail cleavage by either chemical or physical means. Chemically induced separation of heads from tails can be achieved, with varying degrees of damage to one or both structures, by prolonged proteolysis (Miescher, 1878; Pihlaja et al., 1973), exposure to sodium dodecyl sulfate (SDS) (Calvin and Bedford, 1971) or Sarkosyl (Millette et al., 1974), successive treatment with dithiothreitol (DTT) and trypsin (Millette et al., 1973), or incubation with detergents and DTT (Hernández-Montes et al,, 1973). In the last-mentioned preparations, either the heads or the tails are selectively solubilized, depending upon the detergent used in combination with DTT.

II.   Isolation of Sperm Heads and Tails

A.   Separation of Rat Sperm Heads and Tails by Sonication and Equilibrium Density Gradient Centrifugation

1.   MODIFICATIONS OF EARLIER METHOD

A published procedure for the isolation of heads and tails in discontinuous density gradients, developed in our laboratory during studies on the $-SH$ content of rat spermatozoa (Calvin et al., 1973), has since been refined to improve the yields of both fractions and the purity of the tails. The separation of the two major structural components of the sperm depends on the great difference in their sedimentation velocities in media that approach the density of the tails. At lower densities, the tails concentrate too rapidly during centrifugation, thereby trapping many heads. Accordingly, our earlier procedure has been modified by applying the sample to the gradient in 1.80 M instead of 1.60 M sucrose. In addition, the intermediate 2.00 M layer has been fixed more precisely at 2.05 M sucrose, since the tails begin to sediment through 2.00 M sucrose during sufficiently prolonged centrifugation at high speeds.

2.   PREPARATION OF SAMPLES

All operations are performed at $0°-5°C$. Mature spermatozoa are expressed gently with a hemostat from the cauda epididymidis of 400 to 600-gm Sprague-Dawley rats into 0.02 M Na_2HPO_4 (pH 6.0) (Buffer P) and filtered through a double layer of 50-mesh nylon. The heads are cleaved from the tails by sonication until removal of the heads, as monitored with a phase contrast microscope, is > 99% complete. The precise conditions required depend on a variety of factors, including the age and type of equipment employed, and must be defined empirically by each investigator. In our laboratory, the rat sperm suspensions are divided into aliquots of 3–4 ml and sonicated for ca. 2 minutes in 12–ml conical glass tubes with a Bronwill Model II-A 20 kc sonifier, equipped with a 5/32-inch titanium microprobe, at a power setting of 65% (ca. 80 W). The tip of the probe is maintained at approximately the 1.5 ml level, 1/8 inch from the wall of the tube.

The sperm count in the suspensions is determined by applying dilutions ($1.5–2.5 \times 10^6$ cells/ml) to a hemocytometer and scoring heads. For a complete count, it is necessary to focus alternately on the plane of the hemocytometer and on the overlying surface of the cover slip, since a small percentage of heads adhere to the latter.

3. PREPARATION OF GRADIENTS AND CENTRIFUGATION

Because of the viscosity of dense sucrose solutions, they should be prepared gravimetrically when accuracy is required. The gradients to be described include three layers of sucrose at the following concentrations: 1.80, 2.05, and 2.20 M. These solutions are usually made up in Buffer P, using water which has been successively distilled, deionized, and redistilled to minimize contamination by trace metals that oxidize $-SH$ groups. In addition, 10^{-3} M EDTA may be added to the buffer for the same reason (Calvin et al., 1973), unless the metal content of the tails is of interest (see Calvin and Bleau, 1974, and Section II,A,5).

It is particularly important that the 2.05 M sucrose medium be prepared accurately, for it is at the 1.80 M–2.05 M interface that the critical separation of sperm heads and tails must take place. To prepare 1 liter of the 2.05 M sucrose solution ($d_{20}^{20} = 1.263$) gravimetrically, 40 ml of 25 × Buffer P is diluted with water to a total weight of 561.6 gm and mixed with 701.7 gm of sucrose. Similarly, 1 liter of buffered 2.20 M sucrose may be prepared from 528.8 gm of buffer and 753.1 gm of sucrose. Although these solutions are made up at room temperature, they should be maintained at $0°–5°$ C during the density gradient fractionation for reproducible results. The 1.80 M sample layers can also be prepared gravimetrically, but are usually composed with less precision by mixing buffered 2.20 M sucrose with the sperm suspension. The detached heads and tails must be distributed evenly in the sample layer for good separations. Thus it is not advisable to attempt to suspend a pellet of sonicated sperm directly into 1.80 M sucrose.

For resolution of sperm heads and tails, discontinuous gradients are prepared in 2.5 × 8.9 cm (40-ml) nitrocellulose tubes. The bottom two layers of each gradient contain 13 ml each of 2.20 M and 2.05 M sucrose, above which 13 ml of sonicated sperm in 1.80 M sucrose (at concentrations between 2 and 8×10^6/ml) is carefully applied. Following centrifugation at 22,500 rpm ($91,000 g$, r_{max}) for 60 minutes in a Beckman L3–40 ultracentrifuge equipped with swinging-bucket rotor (SW-27), the tail fraction (Fig. 2) appears as a sharp band at the 1.80 M–2.05 M interface, whereas the heads (Fig. 3) are found at the bottom of the gradient, packed in an annular pellet. Spermatozoa that have not been decapitated by sonication migrate to the 2.05 M–2.20 M boundary.

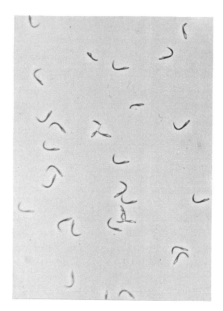

FIG. 2. Fragmented rat sperm tails isolated following sonication. Phase-contrast.
×290.

FIG. 3. Rat sperm heads isolated following sonication. In nearly all of these, the perforatorium remains intact. Phase-contrast.
×290.

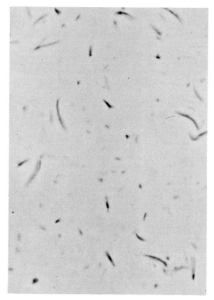

FIG. 4. Rat sperm tails isolated following trypsin-induced cell cleavage. Erosion of
principal piece is indicated by arrow. Phase-
contrast. ×285.

FIG. 5. Fragmented rat perforatoria isolated following sonication of sperm heads in
SDS. Phase-contrast. ×560.

A syringe fitted with a blunt-tipped 18-gauge needle is used to collect the tail band and remove the top two layers of the gradient. The bottom layer is then poured off in one slow continuous motion, in order not to disturb the pellet, the inverted tube is allowed to drain, and the residue on its walls is wiped off with a wet tissue to a level within 2 cm of the bottom of the tube. It is especially important to keep the tube inverted until its walls have been wiped clean, in order to minimize contamination of the head pellet with tail fragments.

4. RECOVERY AND ESTIMATION OF HEADS

Once the heads have been separated from the tails, their loss on the walls of most containers becomes a troublesome problem. Fortunately, it has been observed that the heads adhere much less avidly to nitrocellulose than to other plastics or glass. Consequently, sucrose gradient centrifugation is carried out in nitrocellulose tubes in order to obtain a compact pellet of heads. The pellet, after resuspension in Buffer P, is transferred with a long-tipped disposable pipette into a fresh 5-ml nitrocellulose tube, with care taken never to suck the suspension above the narrow portion of the pipette. The yield of heads is then estimated by diluting the entire suspension to ca. 10×10^6 heads/ml (volume estimated gravimetrically, assuming $d = 1.00$) and counting with a hemocytometer (Neubauer ruling). Aliquots for counting may be diluted to still lower concentrations, without serious under-estimation of heads, provided that the dilutions are prepared in a suitable detergent (e.g., 0.02% SDS). Typical yields range between 60 and 85%. Tail contamination is consistently less than 1%. This can of course be eliminated by recentrifuging the heads through 2.20 M sucrose.

5. RECOVERY OF TAILS AND ESTIMATION OF YIELD

The tail fragments suspended at the 1.80 M–2.05 M interface are transferred to a fresh, preweighed nitrocellulose tube, diluted with at least 3 vol of Buffer P, mixed thoroughly, and recentrifuged at 91,000 g for 20 minutes. After careful aspiration of the resulting supernatant, the flaky tail pellet is resuspended in a suitable volume of Buffer P by stirring, followed by brief sonication at 55 W. In order to estimate the yield at this point, the total volume is determined gravimetrically (assuming that $d =$ ca. 1.00), and the concentration of tails assayed by counting dilutions prepared in 0.02% SDS at concentrations between 1.5 and 2.5 $\times$ 10^6 tails/ml. Using a hemocytometer with Neubauer ruling, the sample is examined within the central, finely ruled area (400 squares/mm^2), and the number of thick tail fragments at least 40 μm in length (i.e., those that contain the midpiece) is estimated. The yield of tails, thus determined, is typically 70–80%, with about 3% contamination by heads. Procedures for further reducing the content of heads in this fraction are discussed in Section II,A,6.

The above scoring method yields slightly low estimates of tails in unfractionated, sonicated sperm suspensions (10–15% lower than the head counts). Its reproducibility depends on the following conditions: (1) use of the minimum degree of sonication necessary to decapitate the sperm, thus avoiding excessive fragmentation of the tails; (2) suspension of the isolated tail pellets at concentrations above 20×10^6 tails/ml for as short a period as possible before counting, in order to avoid significant loss of material on the walls of the container; (3) dilution of the aliquots to be counted in 0.02% SDS, to further minimize wall effects (or even suspension of the final pellet of tails in detergent, when this does not interfere with subsequent procedures); and (4) scoring of tails that stick to the cover slip, as well as those resting on the ruled surface. Failure to observe these precautions will result in serious underestimation of tail content.

An alternative method of calculating the yield of tails depends on the assumption that $> 90\%$ of the total zinc in washed rat sperm is retained within the tail following sonication (Calvin and Bleau, 1974; Calvin, 1975). If EDTA is omitted from the phosphate-buffered media employed during the isolation of tails, recovery of total sperm zinc in the tail fraction is usually 70–80%. This is determined by extraction of both the final tail pellet and an aliquot of sperm [prewashed with 0.02 M Na_2HPO_4 (pH 6.0)] for at least 1 hour in 1 N HC1 and assay of the supernatants by atomic absorption spectrophotometry. It should be kept in mind that the percent zinc recovered in the tail fraction is slightly less than the actual recovery of sonicated tails from the gradient.

6. ALTERNATIVE GRADIENT PROCEDURE FOR PURER TAIL FRACTIONS

The tail fractions obtained by the above procedure usually contain well below 5% heads. Obviously, this contamination can be essentially eliminated by resuspending the material in 1.80 M sucrose, layering it over 2.05 M sucrose, and recentrifuging. Alternatively, the head content can be reduced to about 1% by applying the material to the usual 1.80 M–2.05–2.20 M sucrose gradient in the intermediate 2.05 layer (prepared by weighing out 14.03 gm of 2.50 M sucrose and mixing it with 3.20 ml of sample plus 0.50 ml of 25 $\times$ Buffer P). After centrifugation for 90 minutes at 25,000 rpm (113,000 g, r_{max}) the tail fragments appear as usual at the 1.80–2.05 M interface.

B. Separation of Rat Sperm Heads and Tails by Enzymatic Lysis and Equilibrium Centrifugation

The procedure just outlined for the isolation of heads and tails offers high yields of extremely pure fractions (Figs. 2 and 3). However, it entails the destruction of cell components and the release of enzymes during sonication (see Mohri *et al.*, 1965; Stambaugh and Smith, 1973).

When survival of delicate cell structures such as the acrosome is of importance to the experimenter, harsh physical treatment of the sperm must be avoided. As one alternative, the heads of rat or mouse sperm may be separated from the tails by brief treatment with either trypsin or chymotrypsin to cleave the cells, followed by low-speed (i.e., sedimentation velocity) density gradient centrifugation in continuous sucrose gradients (Edelman and Millette, 1971; Millette *et al.*, 1973). We have tested this method with favorable results, but find the capacity of the gradients to be extremely low. The following modified procedure is therefore proposed, wherein the trypsinized cells are fractionated by equilibrium density gradient centrifugation.

Rat spermatozoa, suspended in 0.05 M Tris–HCl (pH 7.4) (Buffer T) at a concentration of 10–20 $\times$ 10^6 cells/ml, are treated at 25°C with 0.05 mg/ml trypsin (Worthington, TPCK). At 1-minute intervals, the suspension is agitated gently and examined under a phase microscope. After lysis is complete (usually within 5 minutes), 0.5 mg/ml of soybean trypsin inhibitor is added and periodic gentle agitation is continued. A variable degree of sperm agglutination occurs during proteolysis and appears to increase in proportion to contamination of the sperm extract by epididymal cells. This effect is gradually reversed following the addition of inhibitor. The cleaved cells are then collected by centrifugation at 600 g for 10 minutes in a clinical centrifuge, resuspended as homogeneously as possible in Buffer P, chilled to 0°–5°C, applied to the usual 1.80 M–2.05 M–2.20 M sucrose gradients, and centrifuged at 91,000 g for 60 minutes to separate heads and tails. The appearance of the isolated tails is illustrated in Fig. 4. Head contamination (not present in this field) is ca. 5%. Analogous gradients prepared in Buffer T (whose pH is 8.1 at 4°C) yield comparably pure preparations with a more crimped appearance, probably as a result of oxidation of structural $-$SH groups at alkaline pH. Contamination by heads in the tail bands can be reduced by applying the samples in the intermediate 2.05 M sucrose layer (see Section II,A,6).

Since pelleting the heads at high centrifugal force is likely to produce damage to delicate elements such as the acrosome, especially during resuspension of the pellet, isolation of heads by banding in the 30–65% sucrose gradients of Millette *et al.* (1973) may be preferable in some situations. It is suggested that cross-contamination between heads and tails in these gradients can be minimized by keeping the sample concentration below 10 $\times$ 10^6 cells/gradient and centrifuging in a swinging-bucket rotor (Sorvall HB-4) at 1800 g, r_{max} for 30 minutes (three times the force recommended in the original procedure).

We have not yet examined trypsin-treated rat sperm ultrastructurally, but suspect from light microscope observations (Fig. 4) that there is some erosion of the principal piece, even during short incubations. Furthermore, the use of this enzyme and its inhibitor is not suitable for those interested in acrosomal trypsinlike protease (see Section III,A). To overcome the latter objection, the use of chymotrypsin (0.2 mg/ml in Buffer T) is recommended for cell cleavage. When > 95% of the heads are off, the reaction can be stopped with 5 µg/ml of diphenylcarbamyl chloride (Erlanger et al., 1966).

C.   Procedures with Other Species

The procedure described in Section II,A for separation of heads and tails of rat epididymal sperm by sonication and equilibrium density gradient centrifugation has also been applied to the sperm of other eutherian mammals, with modifications. When sperm is collected by ejaculation, as in rabbits and humans, the presence of extraneous soluble and particulate material is far more significant than in rat epididymal sperm suspensions. Therefore, we routinely wash ejaculated rabbit or human sperm several times in Buffer P at 0°–5°C before sonication, centrifuging at 600 g for 10 minutes. (For alternative washing procedures, see Overstreet and Adams, 1971; Harrison and White, 1972.) In addition, more vigorous sonication is required to detach 99% of the sperm heads of rabbits (80 W, 4–6 minutes) or humans (80 W, 6–10 minutes). The thinner, shorter tails in these species are severely fragmented by sonication. Therefore, the gradients used for rat sperm are modified by reducing the concentration of sucrose in the sample layer from 1.80 to 1.60 M, to afford adequate recovery of the finer tail fragments. Because of the smaller dimensions of rabbit and human sperm, as many as 5×10^8 sperm may be loaded on a single gradient.

There is, as yet, no satisfactory procedure for the isolation of intact heads and tails from the sperm of most mammals, since the sensitivity of the head–tail junction in rat and mouse sperm to endopeptidases is exceptional. In our experience, successive treatment with DTT and trypsin, as suggested by Millette et al. (1973), cleaves rabbit and human spermatozoa but severely damages structures in both the heads and tails (see op. cit., Fig. 2c). Swelling of the sperm nucleus is especially evident.

III.   Subfractionation of the Sperm Head

A.   Acrosome and Plasma Membrane

Within the sperm head, the acrosome and the plasma membrane are the two organelles that seem to be of the most critical importance during the initial phases of gamete interaction, entailing the successive penetration by the sperm of the outer layers of the ovum, namely, the cumulus cells, the corona radiata, the zona pellucida, and the vitelline membrane (see Piko, 1969; Bedford, 1970). Because of the malleability inherent in this function during fertilization, both of these organelles are necessarily fragile, hence do not lend themselves readily to isolation as intact structures. However, their delicate nature facilitates selective removal from the sperm by relatively mild treatment.

The following chemical agents have been employed for the selective removal of acrosomal material from mammalian sperm (usually those of the bull, ram, or rabbit): mild base (Austin and Bishop, 1958; Hathaway and Hartree, 1963), detergents (Austin and Bishop, 1958; Hathaway and Hartree, 1963; Hartree and Srivastava, 1965; Srivastava et al., 1970), barbituric acid (Bernstein and Teichman, 1973), acetic acid (Schill, 1973), and hypotonic $MgCl_2$ (Srivastava et al., 1974). Procedures for physical removal include shaking with glass beads (Hathaway and Hartree, 1963; Morton and Lardy, 1967; Multamäki and Niemi, 1972) and freeze-thawing (Pedersen, 1972; Brown and Hartree, 1974). The inevitable contamination of such "acrosomal" extracts by other easily removed cytoplasm is reduced by prewashing the sperm with hypotonic buffer (Allison and Hartree, 1970).

As a further refinement, acrosomal contents have been released in a stepwise manner (Brown and Hartree, 1974; Srivastava et al., 1974). Using this approach, Brown and Hartree (1974) demonstrated that partial removal of ram sperm acrosomes by freeze-thawing, which leaves behind only the inner acrosomal membrane and the posterior region of the acrosome known as the equatorial segment, does not release trypsinlike protease from the sperm. Subsequent treatment with Ca^{2+} or mild acid (pH 3.0) solubilizes this enzyme, which is believed to be required for sperm penetration of the zona pellucida, suggesting that the trypsinlike activity is bound to one or both of the acrosomal components that resist freeze-thawing.

The content of acrosomal extracts is usually evaluated by: (1) light and electron microscopic examination of the denuded sperm, in order to see what has been removed, and (2) assay for marker enzymes, on the assumption that the enzymes in the acrosome are almost exclusively lysosomal hydrolases. Since neither of these criteria is unambiguous, fractionation procedures for the isolation of sperm heads with intact acrosomes or of acrosomes themselves would be welcome supplements to the existing methodology. The recent use of isopycnic sucrose density gradient centrifugation for the isolation of an acrosomal fraction from bull sperm, following release of the anterior portion of the acrosome with Hyamine, is a promising start in this direction. However, the failure of the isolated structures to retain hyaluronidase, long believed to be intrinsic to mammalian acrosomes (Austin and Bishop, 1958), raises the question of the integrity of this preparation, especially since acrosomal vesicles released by Hyamine are not entirely stable in this detergent (Hartree and Srivastava, 1965) and, in any case, do not include the entire acrosome (Brown and Hartree, 1974). Obviously, a more selective reagent than Hyamine is needed for the release of acrosomal material. Thus, it is worthy of mention that the non-ionic detergent, Empilan, has recently been suggested as a milder and more specific substitute for Hyamine in this context (Gombe et al., 1975).

A preparation of boar acrosomal membranes, isolated on sucrose density gradients following Hyamine treatment and sonication (Morré et al., 1974), is, by the above considerations, of doubtful integrity. On the other hand, the same group has also purified boar sperm plasma membranes on similar gradients, following low-intensity sonication of intact sperm, and has identified these by phosphotungstic acid staining (Lunstra et al., 1974; Morré et al., 1974). The use of this mild physical treatment suggests that the isolated vesicles, presumably derived from the entire sperm plasma membrane, are suitable for biochemical analysis. Further fractionation would be needed to study the regional variation in composition of the plasma membrane, which has been delineated in a variety of ultrastructural studies (see Fawcett, 1975) and is also manifested by regional differences in membrane stability (Wooding, 1973).

B.   Nuclear Chromatin

Treatment of rabbit sperm heads with SDS dissolves the acrosome and all membrane vestments, leaving only the fully condensed nucleus and remnants of perinuclear material in the apical, midlateral, and postacrosomal regions of the sperm head (Calvin and Bedford, 1971; Bedford and Calvin, 1974b). Rat sperm heads, which respond similarly to this reagent, retain the nucleus and perforatorium, a very prominent structure bordering the apical and anterior lateral surfaces of the nucleus (see Fig. 1), as well as less prominent remnants of perinuclear material in the midlateral and posterior regions of the sperm head (Calvin and Temple-Smith, unpublished). We have also found that sonication of rat sperm in SDS detaches virtually all of the perinuclear material and that this process is facilitated by preincubation in the presence of DTT, suggesting the following procedures for preparing chromatin.

Rat sperm heads are recentrifuged through 2.20 M sucrose to eliminate traces of tail contamination, suspended at a concentration of 50–100 $\times$ 10^6 ml in 2–3 ml of 0.1% SDS–0.05 M Tris–HCl (pH 8.5)–0.001 M EDTA for

15 minutes at room temperature, cooled to 5° C, and sonicated at 55 W until removal of perforatoria, as monitored with a phase-contrast microscope, is complete (ca. 4–5 minutes). The sonicate is layered over a stepwise gradient of 1.60 M sucrose (30 ml) and 2.20 M sucrose (5 ml), prepared in Buffer P, and centrifuged in an SW-27 rotor at 22,500 rpm (91,000 g, r_{max}) for 20 minutes. Perinuclear material, chiefly the perforatorium (Fig. 5), is found at the upper boundary of the 1.60 M sucrose layer. The pellet consists of pure chromatin, slightly vacuolated but fully condensed, with only trace contamination by perinuclear material (Fig. 6).

As an alternative method, to further reduce the content of perinuclear elements in the chromatin fraction, 1×10^{-4} M DTT is added to the above SDS medium, the sperm heads are preincubated for 30 minutes at room temperature, prior to sonication, and the product centrifuged as above. The isolated chromatin, although slightly decondensed, is essentially free of perinuclear contamination.

We have not attempted the purification of sperm chromatin from species other than the rat. The removal of perinuclear material by the use of detergents is, nevertheless, a common practice (Hartree and Srivastava, 1965; Srivastava et al., 1970) and has recently been applied to the isolation of sperm nuclear chromatin from the bull (Marushige and Marushige, 1974) and mouse (Meistrich et al., 1975). Ultrastructural evidence was furnished for the morphological integrity and high degree of purity of the latter preparation, obtained by treating isolated sperm heads successively with Triton X-100, in the presence of trypsin and deoxyribonuclease, and then with SDS.

C.   Perforatorium

The rat perforatorium may be isolated following sonication of sperm heads in SDS (in the absence of DTT). During centrifugation as described in Section III,B, a band of material whose appearance is illustrated in Fig. 5 forms at the interface between the sample and 1.60 M sucrose layers. The success of this technique requires that no more than 3×10^8 heads be applied to the gradient. Otherwise, a significant fraction of perforatoria (perhaps by interaction with chromatin) will penetrate the 1.60 M sucrose.

As viewed under the phase contrast or electron microscope, the isolated fraction appears to consist almost entirely of fragmented perforatoria. Moreover, solubilization of this preparation and analysis by polyacrylamide gel electrophoresis in 0.1% SDS has revealed only one significant protein component, in support of its homogeneity (unpublished results). Nevertheless, in view of the relatively primitive gradient procedure employed for its isolation, the possibility of significant contamination of the perforatorium fraction by other SDS-insoluble material cannot yet be discounted.

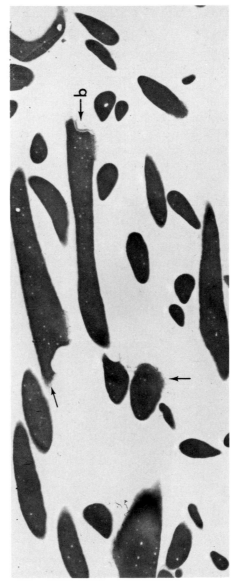

FIG. 6.   Electron micrograph of isolated rat sperm chromatin. Fixed and stained as described by Bedford and Calvin (1974a). Arrows indicate traces of perinuclear material, occasionally including the basal plate (b). × 10,000.

IV. Subfractionation of the Tail

A. Midpieces and Principal Pieces

Subfractionation of the sperm tail for the purpose of enzyme localization has long been of interest to those concerned with the metabolic functions associated with sperm motility. When sperm of bulls (Zittle and O'Dell, 1941) or rabbits (Stambaugh and Buckley, 1969) are sonicated at the minimum intensity required for removal of heads, most tails cleave near the junction between the midpiece and principal piece, with further fragmentation of the latter into segments of varying lengths. Similar division of bull sperm occurs following grinding in the presence of ice (Nelson, 1954). Separation of heads, midpieces, and principal pieces is carried out either by differential centrifugation (Zittle and O'Dell, 1941; Nelson, 1954) or by sedimentation velocity (i.e., nonequilibrium) sucrose density gradient centrifugation (Stambaugh and Buckley, 1969). Although they involve disruption of the plasma membrane and consequent loss of intracellular material, these fractionations have nevertheless enabled the assignment of specific enzyme activities to one or the other of the two major segments of the tail (Zittle and Zitin, 1942; Nelson, 1954; Nelson 1967; Mohri *et al.*, 1965; Stambaugh and Buckley, 1969). It has thus been found that the metabolic profile of the midpiece is rather typical of that of mitochondria (Mohri *et al.*, 1965), which are located in this portion of the sperm tail (Fig. 1).

B. Mitochondria

Isolation of mitochondria from mature sperm in a state suitable for metabolic studies is difficult, in principle, because of the $-S-S-$ linkages which stabilize their association within the sperm tail (Edelman and Millette, 1971; Millette *et al.*, 1973; Bedford and Calvin, 1974a). As an alternative, Machado de Domenech *et al.* (1972) have isolated, on sucrose density gradients, a testicular fraction consisting mainly of presumptive sperm mitochondria, identifiable by their peculiar morphology and high content of the X-isozyme of lactate dehydrogenase. More recently, a short communication has appeared, describing the purification of mature bull sperm mitochondria by isopycnic centrifugation in sucrose, following incubation of the cells in a hypotonic medium and disruption with a French press (Robinson and Forrester, 1974). Although photographs were not furnished, preliminary metabolic studies and unpublished observations with the electron microscope were quoted in support of the integrity of the isolated bodies. The exceptional stability of mammalian sperm mitochondria in media which are hypotonic (Keyhani and Storey, 1973) or hypertonic (Machado de

Domenech *et al.*, 1972), possibly a consequence of $-S-S-$ crosslinking in their outer membranes (Bedford and Calvin, 1974a), is probably not quite absolute (Young and Smithwick, 1975), but nevertheless a factor favoring their isolation as viable organelles by the above procedures.

C.   Keratinlike Structures

Within the tail, the connecting piece, outer mitochondrial membranes, fibrous sheath, and dense fibers resist solution in 1% SDS (Bedford and Calvin, 1974a) and may thus be isolated as an entire keratinlike skeleton by selective solubilization with this reagent. For the isolation of dense fibers and connecting pieces only, the following procedure is suggested.

Sperm tails are prepared in the usual manner and completely freed of head contamination by recentrifuging twice in 1.80 M–2.05 M sucrose gradients (see Section A,5). To minimize the formation of –S–S– links by oxidation during tail isolation, 0.001 M EDTA is included in all media, except in those preparations in which total zinc recovery is desired (Calvin, 1975). Immediately following purification, the tails are incubated overnight in 1% SDS–0.05 M Tris–HCl (pH 7.5)–2 $\times$ 10^{-4} M DTT–0.001 M EDTA. Sonication at 55 W for 2 minutes at 0°–5°C completely releases the fibers from the tail.

Centrifugation of 3-ml aliquots in 11-ml plastic centrifuge tubes at 10,000 g (Sorvall HB-4 rotor, 8000 rpm) pellets all of the sonicated dense fibers and connecting pieces, as well as a small percentage of the surviving remnants of the mitochondrial and fibrous sheaths. The latter are removed by twice resuspending the material in water and recentrifuging at 10,000 g. An electron micrograph of the isolated material appears in Fig. 7. Connecting pieces (not shown) are present as a minor component of this preparation. The numerous electron-dense spots, which appear to be concentrated along the periphery of the fibers, may possibly reflect partial disintegration and recondensation of dense fiber material.

Other workers have reported the isolation and preliminary chemical characterization of dense fibers from sperm of the rat (Price, 1973) and bull (Baccetti *et al.*, 1973), by means of high-intensity sonication and sucrose density gradient fractionation. In some preparations (Baccetti *et al.*, 1973), 0.1% SDS was included in the sonication medium. Unfortunately, the evidence furnished so far does not allow satisfactory evaluation of the purity of the isolated fractions, but it seems likely that they contained both connecting pieces and fibrous sheath fragments, as well as dense fibers.

Pihlaja and Roth (1973) have developed a procedure for dense fiber purification based on the observation that extended treatment of bull sperm with Pronase results in a fragmented preparation of heads and partially de-

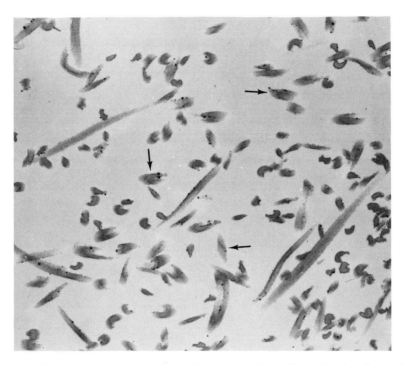

FIG. 7. Electron micrograph of isolated rat sperm dense fibers. Peripherally localized electron-dense spots of varying sizes are evident (arrows). × 14,700.

graded tails (Pihlaja et al., 1973). The tail elements which survive this incubation include a complex of mitochondria and dense fibers, as well as detached, fragmented connecting pieces. Following removal of the heads on sucrose density gradients, isolation of dense fibers from Pronase-degraded tails may be achieved by pressure cavitation in the presence of SDS, which detaches and partially solubilizes the mitochondria, and differential centrifugation. Although the product obtained by this procedure was apparently free of fibrous sheath and mitochondrial elements, no data has been provided concerning: (1) the likely presence of connecting pieces, as well as dense fibers, in this preparation, or (2) the integrity of proteins in the isolated structures, following their prolonged exposure to Pronase.

Yet another approach to the isolation of dense fibers is suggested by the observations of Pihlaja et al. (1973) that proteases selectively attack the principal piece of the tail (see also Fig. 4) and those of Millette et al. (1973) that the integrity of rat sperm midpieces can be disrupted with DTT. We have observed that treatment of rat sperm with trypsin (0.2 mg/ml) for

20 minutes, followed by sonication (55 W, 60 seconds) entirely removes the fibrous sheath. After the addition of trypsin inhibitor, the material is centrifuged (600 g, 10 minutes) and resuspended in 0.05 M Tris–HCl (pH 8.5)– 0.004 M DTT for 30 minutes. Upon vigorous stirring with a vortex mixer, ca. 80% of the dense fibers and connecting pieces are released from the mitochondrial sheath. Complete disruption of the sheath can then be effected by mild sonication (55 W, 30 seconds).

A procedure for the isolation of dense fibers and connecting pieces based on the latter observations has not yet been developed. It is hoped, nevertheless, that they will be of use to those interested in the fractionation of keratinlike tail components, and, in general, that this brief review of methods will aid in the development of improved procedures for the fractionation of mammalian sperm.

ACKNOWLEDGMENTS

The author gratefully acknowledges the skillful technical assistance of Miss Frances Hwang and Miss Herma Wohlrab and would like to thank Drs. J. M. Bedford, G. W. Cooper, and P. Temple-Smith of Cornell University Medical College for their aid in the preparation of Figs. 2–7. This work was supported by Grant No. HD-05316 from the National Institutes of Health.

REFERENCES

Allison, A. C., and Hartree, E. F. (1970). *J. Reprod. Fert.* **21**, 501.
Austin, C. R., and Bishop, M. W. H. (1958). *Proc. Roy. Soc., Ser. B* **149**, 234.
Baccetti, B., Pallini, V., and Burrini, A. G. (1973). *J. Submicrosc. Cytol.* **5**, 237.
Bedford, J. M. (1970). *Biol. Reprod., Suppl.* **2**, 128.
Bedford, J. M., and Calvin, H. I. (1974a). *J. Exp. Zool.* **187**, 181.
Bedford, J. M., and Calvin, H. I. (1974b). *J. Exp. Zool.* **188**, 137.
Bernstein, M. H., and Teichman, R. J. (1973). *J. Reprod. Fert.* **33**, 239.
Brown, C. R., and Hartree, E. F. (1974). *J. Reprod. Fert.* **36**, 195.
Calvin, H. I. (1975). *In* "The Biology of the Male Gamete" (J. G. Duckett and P. A. Racey, eds.), pp. 257–273. Academic Press, New York.
Calvin, H. I., and Bedford, J. M. (1971). *J. Reprod. Fert., Suppl.* **13**, 65.
Calvin, H. I., and Bleau, G. (1974). *Exp. Cell Res.* **86**, 280.
Calvin, H. I., Yu, C. C., and Bedford, J. M. (1973). *Exp. Cell Res.* **81**, 333.
Coelingh, J. P., Rozijn, T. H., and Monfoort, C. H. (1969). *Biochim. Biophys. Acta* **188**, 353.
Edelman, G. M., and Millette, C. F. (1971). *Proc. Nat. Acad. Sci. U.S.* **68**, 436.
Erlanger, B. F., Cooper, A. G., and Cohen, W. (1966). *Biochemistry* **5**, 190.
Fawcett, D. W. (1970). *Biol. Reprod., Suppl.* **2**, 90.
Fawcett, D. W. (1975). *Develop. Biol.* **44**, 394.
Friend, G. F. (1936). *Quart. J. Microsc. Sci.* **78**, 419.
Gombe, S., Norman, C., and Mbogo, D. E. (1975). *J. Reprod. Fert.* **43**, 535.
Harrison, R. A. P., and White, I. G. (1972). *J. Reprod. Fert.* **29**, 271.

Hartree, E. F., and Srivastava, P. N. (1965). *J. Reprod. Fert.* **9**, 47.

Hathaway, R. R., and Hartree, E. F. (1963). *J. Reprod. Fert.* **5**, 225.

Henle, W., Henle, G., and Chambers, L. A. (1938). *J. Exp. Med.* **68**, 335.

Hernández-Montes, H., Iglesias, G., and Mújica, A. (1973). *Exp. Cell Res.* **76**, 437.

Keyhani, E., and Storey, B. T. (1973). *Biochim. Biophys. Acta* **305**, 557.

Koehler, J. K. (1973). *J. Ultrastruct. Res.* **44**, 355.

Lunstra, D. D., Clegg, E. D., and Morré, D. J. (1974). *Prep. Biochem.* **4**, 341.

Machado de Domenech, E., Domenech, C. E., Aoki, A., and Blanco, A. (1972). *Biol. Reprod.* **6**, 136.

McRorie, R. A., and Williams, W. L. (1974). *Annu. Rev. Biochem.* **43**, 777.

Marushige, Y., and Marushige, K. (1974). *Biochim. Biophys. Acta* **340**, 498.

Meistrich, M. L., Reid, B. D., and Barcellona, W. J. (1975) *J. Cell Biol.* **64**, 211.

Miescher, F. (1878). *Verh. Naturforsch. Ges. Basel* **6**, 138.

Millette, C. F., Gall, W. E., and Edelman, G. M. (1974). *Fed. Proc., Fed. Amer. Soc. Exp. Biol.* **33**, 1395. (abstr.).

Millette, C. F., Spear, P. G., Gall, W. E., and Edelman, G. M. (1973). *J. Cell Biol.* **58**, 662.

Mohri, H., Mohri, T., and Ernster, L. (1965). *Exp. Cell Res.* **38**, 217.

Morré, D. J., Clegg, E. D., Lunstra, D. D., and Mollenhauer, E. H. (1974). *Proc. Soc. Exp. Biol. Med.* **145**, 1.

Morton, B. E. (1968). *J. Reprod. Fert.* **15**, 113.

Morton, B. E., and Lardy, H. A. (1967). *Biochemistry* **6**, 50.

Multamäki, S. (1973). *Int. J. Fert.* **18**, 193.

Multamäki, S., and Niemi, M. (1972). *Int. J. Fert.* **17**, 43.

Nelson, L. (1954). *Biochim. Biophys. Acta* **14**, 312.

Nelson, L. (1967). *In* "Fertilization: Comparative Morphology, Biochemistry and Immunology" (C. B. Metz and A. Monroy, eds.), Vol. 1, pp. 27–97. Academic Press, New York.

Overstreet, J. W., and Adams, C. E. (1971). *J. Reprod. Fert.* **26**, 219.

Pedersen, H. (1972). *J. Reprod. Fert.* **31**, 99.

Phillips, D. M. (1970). *J. Ultrastruct. Res.* **33**, 381.

Pihlaja, D. J ., and Roth, L. E. (1973). *J. Ultrastruct. Res.* **44**, 293.

Pihlaja, D. J., Roth, L. E., and Consigli, R. A. (1973). *Biol. Reprod.* **8**, 311.

Pikó, L. (1969). *In* "Fertilization: Comparative Morphology, Biochemistry and Immunology" (C. B. Metz and A. Monroy, eds.), Vol. 2, pp. 325–403. Academic Press, New York.

Price, J. M. (1973). *J. Cell Biol.* **59**, 272a.

Robinson, A. M. and Forrester, I. T. (1974). *Proc. Univ. Otago Med. Schl.* **52**, 51.

Schill, W. -B. (1973). *Arch. Derm. Forsch.* **248**, 257.

Srivastava, P. N., Zaneveld, L. J. D., and Williams, W. L. (1970). *Biochem. Biophys. Res. Commun.* **39**, 575.

Srivastava, P. N., Munnell, J. F., Yang, C. H., and Foley, C. W. (1974). *J. Reprod. Fert.* **36**, 363.

Stambaugh, R., and Buckley, J. (1969). *J. Reprod. Fert.* **19**, 423.

Stambaugh, R., and Smith, M. (1973). *J. Reprod. Fert.* **35**, 127.

Wooding, F. B. P. (1973). *J. Ultrastruct. Res.* **42**, 502.

Woolley, D. M., and Fawcett, D. W. (1973). *Anat. Rec.* **177**, 289.

Young, L. G., and Smithwick, E. B. (1975). *Exp. Cell Res.* **90**, 233.

Zittle, C. A., and O'Dell, R. A. (1941). *J. Biol. Chem.* **140**, 899.

Zittle, C. A., and Zitin, B. (1942). *J. Biol. Chem.* **144**, 99.

Chapter 6

On the Measurement of Tritium in DNA and Its Applications to the Assay of DNA Polymerase Activity

BRUCE K. SCHRIER AND SAMUEL H. WILSON

Behavioral Biology Branch, National Institute of Child Health and Human Development, and Laboratory of Biochemistry, National Cancer Institute, National Institutes of Health, Bethesda, Maryland

I. Introduction 105
II. Evaluation of Methods for Precipitation, Collection, and Counting of
    Tritium-Labeled DNA 106
    A. Optimal Precipitation 106
    B. Collection Methods 107
    C. The Counting of Macromolecules Collected on Filter Discs . . . 109
III. Determination of Optimal Conditions for Collection and Counting of DNA
    Polymerase Assay Products 114
IV. Selection of Methods for a Particular System 117
    A. Guidelines for Evaluation 117
    B. Characteristics of Glass-Fiber Filters 117
    C. Recommendations 118
    References 120

I. Introduction

Extensive use has been made of filter discs as agents for the collection of precipitated radioactive macromolecules contaminated by radioactive molecules of low molecular weight. The methods for such collection suggested by Bollum (1959, 1966) and by Mans and Novelli (1961), as well as other methods in which the filter disc with its adherent macromolecules is immersed directly in scintillation counting solutions, give rise to two-phase counting systems, i.e., the insoluble filter with affixed macromolecules and the scintillation solution. Among the disadvantages of such two-phase systems are (1) difficulty in determining absolute efficiencies of tritium in the solid-phase macromolecules, (2) variations in counting efficiency among samples caused by differences in the total amounts of macromolecules of the filters, and (3) the possible importance of the geometric relationship between the filter and the photomultiplier tube of the scintillation counter.

The importance of solubilization of the macromolecule before counting has been emphasized for glass-fiber and filter-paper discs, respectively, by Birnboim (1970) and by Hodes et al. (1971), and for cellulose acetate filters by Schrecker et al. (1972). These reports, however, recommended different methods for filtration and solubilization of nucleic acids, and Hodes et al. suggested that the procedure recommended by Birnboim may give incomplete recovery of DNA. In addition, Meyer and Keller (1972) have reported that multiple parameters, including amounts both of protein and of DNA, may influence the collection and counting efficiency of tritiated DNA on filter discs. Because of the importance of the filter disc collection technique in DNA polymerase assays and the absence of a clear method of choice in the above reports, we attempted to evaluate systematically: (1) methods for solubilization of nitrocellulose (Millipore) filters, and of ^{3}H-labeled DNA collected on both glass-fiber and nitrocellulose filters; (2) the effect of protein concentration on the collection, solubilization, and counting of ^{3}H-labeled DNA of both high-molecular-weight double-stranded and low-molecular-weight single-stranded forms; (3) methods to reduce the "blank" in DNA polymerase reactions that employ relatively large quantities of ^{3}H-labeled dTTP as substrate along with the filter disc collection procedure for the measurement of reaction products. We here describe the methods used for these evaluations and the results provide a basis for making recommendations about collection and counting of the particular macromolecules that were evaluated (see Table IV).

II. Evaluation of Methods for Precipitation, Collection, and Counting of Tritium-Labeled DNA

A. Optimal Precipitation

We used tritium-labeled DNA isolated from bacteriophage T_7 grown in *Escherichia coli* in the presence of [methyl-^{3}H] thymidine and then purified by isopycnic sedimentation in CsCl (Schrier and Wilson, 1973). This ^{3}H-labeled DNA was indistinguishable from native T_7 DNA with respect to its bouyant density and electrophoretic behavior on polyacrylamide gels; its molecular weight was assumed to be approximately 25×10^6 (Eigner and Doty, 1965). The collection of this high-molecular-weight DNA on nitrocellulose filters after it was treated with 10% trichloracetic acid (TCA) at either $0°-1°C$ or $25°C$, was investigated. Collection was complete during the period from 0.2 to 60 minutes after addition of TCA, and was identical at the two temperatures. Thus, for convenience, a precipitation time of 15 minutes at $0.6°C$ (ice-water bath) was chosen for subsequent experiments. A portion of the DNA-[³H] was sheared in a French pressure cell into single-stranded pieces with average and modal sizes of approximately 185,000 daltons.

B.   Collection Methods

The collection and counting methods for both the low- and high-molecular-weight DNA-[³H] were evaluated in the following experiment. The DNA was collected either on glass-fiber filters (Whatman, GF/A) or on nitrocellulose filters with different amounts of precipitated bovine serum albumin (BSA) (Fig. 1). Glass-fiber filters, after collection of DNA and protein, were either dried and counted in toluene scintillation fluid (designated "insoluble" in Fig. 1) or treated with diluted NCS solubilizer (NCS/3, footnote *l*, Table I) in order to dissolve the DNA and protein, and subsequently counted in a toluene scintillation mixture. Similarly, nitrocellulose filters (Fig. 1A and C) with adsorbed DNA and protein were either dried and counted in toluene scintillation fluid ("insoluble") or treated with ethyl-Cellosolve and counted in Aquasol ("soluble") as shown in Table I. With high-molecular-weight DNA, nitrocellulose filters collected all the DNA even in the absence of the coprecipitant BSA. Increases in the amount of coprecipitant BSA resulted in a marked decrease in counting efficiency in the two-phase system ("insoluble"), whereas solubilization resulted in the same counting efficiency at all levels of BSA. Equivalence between the efficiencies of ³H-labeled DNA and internal standards at 500 μg BSA in solubilized samples confirmed the complete solubilization of DNA from nitrocellulose filters.

With glass-fiber filters and high-molecular-weight DNA (Fig. 1B) collection of ³H-labeled DNA was greater than 90% complete at 200 μg of BSA and above. In the absence of solubilization, glass-fiber and nitrocellulose filters were quite equivalent, except that the glass-fiber filters could accommodate more protein than nitrocellulose filters before reaching impracticable flow rates (note abscissa in Fig. 1A versus 1B). On glass-fiber filters, solubilization of high-molecular-weight ³H-labeled DNA resulted in a much higher counting efficiency than was obtained with insoluble samples.

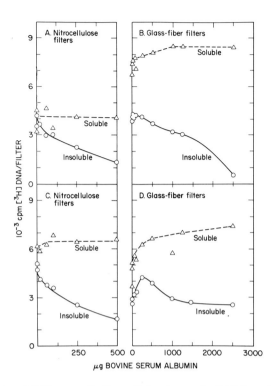

FIG. 1. Effects of BSA on collection efficiencies, and of BSA and solubilization on counting efficiencies of high-molecular-weight and sheared DNA. Native T_7 DNA (20,644 dpm, 0.28 μg) (A and B) or sheared T_7 DNA (24,120 dpm) 0.33 μg, (C and D) was precipitated along with varying amounts of BSA, collected, and washed on either nitrocellulose (A and C) or glass-fiber (B and D) filters as described in Section II,A. Nitrocellulose filters were dissolved in ethyl-Cellosolve (1.0 ml) and counted in 10 ml of Aquasol at 28% counting efficiency ("soluble"), or dried and placed in toluene scintillation mixture ("insoluble"); glass-fiber filters were dried and counted with NCS/3 as described in footnote *l*, Table I at 41.7% efficiency ("soluble"), or dried and counted in toluene scintillation fluid ("insoluble"). Counts were obtained at optimal gain settings. Reproduced by permission from Schrier and Wilson, *Analyt. Biochem.* **56**, 196–207 (1973).

Even without protein in the precipitation, insoluble samples counted at one-half the efficiency of solubilized samples. Complete solubilization of ^{3}H-labeled DNA by our criteria (footnote *b*, Table I) was observed even at 2.5 mg of BSA.

From these results it was clear that, with high-molecular-weight DNA, the glass-fiber filter collection procedure with 200–1000 μg of coprecipitant BSA and then solubilization was superior to any of the other collection and counting methods. However, when a similar experiment was performed with the sheared single-stranded DNA, the data indicated that nitrocellulose filters were clearly superior. Nitrocellulose filters collected essentially all the 185,000-dalton, ³H-labeled DNA (Fig. 1C) without added BSA; BSA additions of up to 500 μg had no effect on counting efficiency in the solubilized system. Increasing the amount of BSA in the insoluble system, as with high-molecular-weight DNA, decreased the counting efficiency markedly. In the case of the glass-fiber filters (Fig. 1D), collection efficiency for the low-molecular-weight DNA was dependent on the amount of albumin used for coprecipitation. With 500 μg of BSA a plateau (solubilized samples) of collection efficiency was reached, but this represented only 66% of the total DNA in the samples. In the presence of 2.5 mg of BSA, where the rate of filtration was too slow to be practicable, only 74.5% of the DNA was collected on the filter. Lack of complete collection was also observed with the insoluble samples, in which, despite decreasing efficiencies due to the BSA (compare to Fig. 1B), count rate increased to 250 μg of BSA and seemed to plateau between 1.0 and 2.5 mg of BSA per filter. These data indicate that the physical properties of the macromolecule to be filtered must be considered before a filtration and counting system is chosen; higher counting efficiencies may not compensate for failure to collect all the radioactive macromolecules.

C.   The Counting of Macromolecules Collected on Filter Discs

The choice of a counting system for tritium-labeled macromolecules adsorbed onto filter discs involved the consideration of several different parameters, some of which cannot be independently measured. Using high-molecular-weight DNA-[³H], we assessed (Table I) a variety of methods for solubilizing and counting under conditions of 100% collection of DNA onto filters; thus the differences in count rate among the various procedures resulted from differences in counting efficiency. Tritiated DNA was considered dissolved when (1) the count rate was stable for several days in a refrigerated spectrophotometer, (2) counting efficiency for the tritiated macromolecule was the same as that of the internal counting standard (hexadecane), and (3) in cases in which the filter was not dissolved, removal of the filter from a counting vial resulted in the removal of less than 10% of the radioactivity.

First, several reagents were tested for their ability to dissolve nitrocellulose filters at 24°C in 30 minutes or less; the reasoning was that such dissolution would at least temporarily suspend the macromolecules in the counting solution. Filters were not dissolved by 2 ml Biosolv (Beckman), isoamyl alcohol, isoamyl acetate, Kodak D19 developer, p-dioxan, absolute ethanol, 36% formaldehyde, isopropanol, methylene chloride, concentrated

TABLE I

COUNTING EFFICIENCIES FOR HIGH-MOLECULAR-WEIGHT ^{3}H-LABELED DNA UNDER VARIOUS CONDITIONS[a]

Solubilization procedure	Solubility[b]		Counting efficiency (%)				
	Filter	DNA	Toluene[c]	Aquasol	Triton–toluene[d]	Bray's[e]	Cellosolve–toluene[f]
Nitrocellulose filters							
None	–	–	22.4	27.0[g,h]	25.2	8.1[h,j]	15.5
Acetone, 1.5 ml	+	–	7.3[k]	8.2	4.3	2.9	—
NCS, 0.5 ml[l]	±	+	2.9[i]	21.3[m]	13.0[i]	14.8[i]	—
Ethyl-Cellosolve, 1 ml	+	+	—	28.0[m]	26.5[i]	—	18.6[i]
Ethyl acetate, 2 ml	+	–	26.9[m]	12.4[i]	4.3[i]	4.3[i]	—
Dimethylformamide, 1 ml	+	+	6.8	26.7[m]	24.0	13.7[i]	—
Protosol, 1 ml	±	–	4.7[i]	7.5[i]	10.3[i]	8.3	—
Protosol, 1 ml[o]	±	–	5.7[i]	—	—	—	—
Biosolv, 0.5 ml	–	–	14.4[i]	—	—	—	—
Hot PCA[p]	–	+	—	23.5	24.5[m]	—	—
Glass-fiber filters							
None	–	–	28.2	20.4	18.9	12.3	—
NCS, 0.5 ml[l]	–	+	41.6[m]	19.3[i]	19.4	—	—
Same, filter not dried	–	+	43.9[m]	—	—	—	—
Protosol, 1 ml	–	+	34.2[m]	16.9[i]	21.9	—	—
Protosol, 0.5 ml[o]	–	–	31.9[i]	—	—	—	—
Same, filter not dried	–	–	22.3[i]	—	—	—	—
Dimethylformamide, 1 ml	–	–	27.0	21.5	26.6	—	—
Biosolv, 0.5 ml[q]	–	–	24.2	—	—	—	—
Hot PCA[p]	–	+	—	16.5	20.8	—	—

110

[a]Native [3H]-labeled DNA from T$_7$ bacteriophage MW 25 × 10^6 daltons; 20,644 dpm; 0.28 μg DNA) was mixed with 20 μg (nitrocellulose filters) or 100 μg (glass-fiber filters) of BSA and precipitated with 1.5 ml of 10% trichloroacetic acid (TCA) for 15 minutes at 0°C, and then filtered as shown and washed with 50 ml cold 10% TCA. Filters to be counted in toluene scintillation cocktail were dried under an infrared lamp. Incubations with solubilizers were 30 minutes at room temperature, except with PCA (see footnote p). Each vial was counted at five different gain settings, and the optimal settings for that vial were used to obtain the data given. Italicized values denote efficiency tested with internal standards of hexadecane-[3H] (120,700 dpm). Reasonable methods for counting high-molecular-weight DNA are in boldface. Reproduced by permission from Schrier and Wilson, *Analyt. Biochem.* **56**, 196–207 (1973).

[b]Solubility of filter by visual inspection; solubility of DNA (when filter was not solubilized) by > 90% of counts remaining behind after removal of filter, or (when filter was solubilized) by stability of counts with time and equivalence of counting efficiencies for DNA and internal standard. No sign indicates not tested.

[c]1000 ml toluene plus 42 ml Liquifluor (New England Nuclear).

[d]2000 ml toluene plus 1000 gm Triton X-100 plus 165 ml Liquifluor.

[e]3800 ml p-dioxan plus 450 ml methanol plus 90 ml ethylene glycol plus 270 gm naphthalene plus 18 gm PPO plus 0.9 gm POPOP.

[f]750 ml toluene plus 240 ml ethyl-Cellosolve plus 42 ml Liquifluor.

[g]Efficiency of internal standard = 31.6%.

[h]Filter dissolved with time in refrigerated scintillation spectrometer; solubility of DNA unknown.

[i]Count rate decreased with time in refrigerated scintillation spectrometer.

[j]Efficiency of internal standard = 21.4%.

[k]Efficiency of internal standard = 11.2%.

[l]Incubation in 0.5 ml of NCS (Amersham-Searle) diluted with 2 vol of toluene scintillation cocktail; nitrocellulose filters gave very yellow solutions.

[m]Counting efficiencies of [3H]-labeled DNA and internal standard agreed.

[n]Efficiency of internal standard = 37.9%.

[o]Incubation in 0.5 ml of Protosol (New England Nuclear) diluted with 2 vol of toluene scintillation cocktail; nitrocellulose filters gave very yellow solutions.

[p]Filter was placed in 0.9 ml of 0.6 N PCA and incubated for 30 minutes at 70°C; then filter was removed and discarded, and 0.1 ml of concentrated NH$_4$OH added, followed by 10 ml of scintillation fluid.

[q]Incubation in 0.5 ml of Biosolv (Beckman Instruments) diluted with 2 vol of toluene scintillation cocktail.

H_2SO_4, benzyl alcohol, 85% H_3PO_4, or Folin reagent (2 N, Fisher Chemicals). Dissolution did occur with 1.5 ml glacial acetic acid, 1.5 ml acetone, 1 ml ethyl-Cellosolve (ethylene glycolmonoethyl ether), 2 ml ethyl acetate, 1 ml dimethylformamide and 0.5–1 ml of either Protosol (New England Nuclear) or NCS (Amersham-Searle) diluted with 2 vol of toluene scintillation fluid. These two treatments gave very yellow solutions (as previously reported for NCS by Schrecker et al., 1972), and glacial acetic acid dissolution was not compatible with toluene scintillation fluid because of the formation of a precipitate.

The data of Table I show that four counting methods with nitrocellulose filters and three methods with glass-fiber filters are clearly superior to all the others (methods presented in boldface in Table I). By far the best counting efficiency for this high-molecular-weight DNA was obtained with NCS treatment of glass fiber-collected DNA counted in toluene scintillation mixture (slightly modified from the procedure of Birnboim, 1970). Undiluted Protosol (glass fiber–filtered DNA) was effective but less efficient; and it did not give solubilization when diluted with a toluene scintillation mixture. None of the methods with nitrocellulose filters equaled those with glass-fiber filters in counting efficiency. Ethyl-Cellosolve and dimethylformamide dissolution of nitrocellulose filters, with counting in xylene-based scintillation solutions, resulted in very stable count rates at relatively good efficiencies that were equivalent to the efficiencies of the internal standards. Treatment of DNA adsorbed on nitrocellulose filters with hot perchloric acid (PCA) (a method of solubilization that dissolves the DNA but not the filter) also appeared to be satisfactory, although this involved extra manipulations. In the last-mentioned three methods the DNA, according to our criteria, was solubilized.

The potential magnitude of the error involved in estimating counting efficiency when the macromolecule is not dissolved is best seen with the data with Bray's solution (nitrocellulose filter—no solubilizer) in Table I. Even though the filter dissolved with time and the count rate appeared to be stable, the internal standard counted with nearly 3-fold greater efficiency than did the DNA. Efficiency determinations by external standardization and channels ratio methods would also have given spurious efficiency values in such a case. Several other methods for solubilizing nitrocellulose filters were unsatisfactory as a result of a high level of chemical quenching (e.g., acetone, Protosol), failure to solubilize the macromolecule, or production of a marked yellow color which resulted in color quenching (NCS, Protosol). Many of the combinations with a Triton–toluene scintillation mixture, including samples of ^{3}H-labeled DNA in 25 μl of water without a filter, BSA, or solubilizer, resulted in falling count rates. Thus the collection and counting system consisting of glass-fiber filters, NCS/3 solubilizer (footnote *l*, of Table I), and toluene scintillation mixture was the most satisfactory for use with high-molecular-weight DNA and 100 μg of coprecipitant BSA. However, for low-molecular-weight DNA, collection on Millipore filters, dissolution with ethyl-Cellosolve, and counting in Aquasol appeared to be the most advantageous system.

Solubilization of the macromolecules on the filters is most likely to be helpful when samples to be compared contain different amounts of protein and/or DNA, and when macromolecules may be caught in the interstices of the filter, such that beta particles are quenched or adsorbed by the filter or by the macromolecule itself before reaching the scintillation solution. Also, the difficulty of exactly reproducing the conditions of filter and macro-molecule that exist in a given sample makes it impossible to obtain counting efficiency determinations in nonsolubilized samples by attempting to duplicate those conditions using molecules of known radioactivity. Thus methods of quench correction such as external standard, channels ratio, or internal standard can be more accurately employed with solubilized samples. When using solubilizing agents, such as NCS, it is important to avoid chemiluminescence whenever possible. In our experience, some lots of NCS are free of significant chemiluminescence, whereas others are not; when a bottle of solubilizer that does not have chemiluminescence is kept uncontaminated from intromittent pipettes and is stored under refrigeration, chemiluminescence does not develop. When macromolecules are solubilized from glass-fiber filters, even though the filters themselves are not solubilized, the transparency of the filters in toluene scintillation fluid allows any of the quench correction methods listed above to be employed. It is important to note, however, that external standard and channels ratio correction methods can apply to only one specific set of counting conditions (volume and type of scintillant and solubilizer, diluent, and amount of macromolecule). In general, internal standard quench correction is applicable to any solubilized sample under any counting conditions. For this purpose we recommend standards of ³H-labeled *n*-hexadecane, a stable and nonvolatile molecule which has negligible quench of its own. When the precise amount of DNA being filtered is known as conditions are calibrated, internal standard quench correction can detect a discrepancy between the counting efficiency of the DNA and that of the internal standard. As we found such discrepancies to occur rather frequently (Table I), we assume this to be the result of precipitation of DNA from the scintillation solution. Precipitation of DNA-[³H] could also account for decreases in count rate upon storage of counting vials at 3°C; this was particularly true with triton–toluene scintillation solutions. As noted in Table I, it is also important to optimize the gain settings for the tritium channels to be used for the particular counting methods employed.

III. Determination of Optimal Conditions for Collection and Counting of DNA Polymerase Assay Products

An important factor in choosing a collection and counting system for radioactive macromolecular reaction products is the magnitude of the "blank" produced by trapped substrate molecules and/or impurities in the substrate. Accordingly, we investigated the effects of BSA content and filter type on the amount of the reaction blank in DNA polymerase assays with dTTP as the tritium-labeled substrate (Fig. 2). The solubilized counting systems previously found to be optimal for both glass-fiber and nitrocellulose filters were used. In both collection systems, blank values were lower when 5 μg of BSA was present than in the absence of added protein;

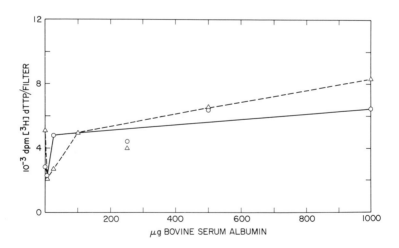

FIG. 2. Effects of BSA and filter type on reaction blanks with ^{3}H-labeled dTTP. Reaction mixtures in silicon-treated tubes containing 7.8 μCi (17.3 $\times$ 10^6 dpm) of ^{3}H-labeled dTTP (17.3 Ci/mmole), and varying amounts of BSA were treated with 10% TCA for 15 minutes at 0.6°C and filtered through nitrocellulose or glass-fiber filters. Nitrocellulose filters (circles) were dissolved in ethyl-Cellosolve and counted in Aquasol, and glass-fiber filters (triangles) were treated with NCS/3 and counted in toluene scintillation mix as described in Table I. Conversion to disintegrations per minute was made with the aid of internal standards. Reproduced by permission from Schrier and Wilson, *Analyt. Biochem.* **56**, 196–207 (1973).

values for the blank increased between 5 and 100 μg of BSA. However, the presence of larger amounts of BSA—up to 1 mg—resulted in only a slight additional increase in the blank. Similar blank values were observed at each of the levels of BSA for nitrocellulose and glass-fiber filters, except at 25 and 100 μg of BSA, where the values using nitrocellulose filters were higher. Flow rates through nitrocellulose filters at 500 and 1000 μg of BSA were so slow as to make the use of these filter techniques impracticable.

Reaction blanks of 2000–8000 dpm, which were observed in the experiments shown in Fig. 2, may be unacceptable for assays of low levels of DNA polymerase activity. Thus an attempt was made to find conditions, for collection and washing of DNA on filters, that would lower the blank values without resulting in a loss of low-molecular-weight single-stranded DNA. Blank values using nitrocellulose and glass-fiber filters were assessed with DNA polymerase reaction mixtures containing 100 and 1000 μg of BSA, respectively, and 4.7 μCi (each) of ³H-labeled dTTP. As shown in Table II, inclusion of 1 mM thymidine and a mixture of sodium ortho- and pyrophosphates (TTPP) in the precipitant did not decrease the blank with nitrocellulose filters, but did effect a significant reduction with glass-fiber filters. Blank values were reduced with both filters when they were washed with chloroform–methanol (A wash), and the combination of this wash with dTTP precipitant resulted in a very significant blank reduction for glass-fiber filters. Some additional decrease in blank values with nitrocellulose filters was accomplished by the addition of washes with HCl and ethanol. It was not possible with any of the treatments tested to lower the blanks as completely with nitrocellulose as with glass-fiber filters. Centrifugation data (footnote f, Table II) are included to show that much of the remaining blank appears to be acid-precipitable or to be trapped in precipitated BSA. Other procedures for reducing blanks may also be useful, such as those described by Oldham (1971).

In similar experiments (Table II) with the sheared single-stranded DNA in place of the labeled dTTP, collection of the DNA on nitrocellulose filters was 82–92% complete and was not affected by any of the wash procedures. In contrast, recoveries of DNA were much lower (as in Fig. 1) on glass-fiber filters, and some washing procedures effective in lowering tritium-labeled dTTP blanks also washed off tritium-labeled DNA. This result confirmed the conclusions from other data that, regardless of obtainable counting efficiencies, glass-fiber filters should not be used for collection of low-molecular-weight DNA. From the data presented in Table II for nitro-cellulose filters, it is clear that significant reductions in blank values can be obtained by combining the dTTP precipitant, A wash, and either washes with HCl and ethanol or the B wash of Waters and Yang (1974).

TABLE II

EFFECTS OF PRECIPITANT SOLUTIONS AND WASHING PROCEDURES ON DNA
POLYMERASE ASSAY BLANK AND RECOVERY OF LOW-MOLECULAR-WEIGHT
[3]H-LABELED DNA[a]

Test number	Precipitant solution	Washing solution(s)	Assay blank (dpm)	^{3}H-labeled DNA recovery	
				Disintegrations per minute	Per-cent
Nitrocellulose filters					
1	10% TCA	10% TCA	3,450	19,740	81.8
2	10% TCA	A wash[b]	1,935	NT[c]	—
3	TTPP[d]	10% TCA	4,860	NT	—
4	TTPP	A wash	2,140	20,410	84.6
5	TTPP	A wash; 1 N HCl; absolute EtOH	1,450	21,210	87.9
6	TTPP	A wash; B wash[e]	1,404	20,360	84.4
7	TTPP	A wash; benzene	9,000	22,120	91.7
8	TTPP	A wash; ligroine	2,517	22,270	92.2
9	TTPP; spun[f]	A wash	972	NT	—
Glass-fiber filters					
1	10% TCA	10% TCA	6,200	12,220	50.6
2	10% TCA	A wash[b]	1,530	NT	—
3	TTPP[d]	10% TCA	1,790	12,960	53.7
4	TTPP	A wash	370	9,590	39.8
5	TTPP	A wash; 1N HCl; absolute EtOH	690	7,150	29.6
6	TTPP	A wash; B wash	4,270	8,300	34.4
7	10% TCA	C wash	5,080	NT	—
8	TTPP	C wash	558	NT	—
9	TTPP; spun[f]	A wash	1,426	NT	—

[a]Each reaction mixture contained 4.7 μCi of ^{3}H-labeled dTTP (17.3 Ci/mmole) or 24,120 dpm of sheared ^{3}H-labeled DNA (see Section II,A), with the indicated amounts of BSA in H_2O at a final volume of 125 μl in siliconized 10 $\times$ 75 mm tubes. Each was treated with the indicated precipitant and washed with TCA as described in footnote a of Table I and other washes (40 ml each) as indicated. Nitrocellulose filters were dissolved in ethyl-Cellosolve and counted in Aquasol as previously. Glass-fiber filters were dried, treated with NCS/3 with toluene scintillation mixture, and counted as described above. Counts were converted to disintegrations per minute to facilitate comparison (counting efficiencies: 28% for nitrocellulose and 41.8% for glass fiber). Reproduced by permission from Schrier and Wilson, *Analyt. Biochem.* **56**, 196–207 (1973).

[b]A wash: first with 10% TCA, then with $CHCl_3$–MeOH (1:1).

[c]NT, Not tested.

[d]TTPP: a 1:1 mixture of saturated solutions of Na_2HPO_4 and $Na_4P_2O_7$ supplemented with solid TCA to 10% (w/v) and thymidine to 1 mM.

[e]B wash: 5% (v/v) glacial acetic acid, 0.7% (v/v) concentrated HCl, and 1% (w/v) NaH_2PO_4 in H_2O (Waters and Yang, 1974).

[f]After 15 minutes at 0°C for precipitation, tubes were centrifuged at 7000 g for 10 minutes at 3°C; pellets were resuspended in 10% ice-cold TCA, filtered, and washed as indicated.

[g]C wash: first with 10% TCA, then with ethanol–diethyl ether (1:1); this treatment destroyed nitrocellulose filters.

IV. Selection of Methods for a Particular System

A. Guidelines for Evaluation

The experiments presented here have evaluated DNA precipitation, collection, and counting techniques as applied to only a limited number of systems, but they offer guidelines by which the optimal assay system for a given experimental design may be determined. Recommendations for handling only two types of DNA can be made using the data presented (see Section IV,C), and these methods may not necessarily apply directly to all other systems. It is important that the optimal conditions for precipitation of the macromolecule be examined under conditions where the easily confused effects of collection and counting efficiencies are minimized. Similarly, in a comparison of collection techniques, constancy of precipitation and counting efficiencies must be ensured, and some method of determining the absolute efficiency of collection should be available and employed. We noted several problems with counting efficiency and several advantages of solubilization; the latter manipulation may perhaps be avoided if constancy of the other parameters and the amount of macromolecular material on the filters is maintained, but counting efficiency must be thoroughly investigated and evaluated frequently in order to ensure reliable results.

B. Characteristics of Glass-Fiber Filters

In this study we evaluated the collection ability of only Whatman GF/A filters; other glass-fiber filters from the same manufacturer or glass-fiber filters offered by other makers were not tested. To provide the reader with a more reasonable method of choosing or excluding a glass-fiber filter, we include Table III which lists nominal retention efficiency [particle size (in micrometers) that is 98% retained], filtration time, and some other characteristics for several glass-fiber filters. Data concerning the amount of a standard macromolecule (such as acid-precipitated BSA) that would result in clogging would be of great value, but are apparently not available. From the data of Table III, it is apparent that the Whatman GF/D filter would be less likely to retain low-molecular-weight single-stranded DNA, whereas the GF/F, and perhaps the 984H, filters might collect such DNA much more efficiently than the GF/A filters used in our experiments. Note that neither Whatman nor Reeve-Angel filters contain any binder. Binders used by other manufacturers may be acrylics, acrylonitrile, or butadiene-styrene. We do not know, and did not find in the literature, the effects of such binder materials on the collection, solubilization, or counting efficiencies of these filters for macromolecules, nor the effect of the binder on the magnitude

TABLE III

SOME PROPERTIES OF GLASS-FIBER FILTERS[a]

Designation	Brand	Nominal retention efficiency[b] (μm)	Thickness (mm)	Weight (gm/m²)	Time to filter[c] (seconds)
GF/A	Whatman[d]	1.6	0.25	52	27
GF/B	Whatman[d]	1.0	0.73	150	56
GF/C	Whatman[d]	1.3	0.26	55	30
GF/D	Whatman[d]	2.6	0.67	100	19
GF/F	Whatman[d]	0.7	0.44	75	55
934 AH	Reeve Angel	(1)[e]	0.30[f]	63[g]	h
984 H	Reeve Angel	(0.75)[e]	0.30[f]	65[g]	h

[a]Both brands are made from borosilicate glass produced by the same manufacturer. Neither brand contains any binder material when received by the investigator. Tests of the glass (Reeve Angel) have shown that boiling for 1 hour in 0.1 N NaOH solubilizes 6.4% of its weight, while boiling for 1 hour in 1 N H_2SO_4 solubilizes 2.3%. In addition, each $8\frac{1}{2} \times 11$ inch sheet of 934 AH has been found to contain 2940 μg of benzene-soluble organic material.

[b]Particle size that is 98% retained during liquid flow.

[c]Time required to filter 100 ml of clean water through a 10-cm² disc at 3 cm Hg.

[d]Made by W. and R. Balston, Ltd., Maidstone, Kent; data for these filters were taken directly or estimated from J. C. Meakin, and M. C. Pratt, in "Manual of Laboratory Filtration" (N. F. Kember, ed.), W & R Blaston, Ltd., Kent, 1972.

[e]These are estimates by Reeve Angel Co., and are based on very limited and very preliminary data. More reliable measurements, although of unknown significance, are the collection efficiencies in air of 99.97% and 99.98%, for 934 AH and 984 H, respectively, of DOP smoke particles 0.3 μm diameter.

[f]Given by Reeve Angel as 0.012 inch.

[g]Given by Reeve Angel as 0.0130 and 0.0133 lb/ft², respectively, for 934 AH and 984 H.

[h]Data not available.

of the reaction blank. We hope that in the near future comparisons of these other filtering supports with each other and with nitrocellulose filters will be made available.

C.   Recommendations

The recommended procedures for collecting and counting ³H-labeled DNA using both types of filters are summarized in Table IV. When the difference in cost is not a factor, and when DNA of unknown size is to be collected, nitrocellulose filters are recommended. When possible, the filter and precipitated macromolecule should be solubilized before counting.

TABLE IV

RECOMMENDATIONS FOR COLLECTION AND COUNTING OF TRITIATED
MACROMOLECULES ON FILTER DISCS[a]

Procedure	Recommendations
Counting	Solubilize filter and macromolecule (nitrocellulose filter) with ethyl-Cellosolve; or solubilize macro-molecule (glass-fiber filter) with NCS/3 or Protosol 1/3
	Xylene-based scintillation cocktail (nitrocellulose filter with ethyl-Cellosolve); or toluene scintillation mix (glass-fiber filter with NCS/3)
	Counting efficiency is about 1.5 times better with glass-fiber than with nitrocellulose filters.
Collection	Nitrocellulose filters with either high- or low-molecular-weight DNA and up to 250 μg protein per filter
	Glass-fiber filters only with high-molecular-weight DNA (90% collection at 500 μg protein) and 250–2500 μg protein per filter. Not recommended for low-molecular-weight DNA
Reduction of DNA polymerase assay blanks	Nitrocellulose filters with low-molecular-weight DNA, using minimum possible protein per filter, TTPP pre-cipitant, A wash, then B wash.
	Glass-fiber filters with high-molecular-weight DNA and > 200 μg protein per filter; use TTPP pre-cipitant and A wash

[a]Reproduced by permission from Schrier and Wilson, *Analyt. Biochem.* **56**, 196–207 (1973).

The xylene-based scintillation cocktail, Aquasol, was found to be excellent for the counting of filters solubilized with ethyl-Cellosolve or dimethyl-formamide. Presumably the cocktail described by Anderson and McClure (1973) would serve as well and be considerably less expensive. In our hands the procedure suggested by Schrecker *et al.* (1972), using cellulose acetate filters (0.5-μm pore size) and dissolution of DNA by NCS and water, gave higher counting efficiencies (34.9%) than we were able to obtain with dis-solved nitrocellulose filters. However, stable count rates could be obtained only when filters were rigorously dried before dissolution. In addition, with these filters, flow rates became impracticably slow when the reaction tubes contained more than 10 μg of BSA.

It is not known to what extent the results and recommendations described here are applicable to macromolecules other than DNA or to DNA mole-cules of sizes other than those evaluated here. Similar phenomena might be expected to occur with RNA, and the behavior of these systems as a function of the amount of BSA included in the precipitate suggests that they may also be valid for collection and counting of radioactive protein. It is also possible that systems to differentiate between DNA and RNA on discs, such as that of Chou *et al.* (1974), may prove valuable on the supports given here, although the possible effects of acids, alkali, and organic solvents on the filters (cf. footnote *a*, Table III) should be evaluated.

REFERENCES

Anderson, L. E., and McClure, W. D. (1973). *Anal. Biochem.* **51**, 173.

Birnboim, H. C. (1970). *Anal. Biochem.* **37**, 178.

Bollum, F. J. (1959). *J. Biol. Chem.* **234**, 2733.

Bollum, F. J. (1966). *In* "Procedures in Nucleic Acid Research" (G. L. Cantoni and D. R. Davies, eds.), Vol. 1, p. 296. Harper, New York.

Chou, S. C., Pan, H. Y. M., Hew, P., and Conklin, X. A. (1974). *Anal. Biochem.* **57**, 62–70.

Eigner, J., and Doty, P. (1965). *J. Mol. Biol.* **12**, 549.

Hodes, M. E., Kaplan, L. A., and Yu, P.-L. (1971). *Anal. Biochem.* **43**, 644.

Mans, R. J., and Novelli, G. D. (1961). *Arch. Biochem. Biophys.* **94**, 48.

Meyer, R. R., and Keller, S. J. (1972). *Anal. Biochem.* **46**, 332.

Oldham, K. G. (1971). *Anal. Biochem.* **44**, 143.

Schrecker, A. W., Sporn, M. B., and Gallo, R. C. (1972). *Cancer Res.* **32**, 1547.

Schrier, B. K., and Wilson, S. H. (1973). *Anal. Biochem.* **56**, 196.

Waters, L. C. and Yang, W.-K. (1974). *Cancer Res.* **34**, 2585.

Chapter 7

The Radioiodination of RNA and DNA to High Specific Activities

WOLF PRENSKY

Molecular Cytology Laboratory,
Memorial Sloan-Kettering Cancer Center,
New York, New York

I.	Introduction	121
II.	Radioiodine	122
	A. Nature of Commercial ^{125}I	123
	B. Physical Properties of ^{125}I	124
	C. Detection of ^{125}I Decay	126
	D. Use of ^{131}I for Iodination of Nucleic Acids	132
III.	Commerford's Reaction: General Description	136
IV.	Equipment Needs	138
	A. Radiation Laboratory	138
	B. Small Equipment and Supplies	139
V.	Radioiodination Reaction	141
	A. Reagents and Solutions	141
	B. Assembly of Reaction Mixture	144
	C. Column Chromatography	145
	D. Specific Iodinations and Results	146
VI.	Comments and Discussion	147
	References	152

I. Introduction

The discovery by Commerford (1971) of an effective method for the chemical iodination of nucleic acids has opened up the possibility of producing eukaryotic nucleic acids of high specific activity with iodine-125 (^{125}I) or iodine-131 (^{131}I). In contrast to nucleic acids obtained from prokaryotes, eukaryotic RNA and DNA can be labeled only to relatively low specific activities, and then at relatively great cost. Nucleic acids obtained from nontissue culture sources often cannot be labeled at all

121

in the course of their biosynthesis and, if labeled, have an even lower specific activity than those obtained from tissue culture cells. Enzymatic synthesis and chemical labeling are therefore the only useful methods when high specific activities are desired.

The preparation of labeled nucleic acids, especially RNA, of various types and specific activity is indispensible for most molecular hybridization studies involving eukaryotic DNA, and in studies of RNA sequence and complexity. The primary result of Commerford's reaction is the replacement of a hydrogen atom at the C-5 position of cytidine by an atom of iodine. The polynucleotide so altered has only slightly different physiochemical properties, and can replace ^{32}P- or ^{3}H-labeled materials in almost all biochemical studies in which a radioactive label is necessary.

In this article we dwell primarily on the problems encountered in the labeling of nucleic acids to a high specific activity and present methods for radioiodination under conditions of relative safety to the investigator. The factors influencing reaction rates and the utility of iodinated RNA and DNA in various areas of investigation were reviewed by G. F. Saunders in Volume IX of this series. Commerford's reaction is a highly reproducible one, and works well over a surprisingly wide range of temperature and iodine and RNA concentrations (Commerford, 1971; Tareba and McCarthy, 1973; W. Prensky, unpublished results). Except for holding the pH of the reaction constant between 4.5 and 5.0, and the thallic ion concentration to about 6 times that of the iodine concentration, we have varied all other reaction factors over a wide range and have generally obtained iodination levels predicted by Commerford's article and by Tareba and McCarthy. "Good" carrier-free ^{125}I can yield iodocytosine levels at least 50% as high as those obtained with ^{127}I, the stable isotope of iodine. The problems we had in reproducing Commerford's iodination levels were due either to lack of knowledge about the nature of the commercial ^{125}I reagent or, on some occasions, to the presence of chemical impurities in the nucleic acid solutions we attempted to iodinate.

A few words of caution are in order. The temperature, pH, and presence of oxidizing agent all conspire to volatilize a fraction of the radioactive iodine present in the reaction mixture. Radioiodination should be attempted only if the investigator is prepared to institute rigid measures of personnel monitoring and protection. Our efforts in this regard are detailed in this chapter.

II.   Radioiodine

Either ^{125}I or ^{131}I can be used to obtain radioactive nucleic acid of high specific activity (Heiniger et al., 1973). When a single chemical label is desired, ^{125}I is much the preferred isotope. It is our belief that ^{131}I labeling should be performed only in studies in which double labeling is absolutely necessary for the success of the investigation. The problems inherent in use of ^{131}I are discussed in Section II,D. Most of the discussion here therefore concerns the properties and use of ^{125}I in the chemical labeling of RNA and DNA.

A. Nature of Commercial ^{125}I

^{125}I is obtained from neutron irradiation of ^{124}Xe and is produced in essentially carrier-free condition in commercial quantities by several companies. A by-product of the reaction is ^{126}I, which has a 13-day half-life and emits gamma rays with energies in the range of 388–1420 keV. The presence of large fractions of ^{126}I therefore require much heavier shielding at all stages of iodination. AEC-Canada produces ^{125}I with very low levels of ^{126}I contamination, and in the United States this product is presently marketed by New England Nuclear Co., Boston, Mass. New England Nuclear also provides a report of independent quality control analyses with each sample.

The production of Na^{125}I at the reactor site entails the solubilization of the ^{125}I in an aqueous medium. The exact process used by each of the manufacturers is regarded as proprietary information, and it is therefore difficult to establish which salts other than Na^{125}I are generally present in the product of each manufacturer. For the purpose of RNA and DNA iodination, some of these salts appear to have an undesired buffering capacity. In our experience "low-pH iodine," with no base added, as marketed by New England Nuclear, has an apparent buffering capacity of about 0.15 M when used in RNA iodinations, and appropriate amounts of acid are added by us to correct for this property. Different amounts of acid are required for ^{125}I from sources other than the Canadian reactor.

The highest possible concentrations of iodine should be purchased when completely carrier-free reactions are to be carried out. Na^{125}I, at a concentration of 200 mCi/ml, is about 0.9 × 10^{-4} M iodine. Since about 0.5 × 10^{-4} M is the desired final concentration of iodine, the reaction mixture is easiest to assemble with concentrations of commercial ^{125}I in the range 400–800 mCi/ml. The obtainable ^{125}I concentration varies from lot to lot, but should not be below 300 mCi/ml for carrier-free iodination of nucleic acids.

The necessity of using high concentrations of ^{125}I in the assembly of Commerford's reaction introduces an additional hazard to the success of the procedure. Typically a commercial lot has a concentration of 400 mCi/ml, and 20 mCi is contained in 0.05 ml of solution. Repeated opening and closing of the isotope vial, aside from presenting the investigator with the hazards of iodine vapor, will result in an appreciable loss of water by evaporation, which changes the concentration of extraneous salts and leads to un-predictable pH fluctuations in the final reaction mixture. It is therefore advisable to purchase concentrated isotope solutions in vials containing not less than 30 mCi and to remove needed aliquots with a special syringe needle.

B.   Physical Properties of ^{125}I

An excellent account of the relevant physical properties of ^{125}I has been published by Ertl et al. (1970), and the data given here are summarized from their article.

^{125}I decays with a half-life 60 days. It emits soft gamma radiation ($E_{max} = 35.4$ keV) and various conversion and Auger electrons with a maximum energy of 34.6 keV. The decay proceeds in two steps: (i) transmutation by electron capture, with a half-life of 60.0 $\pm$ 0.5 days, into ^{125}Te^m; and (2) transition of ^{125}Te^m to stable ^{125}Te, with a half-life of 1.6 $\times$ 10^{-9} seconds, either by internal conversion or by emission of gamma radiation.

The decay scheme is not simple, at least to a nonphysicist. A total of 165.8 gammas are given off per 100 decays of ^{125}I (Table I). Eighty-five percent of all the gamma radiation emitted has a narrow energy spectrum, 27.5–31.7 keV. The remaining activity is divided between 3.7-keV (13.3%) and 35.4-keV (4.2%) photons.

The low energy of the gamma emissions has two practical effects in our context. The first is that relatively thin lead shielding is sufficient for personnel protection when ^{125}I with a low content of ^{126}I is utilized in the laboratory. In practice, 0.8 mm of lead foil reduces gamma-ray exposure from

TABLE I

CALCULATED GAMMA- AND SOFT X-RAY EMISSIONS OBTAINED FROM
^{125}I DECAY[a]

Type of photon emission (keV)	Number of gammas per 100 decays	Relative frequency (%)
3.7	22.0	13.3
27.5	74.3	44.8
27.2	37.9	22.9
31.0	20.1	12.1
31.7	4.5	2.7
35.4	7.0	4.2
Total	165.8	100.0

[a]Ertl et al. (1970).

^{125}I by a factor of more than 1000 times. With adroit use of various shielding materials (lead foil, lead glass, vial and test tube shields) large amounts of ^{125}I can be handled without great risk to personnel from gamma radiation. A second consequence of the soft gamma-ray spectrum of ^{125}I is the fact that it is not efficiently monitored with the Geiger-Muller tubes common to most laboratory survey meters. For adequate monitoring of personnel contamination the acquisition of special detection equipment is therefore a virtual necessity.

The energy spectrum of the electron radiation emitted in the course of ^{125}I decay is shown in Table II. A total of 579 electrons is emitted per 100 decays, over half of which have a maximum energy of 0.5 keV. Ertl *et al.* have calculated that this radiation component of ^{125}I decay has a practical range in water of 0.02 μm. No quantitative studies have been published on the use of ^{125}I in electron microscope autoradiography, but the great abundance and limited range of these electrons should make the use of ^{125}I preferable to that of other isotopes in electron microscope autoradiography. The usefulness of ^{125}I in light microscope autoradiography stems from the fact that, per 100 decays, 35 electrons can be expected to reach a distance of 0.5 μm in water, 20 reach 3.5 μm, 7 reach 10 μm, and 1 reaches 16.8 μm (calculations by Ertl *et al.*). The maximum possible range is 22 μm. Assuming that no image is obtained from the 0.5-keV electrons, 55% of the single grains seen in light microscope autoradiographs localize the source of radiation within 0.5 μm, and 87% of the grains localize it within a radius of 3.5 μm. In dipterans, when exposure is adjusted to give the minimum necessary density of silver grains,

TABLE II

ENERGY AND FREQUENCY DISTRIBUTION OF THE ELECTRON
RADIATION EMITTED IN THE COURSE OF ^{125}I DECAY[a]

Electron energy (keV)	Electrons per 100 transmutations	Relative frequency (%)
0.5	299	51.6
3.0	107	18.5
3.5	54	9.3
3.6	80	13.8
22.7	17	2.9
26.1	9	1.6
31.5	11	1.9
34.2	2	0.3
Total	579	99.9

[a]Adapted from Ertl *et al.* (1970).

the site of molecular hybridization can be localized to a single band of a polytene chromosome (Fig. 6).

C. Detection of ^{125}I Decay

Because of the low level of both the gamma and electron emissions from ^{125}I decay, standard laboratory survey meters will be very insensitive for the detection of ^{125}I. Most of the survey meters equipped with GM tubes have an efficiency of less than 0.05%. The Model 540 Scintillation Meter (Mini Instruments, Ltd., London WC1) sold in the United States by Research Products International, Elk Grove Village, Illinois, is the most satisfactory. The Model 540 has a thin sodium iodide crystal with a very thin aluminum entrance window designed to permit passage of the low energy photons from ^{125}I decay. It is a very sensitive instrument, capable of detecting 10 nCi (22,200 dpm) of ^{125}I radiation or less. It is, at present, the only moderately priced meter with which meaningful surveys of personnel, clothing, and work-area contamination can be achieved by the laboratory investigator.

Lest these injunctions about the need of an adequate low level survey meter for ^{125}I detection be considered a personal idiosyncracy of this author, the following experience should tend to dispel that notion. Early in my involvement with chemical labeling with ^{125}I, I had my thyroid activity determined at several installations and was found to have "normal" thyroid radioactivity. Once I placed a test tube containing 50,000 cpm of ^{125}I in my shirt pocket, and again was found to have no detectable thyroid contamination! Failure to detect this activity was due to the fact that most medical whole-body gamma detectors are constructed for the purpose of detecting ^{131}I or other high-energy gamma emitters, and many nuclear medicine departments do not own large thin-window crystals capable of efficient ^{125}I detection. The same is true of many health physics departments, and the investigator is therefore on his own in this regard.

Aside from the practical difficulties of detecting ^{125}I as an environmental contaminant, it is truly an isotope for all seasons as far as other modes of detection are concerned. Almost all manufacturers of nuclear detection equipment now market gamma-ray counting equipment, usually with thin window crystals designed for the efficient detection of the low-energy photons of ^{125}I decay. The counting efficiency of a good system should be 40–60% with background in the range of 50 cpm. ^{125}I decay can also be detected by liquid scintillation spectrometry (Rhodes, 1965; Bransome and Sharpe, 1972). Bransome and Sharpe described the adjustment of several types of liquid scintillation spectrometers for efficient detection and separation of the energy spectra of ^{125}I and ^{131}I.

The ability to detect two isotopes by spectrometry is especially useful when an enzymatically or biosynthetically labeled polynucleotide is radio-iodinated to increase its specific acitivity. In our experience we could not efficiently separate ^3H radiation from that of ^{125}I by liquid scintillation counting. However, ^{32}P ($E_{max} = 1710$ keV) can be easily separated from the ^{125}I radiation spectra. In addition, ^{32}P can be detected in a liquid scintillation spectrometer by Cerenkov counting, with ^{125}I making no contribution what-soever to the observed rate of radioactive decay. The same vial can sub-sequently be placed in a gamma-ray spectrometer for the evaluation of ^{125}I decay alone. ^{14}C ($E_{max} = 155$ keV) can probably be resolved from ^{125}I by liquid scintillation spectrometry, but we have not attempted to do so.

Since automatic gamma-ray spectrometers are frequently not available, NaI crystal inserts (Bicron Inc., Newbury, Ohio) for the more ubiquitous liquid scintillation spectrometer may provide a measure of counting ver-satility to many investigators. These inserts allow the detection of ^{125}I gamma rays without the use of a scintillation cocktail and worked surprisingly well (efficiency approximately 55%, background about 50 cpm) in the hands of one of our collaborators).

The soft gamma rays of ^{125}I decay efficiently expose many types of medical X-ray film, and four layers of film can be easily exposed at one time to obtain exposures of different intensity in one envelope or cassette. Consequently, autoradiographs of acrylamide gels and various thin layer chromato- and electropherograms can be readily prepared and evaluated (Robertson et al., 1973; Jacobson et al., 1973; Dube, 1973). This property of ^{125}I is especially useful for determining the size distribution of radioiodinated RNA and DNA by acrylamide gel electrophoresis (Fig. 1), for establishing the base sequence complexity of various types of RNA preparations following T_1 or pancreatic RNase digestion (Figs. 2–4), and for determining the types of nucleotides that are iodinated following T_2 RNase digestion (Robertson et al., 1973).

Following the methodology outlined by Gall and Pardue (1971), ^{125}I-labeled RNA was found useful in gene localization by in situ DNA–RNA molecular hybridization (Prensky et al., 1973; Altenburg et al., 1973; Gri-gliatti et al., 1973; Steffensen and Wimber, 1973). It is worthwhile therefore, to examine in some detail the detection of ^{125}I decay in photographic emulsions. Examples of the localization of 5 S rRNA genes in polytene chromosomes of Diptera, in meiotic chromosomes of corn, and in mitotic chromosomes of man are shown in Figs. 5–8. The images seen are due to the emission of electrons, since nuclear track emulsions are relatively insensitive to gamma radiation.

Ada et al. (1966) studied the efficiency of silver grain formation in Ilford L-4, Kodak NTB-2 and AR-10 autoradiographic film emulsions and com-

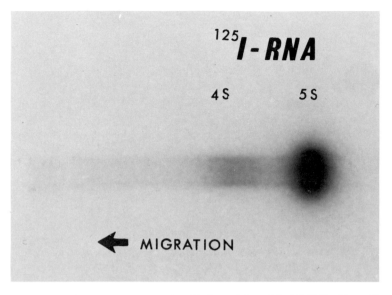

FIG. 1. Acrylamide gel electrophoresis of iodinated *Drosphila* 5 S RNA. X-ray film auto-radiograph of longitudinal section of a cylindrical gel. The major radioactive spot is due to 5 S RNA, and the sample shows an admixture of 4 S RNA.

pared detection efficiencies of electrons from ^{125}I and ^{3}H decay. Two test systems were used, surface-labeled red blood cells and protein films of various thicknesses. The radiation from ^{125}I contains both soft (less than 4 keV) and hard (22–34 keV) electrons, and most are soft (Table II). With protein films of infinite thinness all emergent electrons can affect the emulsion, but with radioactive sources of over 1 μm thickness the latent image in the emulsion is due almost entirely to the hard electrons. Efficiency of silver grain information in photographic emulsion can therefore be described by reference to either one of two physical parameters: grains per disintegration or grains per emergent electron. The detection efficiency of ^{125}I over very thin films was 0.73 grain per disintegration and, as the thickness of the film increased, the number of grains per disintegration fell sharply and reached a plateau value of 0.23 for film thickness above 0.5 μm. The number of emergent electrons also fell sharply but, as thickness increased, the relative efficiency per emergent electron increased from 0.169 to 1.23 grains per electron. In thin films, tritium was detected with 0.116 grains per disintegration. With tritium, as with ^{125}I, efficiency per emergent electron went up as the protein film absorbed a relatively greater proportion of the weak particles, rising from 0.29 grain to over 0.62 grains per emergent electron

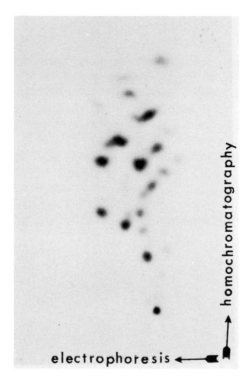

electrophoresis ←

homochromatography →

FIGS. 2–4. X-ray film autoradiographs of RNAse T₁ digests of several iodinated RNA preparations. The number of oligonucleotides increases with the base sequence complexity of the RNA preparations. Methods of preparation are given by Robertson *et al.* (1973). The T₁ digests and autoradiographs were prepared by Elizabeth Dickson and H. D. Robertson.

FIG. 2. 5 S RNA from HeLa cells, specific activity about 40×10^6 dpm/μg. The RNA is 120 bases long (Forget and Weissman, 1967). The unlabeled RNA was a gift of T. Borun.

over thick protein film. The more energetic electrons from ^{125}I were therefore capable of inducing more than 1 grain per electron, and Ada *et al.* report observing tracks of up to 5 grains in their autoradiographs. Possible examples of such tracks are illustrated in Figs. 5–8.

When the isotopes were localized at the surface of red blood cells, 0.44 grain per disintegration were obtained from ^{125}I decay and 0.12 grain from tritium decay. The value for ^{125}I was midway between 0.73 and 0.23 found with protein films of varying thickness, and reflected the fact that half of the ^{125}I decay came from the surface of the cells next to the glass slide. Whereas the greater efficiency of ^{125}I detection is not surprising, it is also clear that the absolute efficiency of ^{125}I detection will vary with dif-

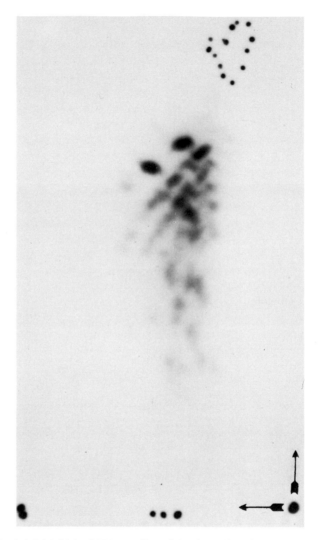

FIG. 3.   Duck 9 S (globin) mRNA, specific activity about 30×10^6 dpm/μg. The RNA was a gift of D. Housman. The sample has a probable complexity of 600 to 1200 nucleotides. When we iodinated and fingerprinted aliquots of rabbit 10 S reticulocyte mRNA, oligonucleotide patterns of greater, equal, and lesser complexity were obtained, depending on the number of steps used in the mRNA purification. Successive iodinations could thus be used to monitor the base sequence complexity of an allegedly pure mRNA preparation.

ferent experimental conditions. In the case of *in situ* hybrids of iodinated RNA, it is highly unlikely that detection efficiency will be less than 0.23 grain per disintegration, and on first approximation should be higher for

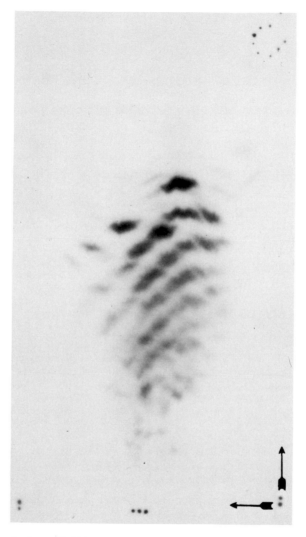

FIG. 4. Bacteriophage f2 RNA, specific activity about 10^7 dpm/μg. This RNA has a complexity of about 3600 nucleotides and was used as a standard for determining the quality of fingerprints from large RNA molecules. This pattern is comparable to the previously established pattern of ^{32}P-labeled f2 RNA (Robertson and Jeppesen, 1972).

RNA bound to metaphase chromosomes than for the thicker polytene chromosomes shown in Figs. 5 and 6. A further complication is introduced by our lack of knowledge as to whether or not efficient DNA–RNA hybrids form in parts of the chromosome bundle away from the surface of the

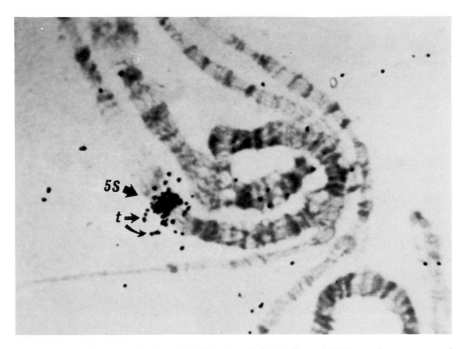

FIGS. 5–8. *In situ* molecular hybridization of 5 S iodinated RNA to chromosomes of different species. Arrow with "5 S" indicates location of strongest and most consistent hybridization signals in autoradiographs; t, position of possible tracks from a single high-energy electron.

FIG. 5. Localization of the genes for 5 S rRNA in *Drosophila* (Wimber and Steffensen, 1970; Prensky *et al.*, 1973). Autoradiographic exposure 1 week, specific activity about 130×10^6 dpm/μg. Prepared by Dr. Paul Szabo.

preparation. The concentration of RNA in solution may be lower in unexposed parts of the chromosome preparation, and most of the hybridization may in fact occur at or near the surface of the chromosome. Detection efficiency would also be higher with thicker autoradiographic film emulsions. However, when there is sufficient signal present, thinner emulsions may yield more precise gene localizations. In fact, because of the great abundance of very low-energy electrons, ultrathin samples and emulsions might yield localizations comparable in precision to those obtained from tritium.

D.   Use of ^{131}I for Iodination of Nucleic Acids

The chemical properties of ^{131}I are identical to those of ^{125}I, and solutions of 200–400 mCi/ml can be purchased from several suppliers. ^{131}I

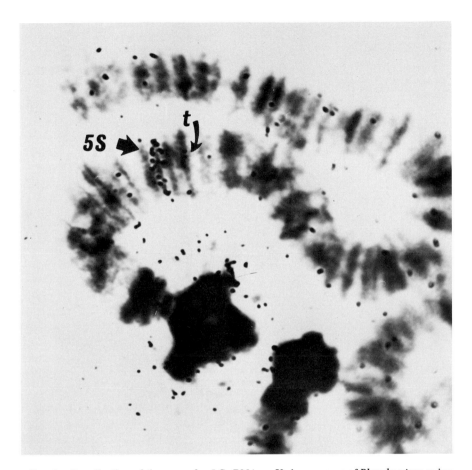

FIG. 6.   Localization of the genes for 5 S rRNA on X chromosome of *Rhynchosciara*, using *Drosophila* 5 S RNA. This hybridization reaction was carried out without the addition of any carrier 18–28 S rRNA. As a consequence the end of the chromosome containing the nucleolar organizer region appears to be labeled as well. Prepared by D. M. Steffensen.

has a half-life of 8.6 days; 79% of the gamma radiation is from a 364.9-keV photon, and about 90% of the β-radiation is in the form of a 608-keV E_{max} particle. These physical characteristics make ^{131}I less than generally useful for most nucleic acid labeling studies. The relatively short half-life may well interfere with the proper utilization of the labeled compunds, although the high efficiency and low background noise with which the gamma or the high-energy beta rays of ^{131}I decay can be detected may make ^{131}I-labeled compounds useful over a period of four to five half-lives.

    The short half-life of ^{131}I theoretically permits the synthesis of materials

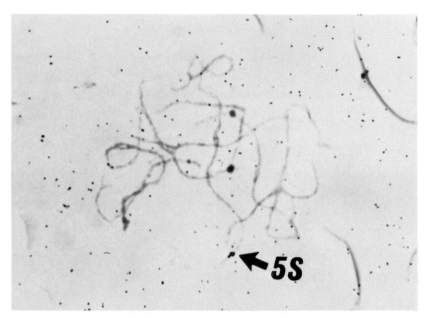

FIG. 7. Localization of 5 S ribosomal genes in pachytene chromosomes of *Zea mays*. Homologous 5 S RNA was used, and the fairly high background is probably due to partial degradation of the iodinated RNA prepared for this experiment. Fingerprint analysis failed to reveal the presence of extraneous species of RNA. Arrow indicates locations of 5 S genes (distal part of long arm of chromosome 2, Wimber *et al.*, 1974). Photograph by D. E. Wimber.

of extremely high specific activity (16,000 Ci/mM for ^{131}I versus 2200 Ci/mM for ^{125}I and 29 Ci/mM for tritium). However, since the Commerford reaction works optimally in the range 0.5×10^{-4} to 1.0×10^{-4} M iodine, it would be more difficult to assemble an efficient reaction mixture containing undiluted ^{131}I isotope. Diluting the isotope with stable iodine would, of course, reduce the specific activity per incorporated iodine atom in the labeled polynucleotide. In practical terms the expectations of extremely high specific activity from the use of ^{131}I are therefore somewhat illusory.

A more general objection to the extensive use of ^{131}I for high-specific-activity labeling of nucleic acids arises from the heavy shielding needed for the handling of large amounts of [131]I. The higher energies of the gamma ray necessitate the use of heavy lead shielding in the course of the relatively large number of steps involved in the synthesis and purification of iodinated nucleic acids. Setting up a minimally safe laboratory for the iodination of

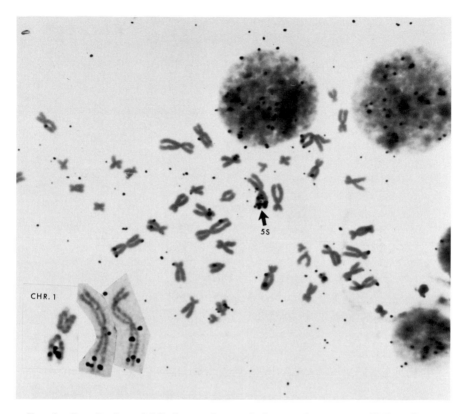

FIG. 8. Localization of 5 S ribosomal genes in human chromosomes. HeLa 5 S RNA, same RNA as in Fig. 2, but iodinated to a specific activity of about 400×10^6 dpm/μg. Localization on long arm of chromosome 1, probably at q32-q43 (Steffensen *et al.*, 1974, 1975; Johnson *et al.*, 1974). Inset shows chromosome 1 from other cells.

nucleic acids with ^{131}I therefore entails many more problems than are posed by the use of ^{125}I, both with respect to shielding as well as remote handling of the reaction mixtures.

^{131}I is appropriate as a label for nucleic acid when double labeling is desired, but labeling in such applications need not be of high specific activity. Its energy spectrum is easily separable from that of ^{125}I in a two-channel spectrometer, either by gamma or liquid scintillation counting (Rhodes, 1965; Bransome and Sharpe, 1972). One species of RNA or DNA can be iodinated with a relatively small amount of ^{131}I, and a second species of nucleic acid can be labeled to a much higher specific activity with ^{125}I. The two species of nucleic acid can then be used in a combined molecular hybridization reaction, and the extent of hybridization of each labeled species determined at the same time.

A second use for ^{131}I is in the labeling of cytidine or deoxycytidine triphosphate. We have recently found that short RNA oligonucleotides and deoxycytidine triphosphate iodinate more efficiently than 4 S or larger molecules (W. Prensky, unpublished observations). The concentration of iodine in such reactions need not be higher than 10^{-5} M, and most likely could be lower. If only one or two species of molecules are to be iodinated repeatedly with ^{131}I, under conditions of fairly low iodine concentration and therefore larger reaction volume, safety procedures can be worked out specific to the iodination and purification of the triphosphate alone. On account of the high energy of the beta particle, enzymatically synthesized RNA-^{131}I and DNA-^{131}I are not as useful for *in situ* molecular hybridization studies as is similar ^{125}I-labeled material. There is no plausible objection to use of such molecules in other types of nucleic acid renaturation studies. ^{131}I-labeled triphosphates may be expected to be relatively more stable than their ^{125}I analogs, because the nature of ^{131}I decay precludes the deposition of much energy in the immediate vicinity of the disintegration site, and secondary damage to other molecules in the solution should therefore be less than with ^{125}I.

III. Commerford's Reaction: General Description

The factors influencing the rate of RNA and DNA iodination have been discussed in Volume IX of this series by G. F. Saunders. We have made no departures from the reaction conditions defined by Commerford (1971), except in so far as they have simplified the assembly and processing of the considerably smaller reaction volumes needed when carrier-free or mildly diluted ^{125}I is used for nucleic acid iodination. So far the only significant departure from Commerford's procedure was given by Orosz and Wetmur (1974). They found that reheating iodinated DNA at near neutral pH and in the absence of thallic ion significantly improves the melting temperature of iodinated DNA–DNA hybrids, probably because of the reduced probability of formation of other than 5-iodo derivatives of cytosine.

Briefly, iodination of cytosine in polynucleotides occurs quite readily under the following range of conditions:

Iodide ion	0.3×10^{-4} to 1.0×10^{-4} M
Cytosine (as RNA or DNA)	0.05×10^{-4} to 1.0×10^{-3} M
Thallic ion	4.0–$10 \times$ the I^- concentration

Sodium acetate (buffer)   0.03–0.2 M
pH   4.0–5.0 (we prefer 4.7)
Temperature   30°–80°C

The iodination is carried out in two steps. The first step involves heating the mixture to 60°C at pH 4.7. The pH of the mixture is then raised to neutrality or higher (we add about 10 vol of pH 7.5 TNE; see Section V,A for buffer formula). It is then heated again for 20–30 minutes at 60°C. The first heating step results in the covalent bonding of iodine to cytosine with the formation of an unstable 5-iodo-6-hydroxydihydropyrimidine. Reheating of the mixture at pH 6.0 or higher causes dehydration of the product and the formation of an iodinated derivative of the nucleic acid whose physiochemical properties closely resemble those of the original polymer.

At 60°C the reaction mixture is heated for 5–20 minutes. With long reaction times iodination is probably terminated by the simultaneous conversion of part of the iodide to iodate. Commerford showed a doubling of the initial reaction rate for every 10°C, and appropriate changes in the reaction time should be used when the reaction temperature is varied. About 20–40% of iodine is bound to RNA or DNA when equimolar amounts of iodine and cytosine are used. Higher percentages of cytosine iodination are observed when the ratio of iodine to cytosine is increased, although lower efficiencies of iodine utilization are then obtained (Commerford, 1971; Tareba and McCarthy, 1973).

The pH of the solution is a critical factor. Since the pK of acetate ion is 4.7, there is no discernible reason for deviating from that pH when RNA is to be iodinated. A somewhat higher pH might be used to prevent the possibility of slight depurination of DNA under the same conditions. At pH 5.5 hardly any iodination takes place.

The reaction is initiated by the additon of the thallic ion to the rest of the reaction mixture. I^- is converted to I_2 and to IOH, and it is the latter that reacts with cytosine. Some uracil is also iodinated, forming 5-iodouracil. Carbon atoms with unsaturated bonds in modified bases of tRNA are also iodinated (Robertson et al., 1973). Analysis of electropherograms of T_2 RNase digest of several natural RNAs has failed to show that any other major unmodified bases (adenine or guanine) are iodinated (Robertson et al., 1973; H. D. Robertson, E. Dickson, P. Model, and W. Prensky, unpublished data). Commercial homopolymers like poly A or poly G may have enough trace cytosine to appear weakly iodinated. Iodination and fingerprint analysis of a commercial preparation of a copolymer of deoxyadenine and deoxythymidine revealed iodocytosine in the digest. Thus Commerford's report of the weak iodination of purine ribopolymers was probably due to contamination of the homopolymer by trace amounts of cytosine.

IV.   Equipment Needs

A.   Radiation Laboratory

Provisions for safe and adequate working facilities are the most costly aspect of the radioiodination of macromolecules. For the most part the investigator has to adapt existing radioactive hood facilities to his own needs, and the comments given here should be viewed in that light.

It is essential to carry out the reaction in a well-ventilated hood, meeting mandatory local or national standards for the containment of volatile radioactive materials. In a big-city environment, filtration of the effluent air through charcoal filters may be required. We found it difficult to work in a hood less than 4 ft (1.2 m) wide. If possible the hood should have provisions for under-the-counter disposal of radioactive waste material. The room containing the hood (the "hot lab") should be off limits to nonessential personnel and, if possible, should be used only for the radioiodination of nucleic acids.

Since small amounts of RNA must be handled under fairly exposed conditions, a clean workbench or counter should be available for the assembly of all nonradioactive reaction components. Shelf space for sterile glassware and disposable plastic ware should also be provided. It is also helpful to have a refrigerator and a freezer in the radiation laboratory for storage of the solutions used in the iodination and purification procedures. A manual gamma-ray counter inside the "hot lab" is of great utility. Both a high- and a low-sensitivity survey meter should be kept on the premises.

Much safer working conditions can be had if the "hot lab" is specifically designed for the purpose. Existing laboratory facilities are more difficult to adapt. In our installation we are constructing a free-standing cinder block wall, facing a hood, which will effectively shield the rest of the laboratory, and on the hood side will support shelves for the storage of most of the supplies needed in the hood.

The space within the hood is divided into two functional parts. One half of the hood is used for all the operations necessary for radioiodination, and the second half is used for separating the iodinated products and for disposal of radioactive wastes. The first half contains a barrier shield (Atomic Products Co., Center Moriches, New York, No. 042–216), which isolates the area where the isotopes are stored and the reaction mixtures assembled. Next to the shield are kept a small water bath (Model 22 Jr., VWR, Rochester, New York), a dry block heater (Cole Parmer, Chicago, Illinois, No. 2090), and two lead-shielded syringe holders (Atomic Products Co., No. 09–220). A 100-μl Eppendorf pipette is kept in one holder, and the other holder contains a series D, 5-μl chromatographic syringe (Pierce Chemical

Co., Rockford, Illinois). A spare syringe and spare needles are also kept on hand. The electrically heated water bath is plugged into a 12-hour electric timer, so that it is automatically turned off after use.

The second half of the hood is taken up by the small tabletop barrier shield, constructed by ourselves, similar to, but much shorter in height than the commercial unit by Atomic Products Co. It is large enough to shield a 25-mm test tube rack, and is used for column chromatography. This half of the hood also houses three waste disposal containers kept behind lead-foil shielding: a large steel pail with plastic liners for dry waste, a 2-liter plastic bottle with gypsum for the solidification of liquid waste, and a disposable plastic container for the containment of potentially sharp objects such as radiocontaminated Pasteur pipettes. An under-the-counter barrel for waste disposal is being incorporated into the design of our hood. The liquid waste bottle is kept in a steel tray containing dry gypsum to provide failsafe containment in case of inadvertent overflow of liquid. Most of the isotope ends up in this bottle, and the bottle is therefore encased in lead foil before being placed in the hood. Plastic bags and bottles, when full, are placed in the cabinet underneath the hood, which is ventilated into the hood exhaust system. A gas mask (Atomic Products Co., No. 120–111, with No. 120–182 iodine cartridge) is worn when waste materials are transferred from the hood for intermediate storage under the hood, or when sealed in a barrel for removal from the premises.

B. Small Equipment and Supplies

The reaction is carried out in 1-ml Kontes Microflex vials (Kontes Glass Co., Vineland, New Jersey). Hollow, cylindrical copper vial holders (19 × 40 mm) were constructed in our machine shop to serve as a radiation shield in the course of the reaction. The copper vial holders have a small Allen set screw on their side to secure the glass vials tightly. Spare aluminum blocks from the Cole Palmer dry-block heater serve as stable supports for the vial holders during all operations. The general policy with respect to any hardware is that, once it is in the hood, it is *never* removed for use or storage elsewhere.

We found that because of the need for frequent handling of the equipment inside the hood, about 40 times more radiation dosage was absorbed by the hands of the operator than elsewhere on his person. We therefore endeavored to construct as much shielding internal to the hood as possible. The test tube rack used for column chromatography was fitted with brass cylinders large enough to hold 16 × 150 mm glass test tubes, which greatly reduced the radiation flux within the hood during column chromatography of the iodination mixtures.

Equipment used outside the hood consists primarily of several Eppendorf or Oxford micropipettes (5–100 μl) and a Drummond automatic micropipette with removable glass barrels (No. BB108, Bolab, Inc., Derry, New Hampshire), which is used for dispensing solutions in volumes of 1–5 μl. Small Eppendorf pipette tips are stored in 50-ml beakers covered with aluminum foil, and autoclaved. They are used for dispensing as well as for partial assembly of the reaction mixtures. Long and short Pasteur pipettes are washed, treated by brief immersion in a 5% solution of dichloro-dimethylsilane (Eastman Kodak, Rochester, New York) in chloroform, air-dried, and then baked overnight in a 130°C oven. The Microflex reaction vials are similarly treated and sterilized, and stored in dust-free containers outside the hood. Additional sterilized screw-cap closures for these vials are kept separately under dust-free conditions.

All column chromatography is performed with 0.7 × 4 cm polypropylene disposable columns (Biorad, Inc., Rockville Center, New York). The columns are loaded with one of three gels, depending on the type of molecule that has to be separated from the iodination mixture. RNA 5 S or larger is isolated on Whatman CF-11 cellulose, using the procedures of Franklin (1966). The cellulose is washed with 1.0 N HCl, H_2O, 0.1N NaOH, and H_2O. It is suspended in H_2O, and only the rapidly settling particles are saved for use in chromatography. For small RNA molecules DEAE-cellulose is used in stepwise salt gradients. For single- or double-stranded DNA, column chromatography grade hydroxylapatite (Biorad) is poured dry directly into the column, to a height of about 1 cm, and 0.05 M phosphate buffer (pH 7.0) is used to wash the column before the iodination mixture is added to it.

All the column materials used operate on the same principle. The iodinated nucleic acid is bound to the column, and the *unreacted* iodine is eluted into shielded 16 × 15 mm test tubes, the content of which is later poured into the liquid waste receptacle. The column is manually moved from tube to tube in the course of the separation. The radioiodinated polynucleotides are eluted into sterile 13-ml screw-cap test tubes (Vanguard International, No. 1200/6). Several 15-ml metal centrifuge tube holders (International Equipment Corp. No. 303) have been calibrated in the gamma counter with respect to their attenuation of ^{125}I gamma rays, which varies from 300 to 500 times with the weight of the holder. These holders permit the evaluation of up to 1 mCi of ^{125}I at a time, without removing small aliquots for counting. They are extremely useful when a manual gamma counting well and crystal are available in the radiation laboratory.

Other equipment includes a generous supply of disposable vinyl rubber gloves. A rubber oversleeve (Fisher Scientific, No. 01–360) serves to keep iodine vapor from getting behind the gloves. Instead of regular laboratory coats, we use disposable surgical gowns because of the added protection they afford the wrist and arms from iodine vapor inside the hood. These added safeguards were adopted after we discovered unacceptable levels of skin contamination on the wrists.

Recently, through the cooperation of Atomic Products, Co., we acquired two pairs of leaded neoprene gloves (No. 116–10), which we cut to elbow length, and placed one pair in each half of the hood. We found that they reduce radiation exposure to the fingers by 95%, and are usable during all operations except the initial dispensing of ^{125}I solution. These gloves also greatly reduced the hazards from iodine adsorption to the skin. We are also arranging for the test of an experimental portable air-decontamination unit for the laboratory, supplied by Atomic Products Company. The effectiveness of this unit in decreasing laboratory air contamination has not yet been determined.

V. Radioiodination Reaction

A. Reagents and Solutions

All chemicals, unless otherwise noted are reagent grade. Double-distilled water used to make solutions is treated with 3 drops per liter of diethyl pyrocarbonate, autoclaved, and stored in the refrigerator. All glassware for liquid storage is acid-washed and baked at 140°C for 40 minutes. One to 5 ml of reagent solutions are dispensed under sterile conditions into 15-ml Falcon (No. 2095) tubes and stored at −20°C unless otherwise indicated.

1. Na iodine (^{125}I), (New England Nuclear Co., Boston, Massachusetts, No. NEZ-033L), 300–800 mCi/ml, concentration as available, stored in a radioactive hood.

2. Sodium acetate and acetic acid, 4 M and 1 M solutions of each, stored in sterile 100-ml milk dilution bottles in a freezer. 4 M sodium acetate solution is filtered through 0.45-μm Millipore or Nalgene disposable 100-ml filter unit before storage.

3. HNO$_3$, 1 N.

4. Thallic(III) nitrate (Ventron Corp., P.O. Box 159, Beverly, Massachusetts, No. 87999) prepared as follows: (a) A 0.1 M stock solution is prepared in 1 N HNO$_3$ and kept in a refrigerator for up to 1 year. (b) The above stock of solution is diluted to 5×10^{-3} M with 1 M acetate buffer (equal parts of 1 M sodium acetate and 1 M acetic acid, adjusted to pH 4.8 with 1 M NaOH). Final pH is about 4.7. One milliliter aliquots are stored in a freezer and used for up to 3 months.

5. KI, 10^{-2} and 10^{-3} M solutions in H_2O.

6. Acidification and iodine dilution reagents. Three solutions are prepared. The purpose of these is to adjust the pH of the commercial ^{125}I reagent and add nonradioactive iodine in desired concentrations when necessary.

a. Carrier-free (CF) solution. 0.3 N HNO_3, 1.50 ml; 0.1 M sodium acetate (pH 4.7), 2.50 ml; H_2O, 2.00 ml.

b. 1× iodine solution: 0.3 N HNO_3, 1.50 ml; 10^{-3} M KI, 0.60 ml; 0.1 M sodium acetate (pH 4.7), 2.50 ml; H_2O, 1.40 ml.

The addition of 11 μl of a 1 × solution adds 10^{-10} moles of stable iodide to the reaction mixture.

c. 10× iodine solution (same as solution 6b, except 10^{-3} M KI is replaced by a 10^{-2} M KI solution). The use of this solution results in a 10-fold dilution of ^{125}I with ^{127}I.

Five to 10 μl of the acidification reagent is used per 5 μl of commercial ^{125}I. Each of these solutions, when used in conjunction with the buffered thallic(III) ion solution, stabilizes the pH of the final reaction mixture to about 4.7. Occasional lots of NEN iodine required more acid than was used in the above reagents for optimal acidification. The ^{125}I of other manufacturers generally required less acid. The degree of isotope dilution can be varied by mixing different amounts of the above solutions. The CF solution is used when no stable iodine is to be incorporated into the polynucleotide.

7. Bromcresyl green, 0.04% solution (Fisher Scientific, Cat. No. SO-1-14), used as pH indicator.

8. TNE solution (0.05 M Tris–HCl, 0.1 M NaCl, 0.001 M EDTA). Because of sterility problems, it is useful to make up a fairly large amount of the sterilized solution, 10 times the necessary strength, and store it frozen as 10-ml aliquots in milk dilution bottles. H_2O is added to 100 ml; 65 ml of TNE is used to prepare a 35:65 ethanol–TNE solution for washing the unreacted iodide from a CF-11 cellulose column, and the remainder of the TNE is used for raising the pH of the iodination reaction and also for eluting RNA off the CF-11 cellulose column according to the procedure of Franklin (1966). When chromatography materials other than CF-11 are used, their sterilized buffer solutions are also stored as frozen concentrated aliquots.

9. Na_2SO_3, 10^{-1} M, made fresh every 2 weeks, used for stopping RNA but not DNA iodination. Tareba and McCarthy (1973) use mercaptoethanol instead of Na_2SO_3.

10. Nucleic acid solutions. The success of the Commerford reaction depends to a great extent on the quality of the RNA and DNA preparations used. In particular protein and excessive salt contamination of the polynucleotide solutions depress the fraction of iodide covalently bound to cytosine. RNA and DNA used for labeling with iodine must not only be

chemically pure, but to have any usefulness in later work must also be prepared with greater than usual care with respect to their homogeneity. Specifically, it is often more difficult to extract from tissue an unlabeled RNA of the required homogeneity than an RNA labeled biosynthetically for a prescribed purpose. For example, pulse-labeling of cells with ^{32}P or uridine-^{3}H followed by rapid extraction of the RNA will primarily result in high specific activity of the rapidly turning over RNA species. Slight or even substantial contamination of such RNA with 18 S or 28 S RNA will not significantly alter the usefulness of the labeled mRNA for most analytical purposes, since the contaminating RNA is of much lower specific activity. However, subsequent chemical labeling with iodine does not afford such protection, since all RNA will be equally labeled, and stringent purification criteria must replace clever experimental protocol which often suffices in biosynthetic labeling procedures. Because large polymers cannot always be expected to survive iodination intact, the need for repurification of the iodinated sample from other RNA or DNA should be avoided if at all possible.

An additional consideration in preparing samples for iodination is the fact that only small amounts of material need to be labeled at one time. Under appropriate conditions an unlabeled sample of RNA or DNA can be stored indefinitely, and labeled with iodine as needed. Careful and thorough preparation of as large an amount of the wanted polynucleotide as possible is therefore both desirable and practical. Once it is properly characterized, its usefulness in immediate as well as future experimentation is greatly enhanced by the repeated iodinations that can be performed. It is beyond the purpose of this article to go into the methodology of RNA and DNA isolation from tissue, but if technology exists to further purify a species of polynucleotide, more often than not that technology ought to be applied before iodination of the sample is attempted. And seldom, if ever, should even a small sample be iodinated in its entirety. We have successfully iodinated RNA for fingerprinting purposes in amounts as small as 0.05 μg or less (Robertson et al., 1973, cytidine-^{14}C-labeled bacteriophage f1 RNA, synthesized in vitro with RNA polymerase from a double-stranded replicative form DNA template), and aliquots of the same preparation could again be iodinated if necessary.

Ideally, RNA prepared for iodination should be dissolved in double-distilled water and microgram quantities aliquoted into sterile 2-ml NUNC vials (No. N1072, Vanguard International), or siliconized glass test tubes and stored in liquid nitrogen. The volume of H_2O should be about 5–10 μl, and the amounts of RNA appropriate for the sample (0.5–20 μg per iodination). The aliquots can be dried if desired, and then taken up in a few microliters of H_2O when needed for iodination.

We have less experience with the handling of DNA. DNA dissolved in

Tris–HCl buffers or in sodium citrate buffer, with or without EDTA, is generally unpredictable with respect to the facility with which it undergoes iodination. Several DNA samples stored in these buffers did not iodinate well in our hands, but were labeled quite readily with carrier-free ^{125}I when first precipitated with alcohol and redissolved in 0.05 M (pH 5.0) sodium acetate. We prefer to prepare DNA in this buffer, although how long it can safely be stored at this pH we do not know. We prefer to adjust the concentration of DNA to about 0.5 mg/ml.

B.   Assembly of Reaction Mixture

For the purpose of this discussion, the concentration of ^{125}I is assumed to be about 420 mCi/ml. The RNA is assumed to be 5 S, at 645 μg/ml in H_2O.

1.   INGREDIENTS

a. 5 μl ^{125}I solution (equal to 9.4×10^{-10} mole ^{125}I)

b. 10 μl acidification reagent (CF, 1×, or 10×, as described in Section A,6)

c. 5 μl H_2O

d. 5 μl RNA solution (equal to 24×10^{-10} mole cytosine)

e. 2 μl buffered thallic ion solution, 3×10^{-3} M

f. 1 drop 0.04% bromcresyl green pH indicator solution

g. 0.5 ml TNE solution, made 10^{-3} M Na_2SO_3

2.   ASSEMBLY AND HEATING OF REACTION MIXTURE

The buffer solution (b above) is dispensed into a 1-ml Microflex vial. The vial is taken to the hood and placed into a small lead shield behind the tabletop barrier. ^{125}I is drawn up into the 5-μl syringe and added to the vial, which is then capped. The water bath and dry-block heater are checked for the desired temperature, usually 60° and 70°C, respectively. The reaction vial shield and a spare screw cap for the vial are preheated on the dry block. The gas mask, two pairs of vinyl rubber gloves, and protective oversleeves are worn during operations within the hood. The gloves are discarded after this step.

A sterile Eppendorf pipette tip is supported in a 10 × 75 mm test tube outside the hood, and a 5-μl drop of H_2O is dispensed into the pipette tip to form a liquid plug, about 1 cm from the distal end. Into this drop are added the RNA and the thallic ion solutions. A drop of the bromcresyl green indicator solution is placed on a piece of wax Parafilm. The test tube containing the pipette tip and the drop of indicator solution are placed inside the hood. The leaded neoprene gloves can be worn for all subsequent operations within the hood.

The tip is picked up with a 100-μl Eppendorf pipette, and the contents are expelled into the Microflex vial. The entire mixture is drawn up and expelled again. The vial is then closed with the prewarmed cap, placed into a pre-warmed shield, secured with an Allen wrench, and placed on the shelf of the water bath. Prewarming prevents condensation of water vapor on the cap, and facilitates rapid attainment of the desired temperature within the reaction vial.

To determine the pH of the reaction mixture, the indicator solution is taken up by the Eppendorf pipette with which the reaction mixture was assembled. Enough of the reaction mixture usually remains within the pipette to change the color of the indicator from blue to green. The pH of the reaction mixture can be determined by reference to the color of a series of pH 4.0–5.4 acetate buffer made up for the purpose. With a little practice, the pH can be determined by rapid visual inspection of the pipette.

Fifteen minutes later the reaction mixture is removed, about 0.5 ml of TNE is added, and the vial is replaced in the water bath for an additional heating period of about 20–30 minutes. At the end of this period the cap is removed, and the vial and its shield are placed in a cool dry-block carrier which is put in the part of the hood used for column chromatography.

C. Column Chromatography

A Biorad column (0.7 × 4.0 cm) is filled with CF-11 cellulose 1 cm in height and washed with 0.1 N NaOH, TNE, and TNE–ethanol (65:35) solutions. The column is supported by a 16 × 150 mm test tube and placed in the shielded test tube rack inside the hood. Two milliliters of TNE–ethanol is added. A test tube containing 1.0 ml ethanol and a Pasteur pipette with a rubber bulb are also placed in the hood. The pipette, with some of the alcohol in it, is used to rapidly transfer the iodinated RNA to the open column; the vial is rinsed out with the rest of the ethanol, and the wash is added to the column buffer and mixed into it. All used glassware, including the Microflex vials, are immediately discarded in the waste receptacles. The column is allowed to run dry and is then transferred to a new test tube; washing is continued with the TNE–ethanol solution. The column is transferred to new test tubes after about every 2 ml of wash, but need not be run dry again. Air bubbles may occasionally be trapped in the column, but do not appear to affect the results. If desired, the activity of each tube may be determined in a manual gamma counter. The column is washed until no more counts are removed from it, and the fourth and successive tubes are therefore evaluated, either in a gamma counter or with a survey meter. The rate of loss of unreacted iodine counts from the CF-11 is rather high, which is one of the reasons we prefer it for the first purification step.

Two or three sterile polycarbonate tubes, properly labeled, are placed in the hood, and 1-ml fractions of 100% TNE eluted into them. These contain the iodinated RNA, and their acitivty is evaluated in the gamma counter, using the International Equipment Corp. No. 309 cup as an attenuator. Several microliters may be removed and counted separately for a more precise determination of specific activity. The remaining column material may be resuspended, removed with a blunted Pastuer pipette into a test tube, and "unelutable" activity determined in the gamma counter with the aid of the attenuator cap. Assuming that 20% of the isotope was incorporated into cytosine, the TNE–ethanol will contain about 1.85×10^9 cpm, and the TNE will contain about 4.6×10^8 cpm of activity as iodinated RNA. The specific activity of the RNA will be about 1.43×10^8 cpm/μg or, assuming a 60% counting efficiency, 2.37×10^8 dpm/μg. The CF-11 cellulose should have an activity of about 30×10^6 cpm. The operation of the column can be improved if the experimental protocol permits the addition of 10–20 μg of an unrelated carrier RNA.

DNA separations, and separations of smaller than 5 S RNA molecules, are carried out on hydroxylapatite and DEAE-cellulose, respectively, using procedures similar to those described above. DNA is loaded in 0.05 M phosphate buffer and eluted with 0.2 and/or 0.4 M phosphate. The unreacted iodine comes off in the loading buffer. 4 S RNA is loaded in TNE on a column containing DEAE-cellulose, washed with 0.25 M NaCl, and taken off the column in 0.6 M NaCl. The type of subsequent purification of the iodinated polynucleotide depends on the purpose for which it will be used.

D.   Specific Iodinations and Results

The sample reaction shown (Section V,B) was presented to illustrate the method of assembly rather than as an invariant recipe for the iodination of 5 S RNA. Higher or lower specific activities may be desired than would be expected from this reaction. For *in situ* hybridization of RNA to the DNA of meiotic or mitotic chromosomes, it might be desirable to obtain the highest possible specific activity. To do so one would need a more concentrated ^{125}I lot than was assumed in the example reaction presented above, or else one would double or triple the amounts of all reagents except the water and the RNA. In designing an actual mixture, it is useful to construct a table listing the total amounts of all reactants, so that a 20% yield of iodine as iodocytosine will yield the desired specific activity. The reaction is then performed, and if it is off by a factor of 2, up or down, the ratio of iodine to cytosine is adjusted appropriately and the reaction run again with the same lot of isotope. If it is down by more than a factor of 2, something is wrong either with the RNA or the commercial iodine solution, unless of course a

blunder had been committed in dispensing the reagents. It is advisable therefore to obtain a large sample of pure RNA for the purpose of standardizing
the iodination reaction, and iodinate that RNA whenever a new lot of ^{125}I
presents any problems.

We often use 5 S rRNA from HeLa cells as our standard RNA sample.
Table III lists the results of four reactions each performed with a different
lot of ^{125}I. Percent of iodine incorporated into RNA varied from 17.6 to 35.6.
The column showing percent iodocytosine in RNA was calculated on the
basis of the total amount of iodine present. In reactions C245 and C256
percentage of iodocytosine in the RNA was kept intentionally low by a
lowering of the iodine/cytosine ratio, and in C224 it was boosted by adding
2 nmoles of the stable isotope of iodine. The lot of ^{125}I used in C245 was over
3 weeks old at the time this iodination took place, which might have ac-
counted for the somewhat lower percentage of iodination obtained.

VI. Comments and Discussion

The procedures and instrumentation described here were chosen for
their effectiveness and relative safety. Effectiveness was judged on the basis
of acceptable levels of cytosine iodination, which now approach those
obtained by Commerford (1971), by Tareba and McCarthy (1973), and by
Orosz and Wetmur (1974), in reactions in which ^{125}I was used only as a trace
label (Table III).

Theoretically, the absence of the stable isotope from the reaction mixture
should not depress the level of cytosine iodination. The use of a solution of
commercial ^{125}I as a large fraction of the final reaction mixture results in the
simultaneous addition of appreciable amounts of unknown salts. These may
or may not have an adverse effect on the pH of the reaction and the final yield
of iodinated polynucleotide. In retrospect, most of our difficulties in defining
conditions for the reproducible iodination of RNA were due to the unknown
components of ^{125}I solutions. High yields of iodocytosine became more or
less routine after we began to titrate the amount of base in ^{125}I solutions by
checking the pH of small acidified aliquots with bromcresyl green indicator
solution. Presently we confine ourselves to checking the pH of the final
reaction mixture. We have further minimized the effect of the extraneous
components of the ^{125}I solution by designing carrier-free reaction mixtures
to be about $0.3-0.5 \times 10^{-4}\ M$ iodine. These are somewhat suboptimal concentrations, but they yield more reproducible results and, for our purposes,
sufficient specific activity in the polynucleotides thus iodinated.

TABLE III

Iodination of 5 S rRNA from HeLa Cells with Different Lots of ^{125}I[a,b]

| Reaction | Commercial ^{125}I Concentration (mCi/ml) | Moles of reagent ($\times 10^{10}$) | | | Results | | |
		^{125}I⁻	^{127}I⁻	Cytosine as RNA	^{125}I incorporated in RNA (%)	Cytosine converted to iodocytosine	Specific activity (dpm/μg)
C210	800	54.0	—	48.4	23.8	26.4	941 $\times$ 10⁶
C224	652	29.0	30.0	29.0	24.0	48.0	877 $\times$ 10⁶
C245	440	9.4	—	48.4	17.6	3.4	125 $\times$ 10⁶
C258	440	18.8	—	48.4	35.6	6.9	246 $\times$ 10⁶

[a] Reaction conditions were as given in Section V, except that iodine and cytosine amounts were varied by varying the colume of the input reagents. For example, C201 has been assembled with 15 μl of ^{125}I and 10 μl of 5S RNA. Thallic(III) ion concentration was 10 times the concentration of iodine. RNA was dissolved at 0.645 mg/ml in water. Reactions C245 and C258 were *not* performed with the same lot of ^{125}I. The formation of 5-iodouracil was ignored in the calculations.

[b] Physical constants: 1 μg RNA = 0.75 $\times$ 10⁻⁹ moles of cytosine; 200 mCi/ml ^{125}I = 0.9 $\times$ 10⁻⁴ M I⁻; 1% cytosine as iodocytosine-^{125}I in RNA = 35.7 $\times$ 10⁶ dpm/μg of RNA.

Even if a lot of carrier-free ^{125}I is not as effective as an equivalent amount of reagent-grade stable ^{127}I, it does not mean that 30 or 40% of the cytosine cannot be radioiodinated with a concentrated (600–800 mCi/ml) solution of ^{125}I. The radioiodination of 1% of the cytosine residues in RNA produces a specific activity of 35.7×10^6 dmp/μg of RNA, and 40% iodination results in specific activity on the order of 1.4×10^9 dpm/μg. Whether such specific activities are desirable is another question, but as Tareba and McCarthy (1973) have already shown, one needs only to reduce the relative concentration of RNA in order to increase its level of iodination. In the case of carrier-free iodination this means the use of more of the isotope solution in relation to a given amount of RNA or DNA.

Commerford used Na_2SO_3 (10^{-2} M) in the second heating step. In the presence of large amounts of radioactive isotope this is primarily a safety step which insures the reduction of volatile components to I$^-$. It is not needed for stopping the reaction, since raising of the pH of the reaction mixture and its dilution with buffer will accomplish that end. There is a danger of iodocytosine deamination in the presence of sulfite (Hayatsu et al., 1970), but by fingerprint and base composition analysis of several iodinated RNA species, we never detected the occurrence of this reaction (Robertson et al., 1973; Jacobsen et al., 1973; unpublished observations). In DNA iodination the sulfite treatment should be omitted (Orosz and Wetmur, 1974).

Products other than iodocytosine have been detected in iodinated DNA (S. L. Commerford, personal communication; Orosz and Wetmur, 1974). Orosz and Wetmur performed the second-step reheating of iodinated DNA at pH 6.0 in the absence of sulfite and all other salts present in the iodination reaction. The melting temperature of this iodinated DNA resembled more closely that of noniodinated control DNA, and less strand scission was observed than when heating was done at pH 9.0. It is not clear whether the improved properties of the iodinated DNA as prepared by Orosz and Wetmur were due only to the reheating at lower pH, or to the removal of the other reagents from the DNA solution. In our laboratory we generally do not separate the iodinated DNA or RNA from the other reaction components before the reheating step. Instead we simply dilute the reaction mixture with relatively large amounts of buffer. With respect to iodinated RNA, in the absence of a critical analysis of its hybridization properties, as distinct from its general usefulness in various molecular hybridization studies (Getz et al., 1972; Prensky et al., 1973; Scherberg and Refetoff, 1973; Grigliatti et al., 1973; Tareba and McCarthy, 1973; Steffensen and Wimber, 1973), the procedures involving the least amount of handling are the ones to be recommended. Robertson et al. (1973) observed no evidence of the formation of side products in RNA labeled as described here, and the inference can be made that Commerford's reaction, as described here, does in fact yield a minimum of products other than 5-iodocytosine and 5-iodo-uridine.

With respect to the separation of RNA and DNA from other components of the iodination reaction, we have adapted methods utilizing a minimum of equipment and resulting in the containment of the ^{125}I within small volumes. As a result we do not use gel-filtration chromatography as we did in our initial experimentation (Prensky *et al.*, 1973). Orosz and Wetmur (1974) have used dialysis to separate iodinated DNA from other reaction components. Reheating of the DNA occurs in the dialysis bag, and any free iodide generated by this step is dialyzed out again. This is an appealing procedure, but it presents difficult waste disposal, quarantine, and temperature-control problems when used with microgram quantities of RNA and millicurie amounts of radioactive iodide. Since we prefer, as far as possible, to remove ^{125}I from the radioactive hood only in macromolecular and therefore nonvolatile form, we do not use dialysis for the initial separation of iodinated polynucleotides.

A useful variation of the Orosz and Wetmur technique, when iodinated DNA is to be prepared for molecular hybridization studies, is to purify the iodinated DNA on hydroxylapatite after the first reaction step, and then follow their dialysis and reheating procedures. Most of the unreacted iodide would then be lost in the course of chromatography, thus minimizing subsequent waste disposal problems. Alternatively, the DNA could be reheated in the phosphate buffer and again passed through hydroxylapatite.

Our aim here was to describe in detail the synthesis and preliminary purification of iodinated RNA and DNA of high specific activity. The final processing of these molecules obviously depends on the nature of the studies they will eventually be subjected to. For enzyme assays of iodinated RNA we generally precipitate the iodinated RNA with 70% ethanol after running a second CF-11 cellulose column and adding appropriate amounts of carrier RNA. For *in situ* molecular hybridization studies we have found it useful to use another means beside CF-11 cellulose to further purify the RNA. 4 or 5 S RNA is generally passed through a DEAE-cellulose or DEAE-Sephadex (A-25) column, and either alcohol-precipitated and brought up in water, or dialyzed against water and partially concentrated by lyophylization. D. M. Steffensen (personal communication) has successfully stored iodinated 5 S RNA in dialysis tubing in 4 × SSC (0.15 M NaCl, 0.015 M sodium citrate) at 4°C for up to 4 weeks before use for *in situ* hybridization. Large rRNA molecules were generally prepared by passage through a second CF-11 column and alcohol precipitation before use. It would be futile to enlarge upon these procedures, since in many instances they might necessarily have to be different in different types of studies. What should not be done, however, is to store iodinated RNA and DNA either as an alcohol precipitate or lyophylized powder, since we have observed rapid breakdown of the polynucleotides under these conditions. There is no doubt that with the increased storage time there is loss of TCA precipitability of the iodinated product, an increase in the proportion of smaller fragments, and that breakdown problems become magnified with increase in specific activity.

As implied elsewhere in this article, the results obtained from Commerford's reaction are highly reproducible and consistent with his observations on the subject. In this connection the need for acquiring a chemically pure sample of RNA or DNA, whose iodination properties are established under a variety of ^{125}I, ^{127}I, and cytosine concentrations, cannot be overemphasized. The quality of a new sample of ^{125}I or of polynucleotide can then be readily ascertained, by the results obtained from control iodinations. When only trace amounts of RNA or DNA are available for iodination, it is a foregone conclusion that their chemical purity cannot always be rigidly controlled. The problem is aggravated by the fact that occasionally very poor lots of ^{125}I are sold by commercial suppliers (David, 1974). Until such time when suppliers are willing to market "RNA-grade" ^{125}I, with iodination potency established by the vendor for each lot of isotope, control RNA or DNA iodinations in the laboratory will remain a virtual necessity when carrier-free ^{125}I is to be used.

The safety problems that arise from the use of large amounts of volatile radioactive materials are also difficult to overemphasize. These problems are especially severe in laboratories where the task is delegated by the senior investigator to other personnel who either are insufficiently trained or lack the authority to examine and solve the safety problems associated with this type of work. When only a few iodinations per year are to be performed, it is wise to fall back on commercial iodination services or on collaborative arrangements with laboratories properly equipped for the purpose. Certainly in the context of larger research institutions the installation of centralized isotope handling facilities for RNA, DNA, and protein iodination would be the most cost-effective means for insuring the safety of personnel engaged in such work.

ACKNOWLEDGMENTS

I would like to thank Ms. Elizabeth Dickson and Dr. Hugh D. Robertson for the preparation of the fingerprints shown in Figs. 2–4, as well as for many other analyses of iodinated RNA which contributed to the rational development of the iodination technique. Thanks are due to Drs. S. L. Commerford and W. L. Hughes for stimulating my interest in the application of iodinated nucleic acids to problems in cell biology, and for their helpful advice with respect to the chemistry of the reaction. Drs. D. M. Steffensen, D. E. Wimber, and Paul Szabo were instrumental in helping me apply the technique to problems of *in situ* hybridization, and Dr. J. G. Wetmur kindly provided me with a preprint of his unpublished report. The 5 S HeLa cell RNA was a gift of Dr. T. W. Borun.

This work was supported in part by NCI grants CA-08748 and CA-16599 and contract number CM-53820.

REFERENCES

Ada, G., Humphrey, J., Askonas, B., McDevitt, H., and Nossal, G. (1966). *Exp. Cell Res.* **41**, 577–582.

Altenburg, L. C., Getz, M. J., Crain, W. R., Saunders, G. F., and Shaw, M. W. (1973). *Proc. Nat. Acad. Sci. U.S.* **70**, 1536–1539.

Bransome, E. D., Jr., and Sharpe, S. E., Jr. (1972). *Anal. Biochem.* **49**, 343–352.

Commerford, S. L. (1971). *Biochemistry* **10**, 1993–1999.

David, G. S. (1974). *Science* **184**, 1381.

Dube, S. K. (1973). *Nature (London)* **246**, 483.

Ertl, H. H., Feinendegen, L. E., and Heiniger, H. J. (1970). *Phys. Med. Biol.* **15**, 447–456.

Forget, B. F., and Weissman, S. M. (1967). *Science* **158**, 1695–1699.

Franklin, R. M. (1966). *Proc. Nat. Acad. Sci. U.S.* **55**, 1504–1511.

Gall, J. G., and Pardue, M. L. (1971). *In* "Methods in Enzymology" (L. Grossman and K. Moldave, eds.) Vol. 21, Part D, pp. 470–480. Academic Press, New York.

Getz, M. J., Altenburg, C. C., and Saunders, G. F. (1972). *Biochim. Biophys. Acta* **287**, 485–494.

Grigliatti, T. A., White, B. N., Tenner, G. M., Kaufman, T. C., Holden, J. J., and Suzuki, D. T. (1973). *Cold Spring Harbor Symp. Quant. Biol.* **38**, 461–474.

Hayatsu, H., Wataya, Y., Kai, K., and Ieda, S. (1970). *Biochemistry* **9**, 209–213.

Heiniger, H. J., Chen, H. W., and Commerford, S. L. (1973). *Int. J. Appl. Radiat. Isotop.* **24**, 425–427.

Jacobson, A., Housman, D., and Prensky, W. (1973). *Mol. Biol. Rep.* **1**, 209–213.

Johnson, L. D., Henderson, A. S., and Atwood, K. C. (1974). *Cytogenet. Cell Genet.* **13**, 103–105.

Orosz, J. M., and Wetmur, J. G. (1974). *Biochemistry* **13**, 5467–5473.

Prensky, W., Steffensen, D. M., and Hughes, W. (1973). *Proc. Nat. Acad. Sci. U.S.* **70**, 1860–1864.

Rhodes, B. A. (1965). *Anal. Chem.* **37**, 995.

Robertson, H. D., and Jeppesen, P. G. N. (1972). *J. Mol. Biol.* **68**, 417–428.

Robertson, H. D., Dickson, E., Model, P., and Prensky, W. (1973). *Proc. Nat. Acad. Sci. U.S.* **70**, 3260–3264.

Scherberg, N. H., and Refetoff, S. (1973). *Nature (London), New Biol.* **242**, 142–145.

Steffensen, D. M., and Wimber, D. E. (1973). *Annu. Rev. Genet.* **7**, 205–223.

Steffensen, D. M., Prensky, W., and Duffey, P. (1974). *Cytogenet. Cell Genet.* **13**, 153–154.

Steffensen, D. M., Duffey, P., and Prensky, W. (1975). *Nature (London)* **252**, 741–743.

Tareba, A., and McCarthy, B. J. (1973). *Biochemistry* **12**, 4675–4679.

Wimber, D. E., and Steffensen, D. M. (1970). *Science* **170**, 639–641.

Wimber, D. E., Duffey, P. A., Steffensen, D. M., and Prensky, W. (1974). *Chromosoma* **47**, 353–359.

Chapter 8

Density Labeling of Proteins

ALOYS HÜTTERMANN AND GERTRUD WENDLBERGER

*Forstbotanisches Institut der University Göttingen,
Göttingen, West Germany*

I. Introduction	153
II. Density Labeling	156
A. With Amino Acid Precursors	156
B. With Heavy Amino Acids	157
C. Problems Regarding Density Labeling	158
III. Equilibrium Density Gradient Sedimentation	158
A. Theoretical Background	158
B. Solvent Systems Used for Isopycnic Banding of Proteins	161
IV. Ultracentrifugation	163
A. In CsCl	163
B. In Metrizamide–D_2O	163
C. Notes	164
V. Use of Density Markers	164
VI. Combination with Other Biochemical Methods	167
A. Combination with Methods for Separation of Isoenzymes	167
B. Combination with Radiotracer Methods	167
VII. Conclusions	169
Appendix: Formulas for Conversion of Refractive Indices to Density Values	170
References	170

I. Introduction

During studies of biological systems it frequently happens that under special conditions (hormone action, differentiation, metabolic stages, certain periods in the cell cycle) pronounced changes can be observed in enzyme activity. This raises the question whether such changes in activity are regulated on the level of protein synthesis.

Changes in activities of enzymes may be due to metabolic interconversions of enzymes, allosteric or covalent modifications, which do not all act at the translational level. Therefore definite evidence is required to determine whether a change in enzymatic activity is the result of *de novo* synthesis of the enzyme protein.

In most reports, only indirect evidence obtained by inhibitor studies is presented for the demonstration of enzyme synthesis; the inducing effect is applied in the presence of an inhibitor of RNA or protein synthesis. If the activity increase is prevented under these conditions, it usually is taken as evidence for a control of this event at the level of enzyme synthesis. However, evidence obtained in this way may not be very solid. Recent reports indicate that the commonly applied inhibitors may have side effects not related to transcription or translation. In addition, metabolic conditions in higher cells are too complex to permit such simple deductions. Therefore it is not surprising, that some recent studies show that interpretations based on inhibitor data may be completely misleading (Yagil *et al.*, 1974; Hizi and Yagil, 1974a,b). A move direct way of attacking such problems is by actual measurement of *de novo* synthesis of the enzyme under study. Usually this is a very time-consuming and tedious project requiring, first, purification of the enzyme to homogeneity, and then a combination of immunological and isotopic methods for the demonstration of enzyme synthesis under conditions in which the increase in enzyme activity is observed.

A simpler way of demonstrating enzyme synthesis—one that does not require prior purification—is the method of density labeling of the enzyme protein with heavy isotopes, followed by equilibrium density gradient sedimentation. This method is based on the same rationale as the classic method of density labeling and isopycnic centrifugation of nucleic acids (Meselson *et al*, 1957), and is the same method by which the semiconservative reduplication of DNA was demonstrated (Meselson and Stahl, 1958). The method was introduced in enzyme chemistry as early as 1962 by the Halvorson group (Hu *et al.*, 1962).

If a protein is centrifuged for sufficient time in a density gradient, it will form a band at the point in the gradient where the density of the gradient corresponds to its own buoyant density. Figure 1 shows the situation for gradients formed during centrifugation. After centrifugation the content of the centrifuge tube is divided into fractions of 3–5 drops. The density slope of the gradient can be conveniently determined by measuring the refractive index in some fractions. The other fractions are assayed for enzymatic activity. Thus a very specific test can be applied for the identification of the position of the enzyme band in the gradient. With this procedure, the

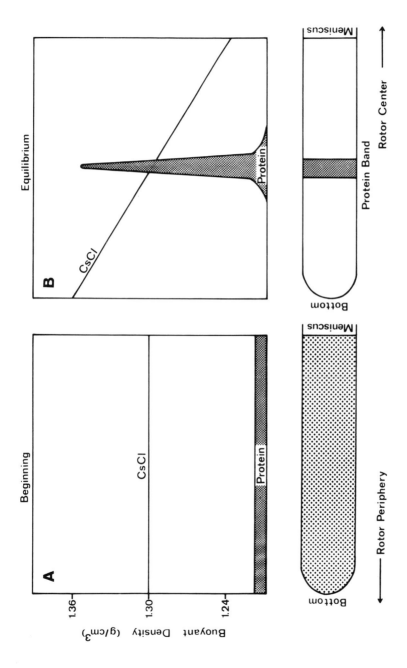

FIG. 1.   Concentration distribution of protein and CsCl in a centrifuge tube at the beginning of centrifugation and after attainment of equilibrium.

buoyant density of an enzyme can be determined with high precision starting with crude extracts.

The density of a protein is a function of its amino acid content. It can be increased experimentally by the incorporation of heavy isotopes such as ^2H, ^{13}C, ^{15}N, and ^{18}O. Hence the experimental design for the study of enzyme synthesis by density labeling is basically very simple; the density of the enzyme under study is determined by using extracts from normal light cultures. Under the conditions in which the increase in enzyme activity is observed, a heavy isotope in suitable form is applied. If after this treatment the enzyme occurs in the subsequent isopycnic sedimentation at a density higher than that of the unlabeled enzyme, this is definite proof that the labeled enzyme was synthesized during the time period of the study.

This method for determining enzyme synthesis has several advantages; no enzyme purification of time-consuming immunological methods are necessary. In addition, the results are always clear-cut, provided proper controls have been carried out. If a shift in density occurs, it must be the result of enzyme synthesis.

Provided the molecular weight of the enzyme being studied is high enough, in many cases a clear separation between "old" and "new" enzyme in the gradients is possible. This differentiation between labeled and unlabeled enzyme in a quantitative relation to each other is impossible to obtain with any other method used for the determination of enzyme synthesis.

II.   Density Labeling

Four heavy isotopes can be used for the incorporation of heavy label into enzymes: ^2H, ^{13}C, ^{15}N, and ^{18}O. The maximal attainable shifts in density of enzymes labeled with these isotopes are given in Table I.

A.   With Amino Acid Precursors

1.   HEAVY WATER (D_2O)

This is the most frequently used method of density labeling. It can be used only with plants and microorganisms that make their own amino acids. If these organisms are grown on D_2O instead of H_2O, deuterium is incorporated into the amino acids. This method of introducing the heavy label has certain disadvantages. Most organisms and plants incubated with heavy water grow significantly slower than those incubated with H_2O. Other

TABLE I

MAXIMAL ATTAINABLE DENSITY SHIFTS IN PROTEINS USING HEAVY
ISOTOPES AS DENSITY LABELS[a]

Isotope	Maximal density shift	Percent
^{15}N	0.013	1
^{13}C	0.05	3.8
^{3}H	0.06	4.5
^{18}O	0.03	2.3
^{15}N, ^{13}C, ^{3}H	0.123	9.5

[a]Values computed from data given by Hu *et al.* (1962).

organisms, e.g., *Physarum polycephalum*, that are able to synthesize amino acids in light water are unable to do so in the presence of D_2O.

Anstine *et al.* (1970) excluded artifacts due to an unspecific exchange of deuterium for hydrogen in already synthesized protein molecules. The normally occurring exchange rate is very slow, even in the presence of 8 *M* urea. No detectable incorporation occurred after 15 hours of incubation of barley amylase with D_2O.

2. HEAVY WATER ($H_2{}^{18}O$)

In plant tissues (e.g., barley embryos) with a low amino acid pool but a high amount of storage protein, storage proteins must be hydrolyzed before new proteins can be synthesized. Here elegant labeling can be done with $H_2{}^{18}O$ (Filner and Varner, 1967).

During protein hydrolysis in the presence of $H_2{}^{18}O$, one ^{18}O atom is incorporated into the carboxyl group of each liberated amino acid. In proteins made from these amino acids, every second peptide bond should contain an ^{18}O atom, enough to result in a detectable density shift.

3. NITRATE-^{15}N AND AMMONIA-^{15}N

In amino acid–synthesizing cells or tissues, an alternative is the application of nitrate-^{15}N or ammonia-^{15}N salts.

B. WITH HEAVY AMINO ACIDS

Probably the best method of density labeling in all organisms is the application of heavy amino acids. With these labels, all detrimental effects on the metabolism of the cells can be excluded. In addition, the label is immediately introduced into the amino acid pool, making any prior amino acid synthesis unnecessary.

Mixtures of 16 deuterated amino acids (98% deuterium), 16 amino-^{13}C acids, and 16 triple-labeled amino acids (^{2}H, ^{13}C, ^{15}N) are available commercially. All amino acid mixtures are made from algal hydrolyzates and are available from Merck, Sharp and Dohme (the triple-labeled mixture is not routinely supplied, but is made on request). ^{15}N-Labeled amino acids are available from several companies that supply biochemicals.

C.   Problems Regarding Density Labeling

Three major problems regarding the introduction of density label should be considered.

The first of these is to check whether the mode of labeling interferes with the biological effect to be studied. D_2O always causes a delay in the growth of plants and may be a source of artifacts. The application of amino acids may interfere with cell metabolism, especially if an enzyme is to be studied during starvation. It should be noted that the amount of amino acids necessary to introduce a heavy label may be higher than in studies with radioactive isotopes. Amino acids of course present no problems as labels, if applied to cells or tissue cultures that contain amino acids in their medium anyway.

A second problem may arise from a carbohydrate moiety of an enzyme. If any active carbohydrate synthesis takes place during the time of labeling, the use of D_2O, or especially of $H_2^{18}O$, requires rather stringent controls, for any shift in density may be the result of a density labeling of the carbohydrate portion of a glycoprotein. If a carbohydrate is synthesized in the presence of $H_2^{18}O$, only 4% of the total glycoprotein mass is sufficient to result in a detectable density shift, thus simulating the occurrence of protein synthesis. This type of labeling, however, allows the study of synthesis of the carbohydrate moiety of glycoprotein independently of the synthesis of their proteins. The complication arising from the carbohydrate moiety has to be considered also in systems with active gluconeogenesis when ^{13}C is used as label.

Another self-evident, but not always recognized, problem in density label studies is that the density label has to be applied at about the same time as the effect being studied takes place. Otherwise, the results are meaningless.

III.   Equilibrium Density Gradient Sedimentation

A.   Theoretical Background

The aim of isopycnic sedimentation of density-labeled proteins is to distinguish between macromolecules of identical molecular weights having

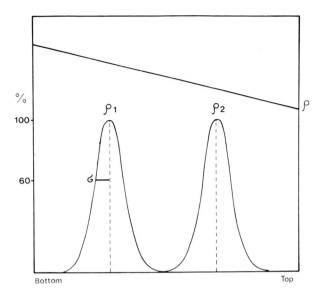

FIG. 2.   Diagram of a density gradient with two macromolecules of identical molecular weight but slightly different densities at their equilibrium position.

small density differences (Fig. 2). For optimal exploitation of ultracentrifugation for this purpose, some theoretical considerations are essential.

1.   BANDING OF MACROMOLECULES IN DENSITY GRADIENTS

The banding of any macromolecule can be described by the equation:

$$\sigma^2 = \frac{RT}{M\bar{v}(d\rho/dr)r\omega^2 r} \tag{1}$$

where σ is the half-width of the Gaussian distribution, M is the molecular weight, R is the gas constant, T is the absolute temperature, $(d\rho/dr)_r$ is the slope of the gradient at the point r, $\omega^2 r$ is the centrifugal field at a given distance from the rotor center, and $\bar{v}$ is the specific inner volume. Equation (1) indicates that the sharpness of the band of a given protein can be increased in two ways, either by increasing the centrifugal speed or the slope of the gradient. Since the experiment should differentiate between proteins having small density differences, the slope of the gradient should not exceed the value of 0.03 gm/cm⁴ (which is the slope of a spontaneous CsCl gradient at 150,000 g). Therefore isopycnic sedimentation of proteins should be carried out using shallow gradients at the highest possible centrifugation speed.

There are two different ways of performing isopycnic sedimentation. Either the sedimentation is carried out on a stable preformed gradient, or the gradient is formed during the centrifugation starting from a homogeneous solution of both the macromolecule and the gradient-forming small molecules.

2. Equilibrium Gradients

After prolonged centrifugation of any small molecule, an equilibrium is attained between sedimentation and diffusion which results in a density gradient along the axis of rotation. The slope of this gradient is defined by (2):

$$\left(\frac{1}{\omega^2 r}\right)\left(\frac{d\rho}{dr}\right) = \frac{M(1 - \bar{v}\rho)}{nRT\dfrac{d\ln a}{d\rho}} \tag{2}$$

The symbols used in this equation are the same as for Eq. (1), except that this time they refer to the small molecule. In addition, $1 - \bar{v}\rho$ is the buoyant factor, n is the amount of ion formed by the molecule, and a is the activity coefficient. Since the right side of Eq. (2) consists only of physicochemical constants for the substance forming the gradient, and these are independent of the conditions of centrifugation, Eq. (2) can be shortened:

$$\left(\frac{1}{\omega^2 r}\right)\left(\frac{d\rho}{dr}\right) = \text{a constant} \tag{3}$$

From Eq. (3) it can be deduced that at equilibrium the steepness of the gradient is a function of the centrifugal speed; the higher the angular velocity, the steeper the gradient. From this it is evident that the demand for the highest centrifugal speeds with shallow gradients is difficult to fulfill in a gradient where the small molecule, which forms the gradient during the sedimentation, has the equilibrium distribution.

3. Preformed Gradients

A way to circumvent the relation between gradient slope and centrifugal speed is to use preformed gradients (Baldwin and Shooter, 1963). If such a gradient is stable for the time necessary for the protein to reach the iso-density point, the steepness of the gradient can be chosen independently of the centrifugal force.

The critical point about preformed gradients is the time during which they remain stable. This is dependent on the speed with whch the equilibrium gradient of the small molecule is formed. The formation of a spontaneous gradient in relation to the centrifugal field is a function of the diffusion coefficient of the small molecule and the geometric dimensions of the centrifuge tube. The time necessary to reach equilibrium distribution over the whole distance of the centrifuge tube is an inverse function of the square of the length of the solvent column, and is directly proportional to the diffusion coefficient. It is independent of the centrifugal speed during centrifugation (Van Holde and Baldwin, 1958). Thus the stability of a preformed gradient increases with the square of the length of the solvent column.

If preformed gradients are used, the sample usually is loaded on top of the gradient. During sedimentation the protein migrates to its isodensity point and does not move after this equilibrium position is reached, provided the gradient does not change. The time necessary to reach this equilibrium is dependent only on the centrifugal speed and not on the length of the solvent column (Baldwin and Shooter, 1963). With preformed gradients, the time necessary for isopycnic runs can be shortened considerably if higher speeds are used (Hüttermann and Guntermann, 1975).

Thus, the use of preformed gradients has several advantages. The slope of the gradient can be adjusted independently of the centrifuge speed, the protein band is sharper, and the time required for a run is shorter than for equilibrium gradients.

B. Solvent Systems Used for Isopycnic Banding of Proteins

An ideal solute for the equilibrium density gradient sedimentation of proteins should have the following properties:

1. It should be inert to proteins and have no influence on the stability and activity of enzymes.

2. It should deliver a density range of about 1.2–1.4 gm/ml [the range of buoyant densities of proteins (Hu et al., 1962)] without significant changes in viscosity.

3. The absolute viscosity in this density range should not exceed 10 cp, since otherwise very long centrifugation times are necessary to equilibrate the proteins.

4. A preformed gradient should be sufficiently stable to allow the protein to reach the equilibrium position without a change in the slope of the gradient.

Currently, three different types of solutes are in use.

1. Solutions of Cs^+ and Rb^+ Salts

These are the most frequently used solutes for isopycnic sedimentation of proteins. Both have the advantage of forming solutions with high densities and low viscosities. However, most proteins do not survive centrifugation in 3 M CsCl or RbCl for 40 hours. Preformed gradients are not stable. Thus the highest speed is about 150,000 g.

2. SUCROSE SOLUTIONS

The most commonly used solute for density gradient sedimentations, in which separation is based on sedimentation coefficients, is sucrose. The great disadvantage of this solute is that in a density range higher than 1.2 it is so viscous that about 100 hours of centrifugation would be required (Hu *et al.*, 1962) to band proteins in their equilibrium position. It is possible, however, to exploit the small increase in sedimentation coefficients of proteins after density labeling for detection of incorporation of the heavy isotope (Hu *et al.*, 1962). Sucrose has been used successfully as a gradient solute for this type of experiment (Hirschberg *et al.*, 1972).

3. METRIZAMIDE–D_2O

Nearly all requirements for an ideal density gradient solute are met by the system metrizamide–D_2O. Metrizamide [2-(3-acetamido-5-*N*-methyl-acetamido-2,4-6-triiodobenzamido)-2-deoxy-D-glucose, supplied by Nygaard and Co., AS, Oslo, Norway] was developed as an inert X-ray contrast medium. Solutions in H_2O have been used successfully as a gradient medium for the fractionation of chromatin (Rickwood *et al.*, 1973) and ribonucleoprotein (Hinton *et al.*, 1974), and for the isolation of nuclei (Mathias and Wynther, 1973), mitochondria, and lysosomes (Aas, 1973). For a recent review, see Rickwood and Birnie (1975).

With protein, only completely reversible interactions have been observed at high metrizamide concentrations. Heavy satellite bands form during the isopycnic banding of proteins in such solutions (Rickwood *et al.*, 1974), the result of van der Waals interactions between aromatic amino acids in the proteins and the benzene ring of metrizamide (C.-C. Gilhuus-Moe, personal communication).

These protein—metrizamide interactions can be largely avoided by supplementing H_2O with D_2O (Hüttermann and Guntermann, 1975). In this solvent system, much less metrizamide is needed to reach the desired density range. Therefore, van der Waals bindings between metrizamide and proteins, which obey the general equilibrium expression, are insufficiently stable to result in such heavy satellite bands. Such bindings have yet to be found in metrizamide–D_2O gradients. Supplementing H_2O with D_2O for the gradients has the additional advantage of lowering the overall viscosity by about a magnitude compared to solutions in light water of the same density, without any significant changes occurring in viscosity in the desired density range.

Preformed metrizamide–D_2O gradients are stable in the main part (about three-fifths) of a 5-cm centrifuge tube for at least 60 hours (Hüttermann and Guntermann, 1975). Thus all advantages of using preformed gradients for the isopycnic banding of proteins can be exploited. The time necessary for an equilibrium banding of proteins can be shortened to 17 hours at 350,000 g.

IV. Ultracentrifugation

A. In CsCl

Preformed gradients of CsCl solutions are not stable, and the slope of the gradient is rather high at high centrifuge speeds. The most commonly used procedure is to start with a homogeneous solution of both CsCl and a protein. Depending on the density of the protein, the refractive index of this solution should have a value of about $n_D^{25} = 1.365$.

Most studies have been performed in 5-cm-long centrifuge tubes in swinging-bucket rotors. If the density of the protein being studied is already known, it is advisable to work with rather short solute columns in order to attain equilibrium distribution of the CsCl rather fast. This will increase the resolution of the system (Hu *et al.*, 1962). In this case, a 1-cm-high artificial bottom layer of fluorocarbon should be applied to the tube first, followed by 1.5- to 2-cm layer of the CsCl solution containing the protein. The density of this solution should be very close to that of the protein being studied. The centrifuge tube then is filled up with paraffin oil and spun at 150,000 g (the centrifugal field at the center of the tube is always given) for about 65 hours at 4°C. After the run, the gradient is divided into fractions of appropriate size (in our laboratory usually 80 μl). The refractive index of every fifth or tenth fraction is determined. This allows easy computation of the density slope in the gradient. The formulas for conversion of the refractive index to the density of CsCl and metrizamide are given in the appendix. The other fractions are assayed for enzymatic activity.

An alternative procedure is to work with fixed-angle rotors (Johnson *et al.*, 1973). Here the gradient is formed during the run along the axis of rotation also. However, the maximal horizontal distance in a fixed-angle rotor is rather short, and so the gradient is formed in a short solvent column. After the run, the gradient is reoriented to a vertical direction. The length of the solvent column in the centrifuge tube is usually much longer than the maximal horizontal distance. Therefore, a flattening of the gradient takes place during the reorientation, resulting in higher resolution.

B. In Metrizamide–D$_2$O

In this system preformed gradients should be used. Depending on the density of the protein to be studied, gradients in the concentration range of either 15–35% or 20–40% metrizamide–D_2O (w/w) should be used. We usually prepare a step gradient (either 15, 20, 25, 30, and 35%, or 20, 25, 30, 35, and 40%) made by pipetting 0.9–0.95 ml of each solution into a 5-ml centrifuge tube. Up to 0.5 ml of the sample solution is layered on top of this gradient, and the run is performed for 40 hours at 150,000 g, 24 hours at 250,000 g, or 17 hours at 350,000 g. The resolution between enzymes of different densities has been found to be identical under all three conditions (Hüttermann and Guntermann, 1975). After the run the gradient is treated in the same way as described for CsCl-gradients.

C. Notes

1. The shape of the gradient should be adjusted so that the protein bands are at least 0.5 cm from both the bottom and the top of the centrifuge tubes. In these regions a change in the hydrostatic pressure occurs during centrifugation, which might lead to misinterpretations.

2. Metrizamide is sensitive to light (near UV, as present in daylight and the light of fluorescent lamps). The required manipulations should therefore be performed without unnecessary exposure to light.

3. After the run, the fractions should be immediately diluted about 2- to 5-fold with either buffer or substrate solution. We found that some enzymes in both metrizamide and CsCl gradient fractions lose their activity within about 60 minutes while standing undiluted on ice. Diluted fractions with the enzymes we have studied can be kept on ice (in the dark in the case of metrizamide) for several hours without significant loss of enzymatic activity.

V. Use of Density Markers

The ideal situation after density labeling would be the complete separation of newly synthesized enzyme from unlabeled enzyme either present in the same sample or added to the gradient as a control. These conditions are only seldom found in density-labeling studies, because in most experiments either too little label is incorporated into newly synthesized protein molecules and/or the molecular weight of the protein under study is too small to give bands sharp enough for a good separation of heavy and light enzymes.

In this case, two methods are available to obtain solid information whether or not the density label has been incorporated into the enzyme molecule. One way is to determine precisely the refractive indices of the light and heavy enzymes in several parallel experiments and to compare the values obtained statistically. If the significance of the density differences can be determined statistically, incorporation of the label can be concluded. A mixture of both light and heavy enzymes centrifuged on the same gradient should then result in a broader band compared to the one obtained for both types of enzymes under identical conditions of centrifugation.

A more direct way of detecting reproducibly small shifts in density is through the use of density markers (Fan, 1966; Jacobsen and Varner, 1967; Anstine et al., 1970; Hüttermann and Volger, 1973; Hoffmann and Hüttermann, 1975; Guntermann et al. 1975).

In this case, about 200 mU of a commercially available enzyme is added to the sample prior to centrifugation. The fractions obtained after the run are divided into two parts. One set of fractions is analyzed for enzyme activity, and the other for the added density marker. If incorporation of the heavy label has taken place, the labeled enzyme should be heavier than the light enzyme, and this shift should be manifested in relation to the internal marker enzyme. This type of experiment needs no absolute density determinations. Only the slope of the gradient should be identical in both runs. For this measurement of density shifts it is desirable to choose a marker enzyme with about the same density as the enzyme to be studied. This should give the best indications for a small shift in density. An example of a density-labeling experiment in which the light enzyme is lighter and the labeled enzyme is heavier than the marker enzyme (β-galactosidase from *Escherichia coli*) is given in Fig. 3.

When CsCl gradients are used, a short distance between the marker enzyme and the enzyme under study is optimal but not mandatory, because the slope of the gradient is defined by the centrifugal speed. In preformed metrizamide gradients, the distance between these two enzymes should be as short as possible, since a small change in the steepness of the gradient may influence the relative distance between the two enzymes. This could lead to the simulation of a density shift that has not really occurred. A variety of enzymes banding at different densities in metrizamide gradients is given in Table II.

It should be noted that the marker enzyme usually should not be present in the extract, because this could lead to complications. If this cannot be avoided, a large excess of the enzyme should be added, and only a small portion of the fractions should be assayed for this activity.

If the synthesis of an enzyme is to be studied that exhibits rather high activity changes under the conditions of induction, a stable constitutive enzyme of the same organism can be used as an additional density marker. In this case, the experimental setup is as follows. The induction is carried

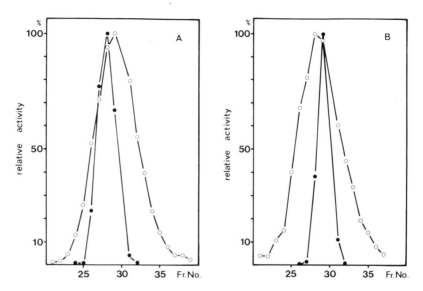

FIG. 3. Use of an external density marker. For the demonstration of a density shift, an externally added protein was used, in this case β-galactosidase from *E. coli*. (A) Equilibrium distribution of aminopeptidase from *P. polycephalum* from extracts of normal cultures. (B) Equilibrium distribution of aminopeptidase in extracts of cultures labeled with deuterated amino acids. Open circles, Aminopeptidase; solid circles, β-galactosidase. (W. Hoffmann and A. Hüttermann, 1975.)

TABLE II

BUOYANT DENSITIES OF SOME ENZYMES IN METRIZAMIDE–D_2O GRADIENTS

Enzyme[a]	Origin	Buoyant density at 25°C	n_D^{25b}
β-Galactosidase	*E. coli*	1.3084	1.3960
Acid phosphatase	Potato	1.2504	1.3770
Phosphodiesterase	Snake venom (*Crotalus*)	1.2489	1.3765
β-Glucosidase	Sweet almond	1.2428	1.3745
α-Glucosidase	Yeast	1.2321	1.3710

[a]All enzymes were obtained from Boehringer-Mannheim, Tutzing, West Germany.
[b]Determined for the peak fraction.

out in one experiment with only light isotopes, and in a second experiment with heavy isotopes. In this case the time period of labeling should be as short as possible. Both types of extracts are then analyzed by isopycnic centrifugation. If the induction is the result of *de novo* synthesis of the induced enzyme to be studied, the relative positions of the two enzymes should be different in the two gradients. The distance between the density marker added externally and the enzyme that serves as an internal density marker should be about the same in both gradients. The density of the induced enzyme, however, should be higher in the heavy extracts made from labeled cells compared to the density of the enzyme in light extracts.

An example of this type of experiment is given in Fig. 4. With this experimental design, preferential protein synthesis, under the conditions of induction, can be clearly demonstrated. Provided the two enzymes are distributed in the same cellular compartments, all complications relating to pool sizes and changes can be avoided, and differential protein synthesis can be studied in a rather simple but conclusive way.

VI.   Combination with Other Biochemical Methods

A.   Combination with Methods for the Separation of Isoenzymes

During morphogenesis of various organisms, a change in the isoenzyme pattern of certain enzymes often occurs. The regulation of this shift can be studied by a combination of density labeling and isopycnic sedimentation and the methods for detection of the isoenzyme pattern. If the isoenzymes can be separated and visualized on starch gel electrophoresis, the elegant and ingenuous method of Quail and Varner (1971) can be used. In this procedure, the extract is first centrifuged in an isopycnic run, and all fractions of the gradient are analyzed together on a multiple starch gel electrophoresis apparatus. The plate is then stained for the enzyme activities being studied, photographs are made, and the prints are analyzed in a densitometer. Thus the density of each individual isoenzyme can be determined simultaneously in a single experiment. If such conditions do not apply, usually the isoenzymes are separated first by isoelectric focusing (Hoffmann and Hüttermann, 1975) or gel electrophoresis (Anstine et al., 1970; Quail and Scandalios, 1971), and then each individual band is analyzed by isopycnic sedimentation. Both types of experiments allow the study of synthesis, degradation, and turnover of individual isoenzyme bands.

B.   Combination with Radiotracer Methods

Density labeling can be combined with radioactive tracer methods. This enables one to determine precisely the turnover of proteins in general or in individual enzymes. Using a triple-label experiment with -^{15}N, -^{14}C, and amino-^{3}H acids, Zielke and Filner (1971) determined the half-life of nitrate reductase in *Nicotiana tabacum*.

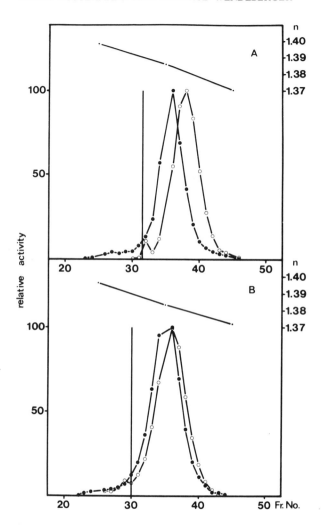

FIG. 4. Use of an internal density marker. For the demonstration that under conditions of induction the preferential synthesis of a given enzyme occurs, a stable constitutive enzyme of the organism under study can be used. In these experiments, α-glucosidase was inducible, β-glucosidase was the stable constitutive enzyme, and β-galactosidase was always added as an external reference. The study was carried out with *Myxobacter* Al–1. (A) Equilibrium distribution of the two enzymes in extracts made from cells induced in the presence of light isotopes only. (B) Equilibrium distribution of the two enzymes from extracts made from cells induced for a short time in the presence of deuterated amino acids. In this gradient, the α-glucosidase is significantly heavier, although the relative position of the β-glucosidase in relation to the external marker is unchanged. Open circles, α-Glucosidase; solid circles, β-glucosidase; —. —, position of the external added β-galactosidase. (U. Gunterman, I. Tan, and A. Hüttermann, 1975.)

VII. Conclusions

Scores of articles have been published in which the synthesis of an enzyme is reported and the interpretation is based only on activity and inhibitor data. Thus the 50 studies in which density labeling was employed for a simple but conclusive demonstration of enzyme synthesis are a surprisingly small fraction. The fact that such a relatively simple method has not been used is even more surprising since it was introduced in protein chemistry as early as 1962 and many details were worked out at that time by the Halvorson group. Three factors may be responsible for this. In the early 1960s, no density-labeled biochemical was available commercially at reasonable cost. The only chemical readily available for this purpose was heavy water. Thus density labeling was used mainly by botanists and microbiologists working with organisms capable of synthesizing their own amino acids in the presence of heavy water. For several years now, deuterated amino acids have been available which offer an easy and inexpensive way of introducing the density label. Thus this method can be applied to all cells and organisms that can be labeled with radioactive amino acids at about the same cost as a radiotracer experiment. Heavy amino acids are especially suitable for use in cell or tissue cultures cultivated in media containing amino acids that can be easily supplemented by deuterated ones without any detrimental effect.

A second possible reason for the limited use of this method may be the fact that only a few enzymes can survive over 40 hours of centrifugation in a 3 M CsCl solution, the most frequently used gradient system. Now that preformed stable metrizamide gradients can be used for protein banding, the time needed for the centrifugation can be reduced to 17 hours. Since metrizamide is inert to all proteins studied thus far, it can be expected that any enzyme that is stable in a refrigerator overnight can be studied with this method.

A third factor is the development of faster centrifuges. Most isopycnic runs were made at 39,000 rpm (150,000 g), the maximal speed available in 1962. Hu *et al.* (1962) predicted that the method could be improved by using higher speeds. This is now possible, and runs can be made at 350,000 g in metrizamide gradients. From these developments it can be expected that this elegant technique will receive the attention it deserves.

ACKNOWLEDGMENTS

The senior author is indebted to Dr. H. O. Halvorson, Brandeis University, for a first introduction to the method of density labeling of proteins.

The work reported in this chapter carried out in the authors' laboratory was supported by grants from the Deutsche Forschungsgemeinschaft, Schwerpunktprogramm "Biochemie der Morphogenese."

Appendix: Formulas for the Conversion of Refractive Indices to Density Values

For CsCl solutions of a density below 1.4:

$$\rho^{25} = 10.2402\, n_D^{25} - 12.6483$$

For metrizamide solutions in D_2O:

$$\rho^{25} = 3.0534\, n_D^{25} - 2.9541$$

REFERENCES

Aas, M. (1973). *Proc. Int. Congr. Biochem. 9th, 1973* Abstract, p. 31.
Anstine, W., Jacobsen, J. V., Scandalios, J. G., and Varner, J. E. (1970). *Plant Physiol.* **45**, 148–152.
Baldwin, R. L., and Shooter, E. M. (1963). *In* "Ultracentrifugal Analysis in Theory and Experiment" (J. W. Williams) pp. 143–168. Academic Press, New York.
Fan, D. P. (1966). *J. Mol. Biol.* **16**, 164–179.
Filner, P., and Varner, J. E. (1967). *Proc. Nat. Acad. Sci. U.S.* **58**, 1520–1526.
Guntermann, U., Tan, I., and Hüttermann, A. (1975). *J. Bacteriol.* **124**(1) (in press).
Hinton, R. H., Mullock, B. M., and Gilhuus-Moe, C.-C. (1974). *In* "Methodological Developments in Biochemistry" (E. Reid, ed.), Vol. 4, pp. 103–110. Longmans, London.
Hirschberg, K., Hübner, G., and Borriss, H. (1972). *Planta* **108**, 333–337.
Hizi, A., and Yagil, G. (1974a). *Eur. J. Biochem.* **45**, 201–209.
Hizi, A., and Yagil, G. (1974b). *Eur. J. Biochem.* **45**, 211–221.
Hoffmann, W. and Hüttermann, A. (1975). *J. Biol. Chem.* **250**, 7420–7427.
Hu, A. S. L., Bock, R. M., and Halvorson, H. O. (1962). *Anal. Biochem.* **4**, 489–504.
Hüttermann, A., and Guntermann, U. (1975). *Anal. Biochem.* **64**, 360–366.
Hüttermann, A., and Volger, C. (1973). *Arch. Mikrobiol.* **93**, 195–204.
Jacobsen, J. V., and Varner, J. E. (1967). *Plant Physiol.* **42**, 1596–1600.
Johnson, C., Attridge, T., and Smith, H. (1973). *Biochim. Biophys. Acta* **317**, 219–230.
Mathias, A. P., and Wynther, C. A. (1973). *FEBS (Fed. Eur. Biochem. Soc.) Lett.* **33**, 18–22.
Meselson, M., and Stahl, F. W. (1958). *Proc. Nat. Acad. Sci. U.S.* **44**, 671–682.
Meselson, M., Stahl. F. W., and Vinograd, J. (1957). *Proc. Nat. Acad. Sci. U.S.* **43**, 581–588.
Quail, P. H., and Scandalios, J. G. (1971). *Proc. Nat. Acad. Sci. U.S.* **68**, 1402–1406.
Quail, P. H., and Varner, J. E. (1971). *Anal. Biochem.* **39**, 344–355.
Rickwood, D., and Birnie, G. D. (1975). *FEBS (Fed. Eur. Biochem. Soc.) Lett.* **50**, 102–110.
Rickwood, D., Hell, A., and Birnie, G. D. (1973). *FEBS (Fed. Eur. Biochem. Soc.) Lett.* **33**, 221–224.
Rickwood, D., Hell, A., Birnie, G. D., and Gilhuus-Moe, C. C. (1974). *Biochim. Biophys. Acta* **342**, 367–371.
Van Holde, K. E., and Baldwin, R. L. (1958). *J. Phys. Chem.* **62**, 734–743.
Yagil, G., Shimron, F., and Hizi, A. (1974). *Eur. J. Biochem.* **45**, 189–200.
Zielke, H. R., and Filner, P. (1971). *J. Biol. Chem.* **246**, 1772–1779.

Chapter 9

Techniques for the Autoradiography of Diffusible Compounds

WALTER E. STUMPF

*Departments of Anatomy and Pharmacology,
University of North Carolina,
Chapel Hill, North Carolina*

I. Introduction	171
II. Dry-Mount Autoradiography	174
A. Preparation of Tissue Mounts	174
B. Freezing of Tissue	175
C. Cryostat Sectioning	178
III. Thaw-Mount Autoradiography	185
IV. Smear-Mount Autoradiography	186
V. Touch-Mount Autoradiography	187
A. Controls and Artifacts	187
B. Examples of Application	188
Appendix: MGP Stain for DNA and RNA as Used for Dry-Mount, Thaw-Mount, Smear-Mount, and Touch-Mount Autoradiograms	192
References	192

I. Introduction

Localization of substances in tissues is essential for the understanding of function, or more specifically, the identification of sites of synthesis, storage, transport, action, metabolism, and excretion. Autoradiography is one of the techniques of choice. As it is based on the use of photographic emulsions and substances labeled with radioisotope, the characteristics of the photographic emulsion and of the radiation of the isotope used are therefore important factors which determine resolution of structure and time of photographic exposure. The compartmental, cellular, or subcellular deposition of radioactively labeled compounds in tissues can be assessed qualitatively and quantitatively. Time-sequence studies and competition experiments with unlabeled analogs or antagonists most effectively provide morphological and functional information simultaneously, which may not be obtainable otherwise.

For the successful application of autoradiography *general prerequisites* must be considered:

1. The range of radiation of the isotopic label and the type of photographic emulsion used must be commensurate with the expected structural resolution.

2. The compound to be studied should be labeled at a high specific activity and with the radioisotope in a chemically stable position(s).

3. The labeled compound should be radiochemically pure at the time of administration.

4. The chemical nature of the radioactivity localized in the autoradiogram should be characterized by (a) autoradiographic studies with competitors, which preferably undergo different metabolism, or (b) radiochemical analysis of tissue extracts.

For the successful application of autoradiography for the localization of diffusible substances, not chemically bound to stable tissue constituents, *special prerequisites* must be considered. During the preparation of the tissue and the autoradiogram, translocation of the radioactively labeled compound and its metabolites must be avoided. This in turn necessitates the prevention of translocation of compounds and tissue constituents due to postmortem changes and the use of fluids, especially solvents, during tissue preparation. Since compounds differ in molecular size and charge, binding, and the degree of immobilizability by histological fixatives, no safe prediction can be made for any of the compounds studied. This even applies to chemically incorporated substances, if absolute values are to be obtained. Therefore stringent prerequisites need to be followed and controls are required in order to render an autoradiographic study meaningful.

Failure to recognize prerequisites mentioned in this chapter and difficulties in establishing acceptable control procedures were obstacles in the field which led to the frequent publication of artifacts until adequate autoradiographic techniques for diffusible compounds were developed and standards established (Stumpf and Roth, 1966; Stumpf, 1968e, 1970b; 1971a,b).

It is the purpose of this chapter to detail existent and tried autoradiographic techniques and to facilitate their application for study of the cellular and subcellular localization of diffusible substances. While many techniques have been proposed in the literature (reviewed in Stumpf, 1971b) only a few fulfill the theoretical requirement that in the preparation of the autoradiogram all steps must be excluded or controlled in which translocation of tissue constituents can occur to the degree of altering or falsifying information.

From our earlier comparative studies of different autoradiographic techniques (Stumpf and Roth, 1966), a critical review of the literature, and own experience over the last 15 years, the following autoradiographic techniques are recommended:

1. Dry-mounting of unembedded freeze-dried frozen sections on emulsion-coated slides (dry autoradiography).

2. Thaw-mounting of frozen sections on emulsion-coated slides.

3. Smear-mounting of cell suspensions on emulsion-coated slides.

4. Touch-mounting of tissue layers or cells on emulsion-coated slides.

All these techniques are based on the exclusion of liquid fixatives and fluids—prior to the end of photographic exposure—and embedding media, as well as the use of photographic emulsion-precoated slides. While the exclusion of fluids and solvents is theoretically advantageous and has proven practically to be so, the possibility of utilizing other techniques that employ liquid fixation and embedding is not excluded. For instance, a specific fixative may be found that immobilizes the compound studied at its original site *in situ*. Also, embedding may not necessarily lead to redistribution, if a suitable fixative is found and perhaps infiltration and hardening occur at short times and reduced temperatures. For instance, glycol methacrylate and Vestopal can be polymerized at room temperature through irradiation with a 10,000-Ci^{137} Cc source within 5 minutes or 1 hour, respectively (W. E. Stumpf and S. Daikoku, unpublished). The photographic emulsion may be applied through a wire loop as described in Volume I of this series by Miller *et al.* (1964), instead of by dipping into a 40°C heated liquid emulsion as is commonly done. Proof of the utility of such wet approaches needs to be provided through proper tests, either with at least one diffusible compound, the localization of which is known, or by comparing the results with those obtained by the dry-mount autoradiographic technique. Since the latter is a dry procedure and has been tested (Stumpf and Roth, 1966), it appears justified to recommend it as a control.

It would be ideal if frozen sections could be dry-mounted on photographic emulsion–coated slides, so that they could remain adherent during subsequent treatment, thus circumventing freeze-drying and avoiding warming and melting of the sections, i.e., excluding recrystallization of tissue ice due to warming and also excluding thawing artifacts of the tissue and diffusion due to melting. In our own experiments in which frozen sections and emulsion-coated slides were kept below −30°C, dry-mounting of the sections was not successful, confirming that frozen sections do not adhere without melting. This is in contrast to the claim of Appleton (1964) that frozen sections can be permanently mounted on photographic emulsions without melting. Appleton, however, recommends that emulsion-coated cover glasses be kept at $-5°C$. It appears that $-5°C$ is close to the eutectic temperature of the tissue and that it implies "brief" melting, when the histological slide or cover glass is transported by hand—as recommended by Appleton—and is pressed against the section on the knife. Therefore this technique is possibly not a dry-mounting but rather a thaw-mounting procedure.

II. Dry-Mount Autoradiography

The need to have histological techniques available that allow the cellular and subcellular localization of diffusible compounds, such as ions or low-molecular-weight substances, has long been recognized. The assumed necessity, expressed by many histologists, for liquid fixation and embedding in order to obtain useful light microscopic tissue preparation, was a major obstacle in the development of a suitable technique, and therefore the problem was considered very difficult or impossible to resolve (Levi, 1969). Many experienced investigators did not succeed (reviewed by Stumpf, 1971b). Progress in this field became possible when it could be demonstrated that (1) thin frozen sections can be cut and (2) freeze-dried without loss of coherence, (3) fixation is not indispensable in order to preserve the tissue structure (but liquid fixation may be adverse to the goal of immobilization of low-molecular-weight compounds), and (4) dry-mounting of sections on desiccated photographic emulsions is possible (Stumpf and Roth, 1964).

The steps of the dry-mounting of freeze-dried section autoradiography are depicted in Figs. 1–5.

A. Preparation of Tissue Mounts (Fig. 1)

Different types and sizes of brass tissue mounts should be available and selected according to specimen size. Prior to the experiment, tissue mounts are numbered and placed in a holder which is put on ice. Liver from a nonexperimental animal is minced with scissors or razor blade within a petri dish and is also kept in an ice tray which contains one or two additional petri dishes for specimen preparation. Some finely minced liver, serving to adhere the tissue to the mounts, is placed on top of the mounts (see Fig. 1, inset). Too little liver may impair cutting of the base of the

PREPARATION
of tissue mounts

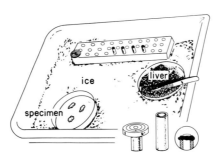

FIG. 1. Various steps of the dry-mount autoradiographic procedure which include preparation of the tissue mounts.

tissue; too much liver may result in slow freezing or cause cracking during rapid freezing. Tissue from the experimental animal may be excised after decapitation or under anesthesia and prepared for mounting on a brass holder in a petri dish in the ice tray. The specimen is positioned on the brass mount for later convenient cutting. The base of the specimen may be slightly embedded in liver.

B.   Freezing of Tissue (Fig. 2)

Prior to the experiment, the coolant for tissue freezing is prepared. Rapid freezing for minimizing ice crystal formation may be accomplished by using fluids characterized by a large temperature span between freezing and boiling points, unlike liquid nitrogen or liquid helium, in order to keep minimal the vapor insulation (Leyden-Frost phenomenon) during freezing. Optimal thermoconductivity may be accomplished by using liquid polyhydrocarbons, such as propane, Freon, or 2-methylbutane, cooled by liquid nitrogen close to their freezing point.

Propane gas (100–150 ml) is liquefied under a hood by prior cooling in liquid nitrogen of a 500-ml round flask with a stopper with a gas inlet and pressure outlet. Liquid propane is then transferred to a precooled beaker which is kept partially inserted in liquid nitrogen. Adjustment of insertion of the beaker, as well as occasional stirring of the propane, are necessary to maintain optimal low-temperature freezing conditions, for propane approximately $-180°$C.

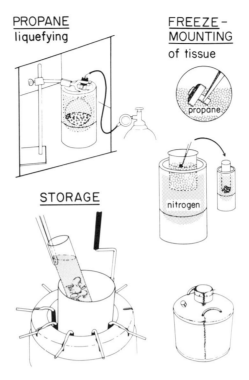

FIG. 2. Propane liquefying, freeze-mounting, and tissue storage in liquid nitrogen.

1.   FREEZE-MOUNTING

The tissue is frozen and mounted *simultaneously*. This is important, since separate freezing and mounting may cause tissue damage as the result of a rise in temperature. With a long forceps the brass holder with the tissue is inserted into the propane in a fashion that minimizes ice crystal formation due to freezing too slowly and cracking of tissue due to freezing too quickly. Therefore the speed and mode of freezing must be varied depending on specimen size. Three procedures have been developed in our laboratory:

1. *One-step freezing*: Small specimens, up to 1–1.5 mm³, may be plunged instantaneously into the coolant without fracturing.

2. *Two-step freezing*: Specimens larger than approximately 1.5 mm³ are likely to crack during sudden immersion into liquid propane at about $-180°C$. This can be prevented by two-step freezing. The tissue with the holder is immersed to about two-thirds its length with only the upper end of the tissue, or of the liver adhesive, remaining above the coolant for 1 or 2 seconds before total immersion. This allows for expansion during cooling,

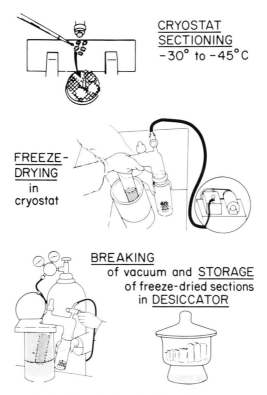

CRYOSTAT
SECTIONING
-30° to -45°C

FREEZE-
DRYING
in
cryostat

BREAKING
of vacuum and STORAGE
of freeze-dried sections
in DESICCATOR

FIG. 3.   Cryostat sectioning, freeze-drying, breaking of the vacuum, and storage of freeze-dried sections in a desiccator.

and the development of a more-or-less pronounced protrusion of the tissue ("terminal spine"). In this two-step freezing the important part of the tissue can be frozen rapidly with minimal ice formation and good preservation of structure, while ice crystal formation may become apparent in the more slowly frozen part.

3. *Successive freezing*: Specimens larger than about 0.5 cm³ may require a slower freezing process in order to avoid strain and cracking due to expansion-contraction during cooling. The mount with the specimen is immersed first either at its base or at its lateral rim. As part of the liver base and tissue starts to freeze, the specimen is gradually lowered into the coolant, allowing the coolant to advance as the freezing front advances throughout the tissue. Lowering the specimen mount too fast may result in fracturing of the tissue, while lowering it too slowly causes ice crystal growth and tissue disruption. At the present state of the art it is uncertain what the size limit is for tissue blocks that can be frozen, if useful microscopic resolu-

DIPPING
in liquid emulsion

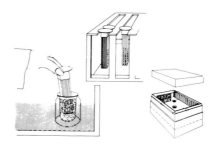

DRY - MOUNTING
of freeze - dried sections

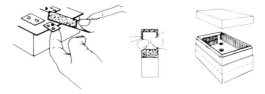

FIG. 4.   Dipping in liquid emulsion and dry-mounting of freeze-dried sections.

tion is to be obtained. We have frozen whole rat brain and cubes of squirrel monkey brain with successive freezing for the screening of hormone and drug distribution at the light microscopic level.

2. STORAGE

The freeze-mounted specimen which is immersed and kept in liquid propane for several seconds for freezing is then transferred rapidly and over the shortest possible distance to a plastic test tube filled with and kept in liquid nitrogen. Different sizes of test tubes are used, depending on the size and number of tissue mounts. At the end of the experiment, the test tubes are stored in a liquid-nitrogen tissue storage container until cryostat sectioning.

C.   Cryostat Sectioning (Fig. 3)

The test tube containing the specimen in liquid nitrogen is transferred in a Dewar storage container to the cryostat. With a precooled forceps the specimen mount is attached to the microtome chuck. Proper adjustment of specimen position will minimize the need for trimming. The optimal

THAW-MOUNTING

of cryostat sections on emulsion
coated slides

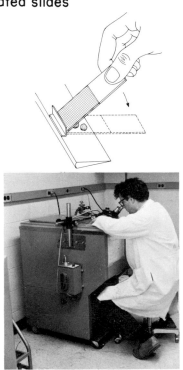

FIG. 5. The thaw-mounting of frozen sections on emulsion-precoated slides (top) and the Wide Range cryostat as used for the dry-mount and thaw-mount autoradiographic techniques, with fiber-optics illumination and disecting microscope (bottom). Although an open-top-type cryostat with convenient access, it has only a narrow opening during operation, keeping cutting conditions relatively constant. (Lower half of Fig. 5 reprinted with permission of Academic Press, Stumpf and Sar, 1975.)

cutting angle of the knife depends on temperature, section thickness, and type of tissue, as well as the included angle of the cutting edge (Stumpf and Roth, 1965); it must therefore be determined by experience. An inter-relationship exists between temperature and section thickness for optimal cutting. Accordingly, at $-20°C$ sections of 8 μm can easily be cut, while frozen sections of 2 or 1 μm require a temperature of $-40°C$ or lower (Stumpf and Roth, 1965). Optimal preservation of tissue is observed at lower temperatures, which also permit thinner sectioning and thus higher resolution. Reduced staining contrast, however, may necessitate the use of 2 μm sections instead of 0.5- or 1-μm sections. Routinely we cut 2- to 4-μm sections which, in general, provide good light microscopic resolution as well as good stainability with the silver grains and histology being in the same focal plane. It is important to use a cryostat (Fig. 6, bottom) that provides temperatures between $-20°$ and $-50°$C and constant temperature conditions by having a large freezing compartment (Wide Range Cryostat, Harris Manufacturing Co., North Billerica, Massachusetts). Such a compartment has the advantages of easy accessibility, space for safe tissue handling, and a large low-temperature air cushion which minimizes the danger of inadvertent warming of the specimen or cut sections due to convection of air or body heat.

The cutting of 2- to 4-μm thin, flat frozen sections is facilitated by observation through a dissection microscope equipped with glass-fiber cold light and homoiothermic conditions, i.e., knife, specimen, and surrounding air have equal or similar temperatures. Therefore the top opening of the recommended cryostat is kept small (Fig. 6). There is no need for a glove box. Rolling up of sections occurs more easily when a considerable temperature gradient exists above the knife level within the cryostat, as is the case in smaller, open-top microtome cryostats. During cutting, the section is held down with the bristles of a fine camel's hair brush. The cutting stroke may begin slowly, advance faster in the middle of the section and, before detachment of the section from the tissue block, be halted for a second, which reduces the tendency for the section to roll up. A well-sharpened knife is essential. The section is transferred with a fine brush to a vial located near and below knife level. During transfer of the section and, more generally, during handling of the frozen tissue, care must be taken not to raise the temperature, in order to prevent tissue disruption due to temperature-dependent recrystallization and ice crystal growth. A punched-out mesh on top of the vial carrier permits wiping off of the brush and retention of sections in the Polyvials during freeze-drying. Depending on the size of sections and vials, 10, 20, or more sections may be placed in one vial. After sectioning, the vial carrier is transferred with a precooled long forceps to the precooled specimen chamber of the vacuum system within the cryostat, the specimen temperature again being kept low at all times. Inadvertent warming of sections, even for a fraction of a second, may result in partial melting, diffusion, and shrinkage.

1. Freeze-Drying

While different vacuum systems may be utilized a compact cryosorption pump is conveniently used. A running fore pump is connected to the

Cryopump (Thermovac Industries, Copiague, L.I., New York), which is kept within the cryostat. The specimen chamber is attached to the Cryopump with the O-ring joint seating itself as a result of suction of the mechanical fore pump. The assembled Cryopump is then preevacuated for about 10 minutes with the adsorbant, Molecular Sieve (Linde A5), being activated. After a vacuum of about 50 torr is reached, the stopcock on top of the pump is closed and the forepump disconnected. The cryosorption chamber of the Cryopump, containing the Molecular Sieve, is kept in an empty Dewar during preevacuation. After the forepump is disconnected, the Dewar is filled with liquid nitrogen for the cooling of the absorbant, while the specimen chamber of the Cryopump is kept cool by the cryostat at about $-40°C$. A vacuum between 10^{-5} and 10^{-6} torr is created.

In general, it is important that a vacuum better than 10^{-4} torr be obtained, that the temperature of the specimen be kept low until the end of freeze-drying, and that termination of the vacuum and a rise in temperature be avoided before the end of drying. The minimal freeze-drying time required may be determined by experience, since the type of tissue, section thickness, temperature of the specimen, vacuum, dimensions of the vacuum chamber, and other factors influence performance.

Freeze-drying of the sections may be done overnight and terminated after about 20 hours by removing the Dewar with the Cryopump from the cryostat and placing it on a laboratory bench (Stumpf and Roth, 1967).

2. Breaking the Vacuum

One and one-half or 2 hours should be allowed for the specimen chamber, still under vacuum, to warm to room temperature. During this time, the portion of the pump containing the vapor trap must remain cooled by liquid nitrogen. The vacuum is broken by attaching a hose which is connected to a source of dry gas to the stopcock and by opening the stopcock to permit minimal flow in order to avoid turbulence. The procedure is facilitated by the use of a breathing bag, inserted between the gas cylinder and the vacuum system (Fig. 3, bottom left). After the vacuum is broken, the specimen chamber is removed from the vacuum system, and with a long forceps the specimen is transferred to a desiccator for storage. Exposure to high relative humidity may cause damage to the hygroscopic freeze-dried tissue as a result of enzymatic autolysis; also, diffusion of substances may occur.

3. Liquid Emulsion Coating of Slides (Fig. 4, top)

In order to have desiccated emulsion–coated slides available for the dry-mounting of freeze-dried sections, thaw-mounting of frozen sections, or smear-mounting of cell suspensions, a day or more before the section

mounting, the photographic emulsion (e.g., Kodak NTB3 or NTB2) is liquefied in a water bath at 40°–45°C in a light-proof container. After about
$\frac{1}{2}$–1 hour a portion of the liquefied emulsion is poured, under a 2-feet-
remote safe light, into a small wide-mouth container of about 100–200 ml,
permitting five slides to be dipped at a time. The liquefied emulsion can be
used undiluted or may be diluted with distilled water. The wide-mouth
beaker is placed into a light tight container in the water bath and allowed
to stand for another 1–2 hours in order to permit air bubbles to escape
and to obtain smooth emulsion-coated slides. Histological slides (those
with a frosted end are preferred) are prepared for dipping by cleaning them
in absolute alcohol, placing them into slide grips, holding five slides each,
and hanging them up on a slide-grip holder. The rack with the slide grips
is placed into an oven for warming to about 45°C. Dipping may be done, in
darkness or with a remote safe light on, by gentle insertion of a slide grip,
keeping the slides for 1 or 2 seconds in the emulsion, and gentle withdrawing,
allowing liquid emulsion to drip off before placing the slide grip into a
second rack. The slides may be air-dried on the darkroom table in a vertical
position. Arrangements for horizontal drying, if required, can easily be
made. No forced heated air should be applied, since reticulation of the
emulsion may occur.

The air-dried slides are removed, preferably in darkness, from the slide
grips and placed into lightproof desiccator boxes for storage in the refrigerator. One slide is kept for immediate photographic processing in order
to assess the silver grain background under the microscope with a calibrated
eyepiece reticule at 1000 × magnification. Less than 1 grain/1000 μm^2 is
usually counted with new photographic emulsions. Silver grain counts above
4–6 per 1000 μm^2 may impair the evaluation of the autoradiogram. Such
counts may be encountered with photographic emulsion or coated slides
older than 2 months or those exposed to light. Silver grain background
counts are also recommended prior to dipping a large number of slides,
using a test slide, or before utilization of stored coated slides.

4.   DRY-MOUNTING (Fig. 4, bottom)

One of the prerequisites for obtaining cellular and subcellular resolution
is close contact between sections and photographic emulsion. This is not
always assured in the apposition techniques, but has been optimally resolved
in the liquid emulsion coating (Joftes and Warren, 1955), stripping film
(Pelc, 1947), and imbibition (Siess and Seybold, 1957) procedures. The
use of wet film, however, is adverse to the preservation of diffusible substances. Dry-mounting, i.e., attainment of permanent close contact of
sections with desiccated emulsion–coated slides under conditions of low
relative humidity (20–40%), has become possible (Stumpf and Roth, 1964),

probably because photographic emulsions contain plasticizer, unspecified by the manufacturer, so that pressure impresses the sections slightly into the emulsion.

In the darkroom and under conditions of relative humidity between 20 and 40% (if the humidity is too low, electrostatic discharge may fog the emulsion and electrically charged sections may be difficult to handle and, if the humidity is too high, diffusion of radioactive material and autolysis of sections may occur) sections are placed on clean, smooth-surfaced Teflon pieces about $1\frac{1}{2}$–$2\frac{1}{2}$ cm long and 1.5 mm thick (sheets of Teflon, 12 × 12 inches and $\frac{1}{16}$-inch thick, may be obtained from Crane Packing Co., Morton Grove, Illinois). The sections are transferred from a Polyvial with a fine forceps and placed on the Teflon supports, the folds being removed. While larger sections can be handled with the unaided eye, small sections may be examined and positioned under a dissection microscope. Ten or more Teflon pieces may be prepared at one time. The light is then turned off and, under a safe light (Wratten No. 2), an emulsion-precoated slide is removed from its box and laid over a Teflon piece with the sections being pressed on by forefingers and thumb, pressing Teflon and slide together, without shifting movement. When the pressure is released, the Teflon falls off, and the slide with the section(s) is placed into a desiccator black box for exposure. From a given tissue 8 to 10 slides may be prepared in order to account for different lengths of exposure, different staining, and reproducibility. Conveniently, the red light is placed so that, by being placed low, the Teflon pieces can be observed against the background of the red light. The exposure of the emulsion-coated slide(s) to the red light must be kept minimal, and the open slide boxes should be shielded from it.

The boxes containing the section-mounted slides are sealed and placed in a freezer or refrigerator for exposure. A freezer, kept at $-15°C$, is used in our laboratory. The length of exposure needs to be determined by experience. It depends on the specific activity of the compound, type of radiation and its energy and half-life, specific concentration of the compound in the tissue, and other factors. It is important that during exposure the humidity be kept low, since the rate of latent image fading is exponentially related to relative humidity and the amount of moisture retained by the gelatin (Albouy and Faraggi, 1949).

5. PHOTOGRAPHIC PROCESSING

The lightproof desiccator box, containing the slides, is removed from the freezer or refrigerator and allowed to warm to room temperature before opening, since precipitation of moisture could damage the sections or cause chemographic artifacts. Under a safe light, a slide is removed and breathed on so that the brief moisture of the breath secures the contact between sections and the photographic emulsion, and loss or shifting of sections during subsequent development and fixation is avoided or minimized. This very brief application of moisture prior to development does not seem to have adverse effects, and is simpler and probably less damaging than the use of a gelatin solution as recommended by others in the literature.

Developer (Kodak D19—undiluted or diluted with water 1:1), fixer (Kodak), and water for rinsing need to be prepared in advance in order to permit adaptation to a uniform temperature for all the fluids. The rinsing water, the temperature of which is set by a water mixer, may serve as a water bath for the developer and fixer. Differences in temperature may cause reticulation of the emulsion and must therefore be kept minimal. In our laboratory the temperature is set at 15°C. This requires mechanical cooling of the cold water during the summer months.

The slides are kept in the 1:1 diluted developer for about 2 minutes without movement. The developing time may have to be altered, depending on the optimal size of the silver grains desired, the water temperature used, and whether or not the developer is diluted. The slide holder is then dipped for a few seconds into gently flowing tap water and placed in the fixer for 4 or 5 minutes. The slides are then rinsed for 5–10 minutes. Vigorous movement of water or handling of the slides must be avoided, since sections may be lost or portions of them shifted.

After photographic processing the sections may be stained while still wet. Intermittent air-drying, storage of the slides, and later staining is also possible. We prefer to stain the still wet slides, for instance, with methyl green–Pyronine (MGP) (see recipe in Appendix, p. 192) for DNA and RNA, which is a single-step stain and provides good visibility of silver grains, cell nuclei, and cytoplasm without noticeably staining the photographic emulsion. After the staining for a few seconds and brief removal of excess dye, the slides are permitted to air-dry and are then mounted with Permount and a cover glass. Alcohol dehydration may be used instead of air-drying. Histological fixation may be employed prior to staining or prior to photographic processing. Controls to prevent possible alteration of the latent image or the developed grains, however, are necessary. For instance, 10% formalin fixation after exposure and prior to photographic development may result in fogging. In general, we did not find disadvantages in using nonfixed sections. According to the principle of minimal tissue treatment and following the maxim of studying the unmolested tissue (Stumpf, 1970), we have excluded all unnecessary and potentially complicating steps. The combination of autoradiography with other histochemical reactions may require brief histological fixation.

III. Thaw-Mount Autoradiography (Fig. 5)

This procedure follows the steps described for the dry-mount technique as far as cryostat sectioning. It circumvents freeze-drying, but includes warming and melting of the frozen sections. Thus it is not a dry procedure. Thin sections between 2 and 4 μm give optimal results for cellular and subcellular light microscope resolution. The amount of diffusible fluid in thin sections is minimal. The sections are well stained, and tissue structure and silver grains can be viewed in the same focal plane. Superimposition of structures is minimal at this thickness. Since it is a wet technique, although the moist state is brief, controls to prevent possible artifacts due to negative and positive chemography are required. The dry-mount technique, described above, appears to be an adequate control, and the dry-mount and thaw-mount autoradiographic techniques are combined to advantage in our laboratory. The thaw-mount technique is superior in providing quick surveys and preserving better the integrity of large sections of tissues with different density components, as well as tissues with high water content which may disintegrate after freeze-drying. In the thaw-mount technique, tissue damage and diffusion can be kept minimal when the mounting is done skillfully and thin sections are used. It appears that, even under optimal conditions, subcellular resolution is inferior to that of the dry-mount technique, and tissues with small cells and high cell density, such as the pituitary, lymph nodes, and even liver, may not yield satisfactory resolution with thaw-mounting. Efforts to accomplish dry-mounting of frozen sections without melting, by using precooled stripping film-coated cover glasses (Appleton, 1964) or emulsion-coated slides, have been made by several investigators. While this would be an ideal approach, frozen-section mounting with permanent close contact between section and emulsion without melting has been claimed to be feasible by some but discounted by other investigators. In our laboratory earlier efforts at frozen-section dry-mounting were not successful. Cellophane tape has been used as an aid to accomplish this (Hammarström et al., 1965); the published results as judged by the autoradiographic resolution appear to be inferior.

For thaw-mounting, cutting of the frozen tissue may be performed in a darkroom under normal light, conveniently with the fiber-optics illumination of the Wide Range Cryostat (Fig. 5, bottom). After a few sections have been cut, the light is turned off and red light, positioned above the knife on top of the cryostat, is turned on. A photographic emulsion–coated slide is removed from a desiccator blackbox and placed over the sections on the knife. This picking up of frozen sections requires the development of skill, since distortion of the tissue may occur if the slide approaches the sections too slowly and also if a warm slide is pressed onto the melting section(s). A stop (e.g., provided by a clamp) at the knife base may serve as a convenient pivot for the slide so that it can be moved rapidly down and up for section pickup. The slide with sections mounted is placed into a light-proof desiccator box. The boxes containing the emulsion-coated slides are covered, the light is turned on, and the procedure is repeated. At the end, the desiccator box with the section-mounted slides is sealed and stored in the freezer or refrigerator for exposure. In order to increase efficiency, a few investigators have been able to cut under red light (Wratten No. 2), avoiding repeated light-dark adaptations and lightproofing of slide boxes.

After exposure, photographic processing is performed as described for the dry-mount technique, except that the brief application of breath moisture at the end of exposure is not required.

IV.   Smear-Mount Autoradiography (Fig. 6, top)

This is an autoradiographic technique which can easily be applied in combination with biochemical or *in vitro* studies (Stumpf, 1968e). Cellular or

SMEAR-MOUNTING

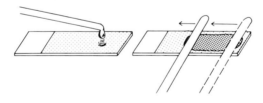

TOUCH-MOUNTING

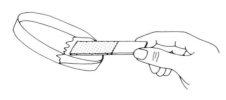

FIG. 6.   Smear-mounting (top) and the touch-mounting (bottom) autoradiographic techniques, using photographic emulsion-precoated slides.

subcellular fractions as obtained after tissue homogenization and centrifugation, or cell suspensions as obtained after enzymatic cell separation of organ pieces or from tissue culture, may be used. The radioactively labeled compound may be added *in vivo*, prior to tissue fragmentation, or *in vitro*. Attention must be given to proper concentration of the cellular or subcellular elements to be studied, generally after washing, centrifugation, and resuspension. Under a safelight, one or a few drops of the suspension are placed at one end of a desiccated emulsion–coated slide. With a glass rod or pipette, the fluid is gently spread over the slide, which is then allowed to air-dry and is placed into a desiccator box for exposure. Exposure, photographic processing, and staining are performed as described for the thaw-mount technique.

In evaluating the autoradiograms of cell suspension, superimposition of structures must be considered. Also, controls to prevent positive or negative chemography are required, since interaction between the wet specimen and the photographic emulsion can take place. If required, histological fixation can be included before or after exposure, however, not without prior control to avoid possible fixative-related aritifacts.

V.   Touch-Mount Autoradiography (Fig. 6, bottom)

This procedure resembles the smear-mount method; the only difference is that the surface of the photographic emulsion–precoated slide is brought in contact with the surface of the specimen, which may be a cell suspension, tissue culture fluid, or the cut surface of bone marrow or other tissue.

A.   Controls and Artifacts

Controls to prevent positive and negative chemography are required. This is a general prerequisite in autoradiography, no matter what technique is used.

1.   POSITIVE CHEMOGRAPHY

It is possible that the silver grains visible in autoradiograms are derived not from the radioisotope but rather from various chemical interactions between tissue components and the photographic emulsion. A case of *specific positive chemography* could exist if the unlabeled compound under investigation ionized silver halide crystals, resulting in specific silver grain formation on photographic development. Components of the tissue may cause "unspecific" chemography. Positive chemography may also be caused by mechanical pressure, by fixatives, e.g., higher concentrations of formaldehyde, and by dyes (Oehlert et al., 1962). Sections from tissues that do not contain radioactive material but the unlabeled compound or no compound, with or without fixative or dye, may serve as control to prevent positive chemography.

2. NEGATIVE CHEMOGRAPHY

The latent image, formed by the radiation of the isotope, or the developed silver grains can be partially or totally eliminated by components of the tissue, e.g., enzymes, or fixatives and dyes, especially those of high acidity. This can be tested by using a lightfogged slide for the autoradiographic exposure of the sections. If negative chemography is produced, the developed slide will show blanched areas where the latent image has been altered.

B.   Examples of Application

Figures 7–9 depict applications of the dry-mount (Fig. 7), thaw-mount (Figs. 8 and 10), and smear-mount (Fig. 9) autoradiography. Numerous other applications have been published in the literature from this laboratory (e.g., Stumpf, 1968a–d, 1969, 1970; Sar and Stumpf, 1973; Keefer et al., 1973; Stumpf and Sar, 1973; 1974a; Stumpf et al., 1974) and other laboratories (e.g., Thomson et al., 1971; Tachi et al., 1972; Knigge et al., 1972; Nakai et al., 1972; Kennedy and Little, 1974), using radioactively labeled steroid hormones, extracellular space indicators, drugs, and polypeptides.

The autoradiographic techniques detailed here are practical, simple—if this is a criterion for consideration—and tried. Ample demonstration of applications has been provided. The investigator planning to select one of the many advertised techniques for the autoradiography of water-soluble or diffusible substances is advised to look for the evidence. It is important to reemphasize here that the published autoradiogram is the evidence, and that a technique must be judged by the demonstrated cellular and subcellular localization of diffusible compounds known to be localized. This was done with the dry-mount autoradiographic technique *before* it was recommended (Stumpf and Roth, 1966); also, the histological resolution obtainable with unfixed, unembedded, freeze-dried frozen sections has been demonstrated in high-power phase-contrast micrographs (Stumpf and Roth, 1967). Unfortunately, descriptions of autoradiographic techniques and claims for their utility have often been accepted for publication *without* such evidence, but instead with sometimes elaborate statistics on silver grain distribution, which may be deceiving unless supported and verified by a theoretically sound procedure and convincing demonstration of authentic autoradiograms.

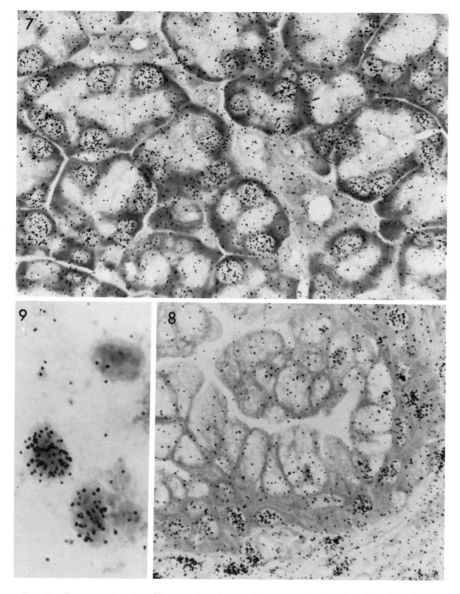

FIG. 7. Dry-mount autoradiogram showing nuclear concentration of radioactivity in acini of rat parotid gland, prepared 3 hours after the injection of 1,2-testosterone-^{3}H. Exposure time 90 days. 4 μm. Stained with methyl green–Pyronine. 610 ×.

FIG. 8. Thaw-mount autoradiogram showing nuclear concentration of radioactivity in basal cells of cervical gland and stromal cells of estrogen-primed guinea pig uterus, prepared 15 minutes after injection of 1,2,6,7-progesterone-^{3}H. Exposure time 42 days. 4 μm. Stained with methyl green–Pyronine. 320 ×.

FIG. 9. Smear-mount autoradiogram, showing concentration of radioactivity in certain nuclei of a suspension made from a nuclear fraction of estrogen-primed guinea pig uterus, prepared 15 minutes after injection of 1,2,6,7-progesterone-^{3}H. Exposure time 11 days. Stained with methyl green–Pyronine. 1100 ×. (M. Sar, T. Guerigian, and W. E. Stumpf, unpublished).

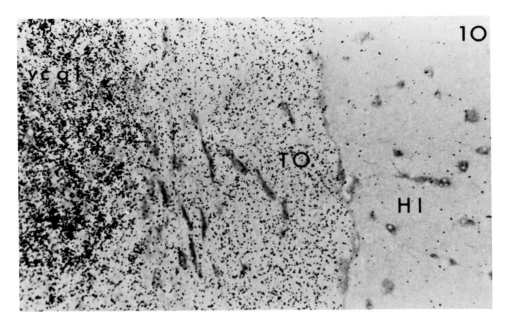

FIG. 10. For nerve pathway tracing, thaw-mount autoradiography is a sensitive and rapid procedure, superior to the paraffin embedding techniques. After multiple injections of precursor cocktail into the vitreous body of the eye (tree shrew), characteristic labeling of axon terminals in the nucleus ventralis corporis geniculati lateralis (vcgl) and of passing fibers in the tractus opticus (TO) can be seen, while the unlabeled adjacent hippocampus (Hi) has only low tissue background of radioactivity. Exposure time 3 months. 4 μm. Stained with methyl green–Pyronine. 530 ×. (from Conrad and Stumpf, 1975).

At the time the techniques described here were introduced, frequently and almost as a rule, the results obtained with them were in disagreement with those obtained with more classic histological approaches published in the literature. Standards for the autoradiography of diffusible compounds have since been set.

Many areas of research benefit from combined functional-morphological studies, and the application of the dry-mount and thaw-mount techniques will expand in the future, since adequate equipment has become available. A recent effective application of the thaw-mount technique is for the tracing of neural pathways (Fig. 10), which appears to be superior to the use of paraffin-embedded sections (Conrad and Stumpf, 1974). Also promising is the extension of these techniques to electron microscopic resolution, still to be accomplished, and, perhaps more so, the combined use of the dry-mount technique with fluorescence and immunohistochemistry. Localiza-

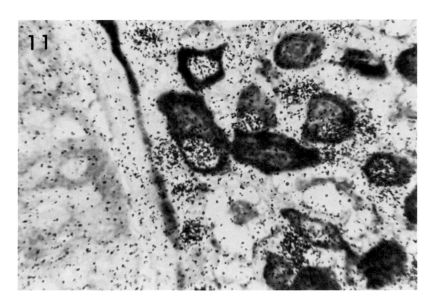

FIG. 11. Immunoautoradiogram of rat pituitary showing simultaneous localization of peroxidase-labeled anti-HCG and estradiol-^{3}H. Estrogen target cells are characterized by the accumulation of silver grains, especially over nuclei. Gonadotropes show heavy cytoplasmic immunostaining. Freeze-dried frozen section, dry mounted, used in combination with postexposure fixation. The autoradiogram was stained according to the immunoglobulin-enzyme bridge technique.

tion of catecholamines and estradiol-^{3}H has been demonstrated simultaneously in identical and in subsequent sections (Grant and Stumpf, 1975). An immunoautoradiogram (Fig. 11) shows the simultaneous localization of estradiol-^{3}H and peroxidase-labeled antihuman chorionic gonadotropin (anti-HCG) in rat pituitary recently accomplished in our laboratory (D. A. Keefer, W. E. Stumpf, M. Sar, and Petrusz, unpublished), utilizing freeze-dried section dry-mount preparation with post-exposure fixation, and staining after the immunoglobulin-enzyme bridge technique (Nakane and Pierce, 1967; Mason et al., 1969).

ACKNOWLEDGMENTS

During the development of the tissue freezing procedure, Dr. Don Keefer introduced what is described as "successive freezing." Dr. Peter Sheridan has used the slide stop at the microtome knife as a pivot for the emulsion-coated slide in the thaw-mounting procedure. Both worked as postdoctoral fellows in our laboratory during this time. Supported in part by PHS grant RO1 NSO9914.

Appendix: Methyl green–Pyronine (MGP) Stain for DNA and RNA as Used for Dry-Mount, Thaw-Mount, Smear-Mount, and Touch-Mount Autoradiograms

A.   Solutions

Pyronine Y (C.I. 45005)	0.25 gm
Methyl green (C.I. 42590)	0.75 gm
Phosphate buffer (pH 5.3)	100.0 ml
Solution A: 0.2 M Na$_2$HPO$_4$	52.5 ml
Solution B: 0.1 M citric acid	47.5 ml

Dissolve both substances in 25% methyl alcohol to prevent mold growth; correct for pH 5.3. Stock solutions are stable.

Solution A: To make 1 liter 0.2 M Na$_2$HPO$_4$ (anhydrous, MW 141.98), weigh out 28.396 gm Na$_2$HPO$_4$ and bring to 1 liter (750 ml H$_2$O + 250 ml methyl alcohol).

Solution B: To make 1 liter of 0.1 M citric acid monohydrate (MW 210.4), weigh out 21.015 gm citric acid and bring to 1 liter (750 ml H$_2$O + 250 ml methyl alcohol).

To this add:

0.5% Phenol solution	0.5 ml
1.0% Resorcinol solution (fresh)	2.5 ml

The above MGP reagent requires 2–3 days to age and remains stable up to 6 months at room temperature. It can be reused; filter before use.

B.   Procedure

1. The slides with the sections (in autoradiography after photographic processing) are stained in MGP reagent for approximately 1–20 seconds, depending on the type of tissue and section thickness. No fixation is required.
2. Wash gently in two changes of tap or distilled water.
3. Air-dry and mount cover slip with synthetic resin.

C.   Results

Cytoplasmic and nucleolar RNA red; DNA blue (the blue color is said to be due to the presence of methyl violet in commercial methyl green). This stain is advantageous, since it is a simple single-step stain, is not retained in the photographic emulsion, and permits good visibility of silver grains.

REFERENCES

Albouy, G., and Faraggi, H. (1949). *J. Phys. Radium* **10**, 105.

Appleton, T. C. (1964). *J. Roy Microsc. Soc.* **83**, 277.

Conrad, C. D., and Stumpf, W. E. (1974). *Cell Tiss. Res.* **155**, 283.

Grant, L. D., and Stumpf, W. E. (1975). *In* "Anatomical Neuroendocrinology" (W. E. Stumpf and L. D. Grant, eds.). Karger, Basel.

Hammarström, L., Appelgren, L. E., and Ullberg, S. (1965). *Exp. Lab. Res.* **37**, 308.

Joftes, D. L., and Warren, S. (1955). *J. Biol. Photogr. Ass.* **23**, 145.

Keefer, D. A., Stumpf, W. E., and Sar, M. (1973). *Proc. Soc. Exp. Biol. Med.* **143**, 414.

Kennedy, A. R., and Little, J. B. (1974). *J. Histochem. Cytochem.* **22**, 361–367.

Knigge, K. M., and Joseph, S. A., Silverman, A. J., and Vaala, S. (1972). *Progr. Brain Res.* **39**, 7–20.

Levi, H. (1969). *In* "Autoradiography of Diffusible Substances" (L. J. Roth and W. E. Stumpf, eds.), pp. 113–120. Academic Press, New York.

Mason, T. E., Phifer, R. F.; Spicer, S. S., Swallow, R. A., and Dreskin, R. B. (1969). *J. Histochem. Cytochem.* **17**, 563–569.

Miller, O. L., Jr., Stone, G. E., and Prescott, D. M. (1964). *In* "Methods in Cell Physiology" (D. M. Prescott, ed.), Vol. 1, pp. 371–385. Academic Press, New York.

Nakai, T., Sakamoto, S., Kigawa, T., and Shigematsu, A. (1972). *Endocrinol. Jap.* **19**, 47.

Nakane, P. L., and Pierce, G. B. (1967). *J. Cell Biol.* **33**, 307–318.

Oehlert, W., Nettesheim, P., and Machemer, R. (1962). *Histochemie* **3**, 99.

Pelc, S. R. (1947). *Nature (London)* **160**, 749.

Sar, M., and Stumpf, W. E. (1973). *Science* **179**, 389.

Siess, M., and Seybold, G. (1957). *Z. Wiss. Mikrosk.* **63**, 156.

Stumpf, W. E. (1968a). *Endocrinology* **83**, 777.

Stumpf, W. E. (1968b). *Endocrinology* **85**, 31.

Stumpf, W. E. (1968c). *Science* **162**, 1001.

Stumpf, W. E. (1968d). *Z. Zellforsch. Mikrosk. Anat.* **92**, 23.

Stumpf, W. E. (1968e). *In* "Radioisotopes in Medicine: In Vitro Studies" (R. L. Hayes, R. A. Goswitz, and B. E. P. Murphy, eds.), AEC Symp. Ser. No. 13 (CONF-671111), p. 633. U.S. At. Energy Comm., Oak Ridge, Tennessee.

Stumpf, W. E. (1970a). *Amer. J. Anat.* **129**, 207.

Stumpf, W. E. (1970b). *J. Histochem. Cytochem.* **18**, 21.

Stumpf, W. E. (1971a). *Acta Endocrinol. (Copenhagen), Suppl.* **153**, 205.

Stumpf, W. E. (1971b). *Amer. Zool.* **11**, 725.

Stumpf, W. E., and Roth, L. J. (1964). *Stain Technol.* **39**, 219.

Stumpf, W. E., and Roth, L. J. (1965). *Nature (London)* **205**, 712.

Stumpf, W. E., and Roth, L. J. (1966). *J. Histochem. Cytochem.* **14**, 274.

Stumpf, W. E., and Roth, L. J. (1967). *J. Histochem. Cytochem.* **15**, 243.

Stumpf, W. E., and Sar, M. (1973). *J. Steroid Biochem.* **4**, 1.

Stumpf, W. E., and Sar, M. (1974). *Endocrinology* **94**, 1116.

Stumpf, W. E., and Sar, M. (1975). *In* "Methods in Enzymology" (B. W. O'Malley and J. G. Hardman, eds.), Vol. 36, pp. 135–156. Academic Press, New York.

Stumpf, W. E., Sar, M., and Joshi, S. G. (1974). *Experientia* **30**, 196.

Tachi, C., Tachi, S., and Lindner, H. R. (1972). *J. Reprod. Fert.* **31**, 59–76.

Thomson, J. A., Rogers, D. C., Ginson, M. M., and Horn, D. H. (1971). *Cytobios* **2**, 78–88.

Chapter 10

Characterization of Estrogen-Binding Proteins in Sex Steroid Target Cells Growing in Long-Term Culture

A. M. SOTO, A. L. ROSNER,[1] R. FAROOKHI,
AND C. SONNENSCHEIN

Tufts University School of Medicine
Tufts Cancer Research Center,
Boston, Massachusetts

I.	Introduction	195
II.	Materials and Methods	196
	A. Nomenclature	196
	B. Cell Lines	197
	C. Growth Conditions of Cells in Culture	198
	D. Cell Fractionation	199
	E. Preparation of Nuclear Extract	200
	F. Protein Assays	201
	G. ER Assays	202
III.	Discussion	207
	References	210

I. Introduction

The presence of cytoplasmic proteins capable of binding certain estrogens to saturation with high affinity, high specificity, and low capacity has become accepted as a criterion by which certain cells are defined as targets for these estrogens (Talwar *et al.*, 1964; Toft and Gorski, 1966). It has been suggested that, at least for the uterus, estradiol-17β (E_2) is the most active of the natural estrogens (Jensen and Jacobson, 1962). Three to ten times more estrone (E_1) or estriol (E_3) is required to produce effects similar to those of E_2. Evidence for the metabolic interconversion of these estrogens

---

[1] Present Address: Department of Pathology, Beth Israel Hospital, Boston, Massachusetts 02215.

195

has been presented (Martin and Stone, 1965; Tseng and Gurpide, 1972).

Reviews of the physicochemical characteristics and biological aspects of the association between estrogens and the specific proteins that bind them have been published (Raspé, 1971; O'Malley and Means, 1973; King and Mainwaring, 1974).

In exploring the mechanism of the action of estrogens, our use of cell cultures has distinct advantages over investigations of hormone action involving entire organs. Primarily, cell culture systems bypass the homeostatic mechanisms that persist in studies performed in whole animals. In addition, the homogeneous population of target cells growing in a long-term culture permits the study of hormone action to be addressed to genetic events. This goal is realized by cloning cells which display variations of the specific steps involved in the mechanism of action of these steroid hormones (Sonnenschein *et al.*, 1974b). A similar approach in the study of the mechanism of action of glucocorticoids has been reported (Sibley and Tomkins, 1974).

The purpose of this article is to evaluate the methods we have used to characterize estrogen receptors (ER) in cells growing in long-term culture.

II.   Materials and Methods

A.   Nomenclature

1.   HORMONES

We have used the following hormones: estrone (3-hydroxyestra-1,3,5(10)-trien-17-one) (E_1); estradiol-17β (estra-1,3,5(10)-triene-3,17β-diol) (E_2); estriol (estra-1,3,5(10)-triene-3,16α,17β-triol); testosterone (17β-hydroxyandrost-4-en-3-one); and progesterone (pregn-4-ene-3, 20-dione).

2.   RECEPTOR

We invoke this word as a synonym for estrogen-binding proteins, which are found in the cytosol and which eventually may be transferred into the nucleus of the target cell. The main characteristics of this receptor are its saturability, high specificity, high affinity, and low capacity with respect to the interacting steroid hormone. Because we sometimes find ER permanently associated with the nucleus of cultured cells (in addition to a translocated cytosol receptor observed in the nucleus of E_2 target cells), it is expedient to designate the cytosol ER as receptor I and the "indigenous" nuclear receptor as receptor II. At this time, we cannot ascribe any definite functions to these receptors beyond the physical properties mentioned above.

Reports of resident sex steroid nuclear receptors (translocation from the cytoplasm not being evident) have been published (Lebeau *et al.*, 1973; Best-Belpomme *et al.*, 1975; Shao *et al.*, 1975).

B.   Cell Lines

From all the cell lines established in long-term cultures in this laboratory, two groups can be distinguished by their ability to bind E_2 specifically to subcellular fractions. This property is coincident with the presence or absence of ER in the cellular extracts.

1.   ESTABLISHED CELL LINES CARRYING ER

a. Rat Pituitary Cell Lines.   1. GH_3. These cells were derived from a rat pituitary tumor called MtT/W5 which was originally described by J. Furth (Yokoro *et al.*, 1961). This transplantable tumor was adapted to grow in long-term culture, and several clones were isolated (Yasumura *et al.*, 1966). Some of these clones are sensitive to hypothalamic releasing hormones (Tashjian and Hoyt, 1971) and thyroid hormones (Samuels *et al.*, 1974); in addition, some secrete pituitary hormones (Tashjian *et al.*, 1968; Son-nenschein *et al.*, 1970) and carry ER in their subcellular fractions (Mester *et al.*, 1973).

2. RAP. These cell lines were established from rat pituitary tumors in-duced in female Wistar/Furth rats by the administration of hyperphysio-logical doses of estradiol valerate (Delestrogen, E. R. Squibb and Sons, New York. N.Y.). Clones were isolated from four cell lines, each derived from a discrete tumor carried by a separate animal (Sonnenschein *et al.*, 1975).

3. FPG. This cell line has been established from the rat pituitary trans-plantable tumor MtT/F_4, which grows in Fisher rats, and has the ability to secrete somatotropin, prolactin, and adrenocorticotropin (Furth *et al.*, 1956). Clones of this cell line have been isolated (Sonnenschein *et al.*, 1975). The hormone secretory properties of the cells in culture have not been tested.

b. Rat Endometrial Cell lines.   A cell line has been established from normal rat endometrial cells explanted from a 10-day-old Wistar/Furth female rat. When these cells were injected into a homologous host, an adenocarcinoma of the endometrium developed. Several clones have been characterized from this cell line which is designated U15 (Sonnenschein *et al.*, 1974a).

2.   ESTABLISHED CELL LINES NOT CARRYING ER

a. HeLa.   This is a human cell line established originally from an adeno-carcinoma of the uterine cervix (Gey *et al.*, 1952).

b. Mouse Myeloma Cells. MPC_{11} is a mouse cell line of immuno-globulin-secreting cells growing in suspension. When injected into mice, they develop into tumors whose cells secrete immunoglobulins (Laskov and Scharff, 1970).

For more specific details of the characteristics of these cell lines we suggest consultation of the references cited above.

C. Growth Conditions of Cells in Culture

The cell lines listed above are routinely grown in media consisting of Dulbecco modified Eagle's medium (DME) plus 15% horse serum and 2.5% calf serum. A mixture of penicillin (final concentration 240 units/ml, E. R. Squibb and Sons, New York, New York), Fungizone (final concentration 6×10^{-3} mg/ml, E. R. Squibb and Sons, New York, New York), strepto-mycin (final concentration 0.12 mg/ml, Pfizer Co., New York, New York), and tylocine (final concentration 72 μg/ml, Gibco, Grand Island, New York) is included to minimize the risk of contamination by bacteria, fungi, and PPLO. A special medium lacking estrogen and consisting of DME plus 10% castrated and adrenalectomized calf serum (Iowa State University, College of Veterinary Medicine, Ames, Iowa) is used for the varying periods of time (hours to months) we wish to assure that no steroid hormones have been made available to the cells.

Cells are cultured at 37°C in glass roller bottles (Bellco Glass Co., Vine-land, N.J.) on a Rollacell apparatus (New Brunswick Scientific, New Bruns-wick, New Jersey) or in plastic flasks (Falcon Plastics, Oxnard, California) in a CO_2 (CO_2, 5%; air, 95%) incubator for routine maintenance. Subcultures of cells are developed by:

1. Transplanting only free-floating cells from stationary cultures or roller bottles.

2. Agitating media-containing bottles in which the cells are loosely attached to the growing surface.

3. Briefly applying trypsin (final concentration 1.3 mg/ml, Gibco, Grand Island, New York) or Viokase (final concentration 25 mg/ml Gibco, Grand Island, New York) to the more firmly attached cells until the latter become free-floating.

Cells growing in suspension whose ER is to be characterized are harvested by centrifugation and thoroughly washed three times with serumless medium or with Hanks' balanced salt solution. In order to determine the presence, molar concentration, and number of binding sites of ER in the cytosol, we use at least 0.2 ml of packed (2×10^8) cells; when the properties of nuclear ER are to be ascertained, we require 0.5 ml of packed (5×10^8) cells. Following this technique, we routinely harvest up to 3 ml of packed estrogen target cells.

D.   Cell Fractionation

1.   CELL DISRUPTION

After three washes with DME, the packed cells are resuspended in 3–4 vol. of buffer A [2 mM MgCl$_2$, 10 mM Tris–HCl(pH 7.4) at 4°C], allowed to swell for 5 minutes at 4°C, and driven through a syringe fitted with a 22-gauge needle. The last mentioned step should be repeated until all the cells have been disrupted. Crude nuclear and cytoplasmic fractions are obtained after centrifugation at 700 g for 10 minutes.

2.   PREPARATION OF CYTOSOL

The supernatant obtained from the disrupted cells contains cellular sap, mitochondria, and endoplasmic reticulum. To remove the organelles, crude cytosol is centrifuged at 129,000 g for 20 minutes in a T-65 rotor (Beckman-Spinco). The pellet represents membranes and whole mitochondria, while the supernatant is called the cytosol.

3.   PURIFICATION OF NUCLEI

The 700-g pellet (obtained above) contains nuclei with relatively little attached cytoplasmic material. Although cytosol and nuclei can be obtained with buffer B [1.5 mM EDTA, 10 mM Tris–HCl (pH 7.4) at 4°C], as well as with buffer A with the same kinetic results (Fig. 1), nuclei obtained in the Tris buffer containing MgCl$_2$ appear more intact than those recovered in the presence of EDTA.

Treatment of the crude nuclei with 0.1% Triton X-100 in buffer C [0.32 M sucrose, 2 mM MgCl$_2$, 10 mM Tris–HCl (pH 7.4) at 4°C] decreases cytoplasmic contamination by removing the outer nuclear membrane. The procedure is as follows: the pellet is resuspended in 6 vol of Triton X-100 solution, incubated 5–10 minutes at 4°C, diluted with 6 vol of buffer C, centrifuged at 800 g for 10 minutes, and washed twice with 12 vol of buffer C. The nuclei obtained by this procedure can be easily resuspended without clumping, and a recovery of 85–100% can be anticipated.

A further purification is achieved using a modification of the method of Chauveau (Chauveau et al., 1956). After cell disruption, nuclei are resuspended in 25 ml of buffer D [2.2 M sucrose, 2 mM MgCl$_2$, 10 mM Tris–HCl (pH 7.4) at 4°C], transferred to a Teflon-to-glass homogenizer (with a clearance of 0.15–0.23 mm), and homogenized. The high shear forces produced in the viscous suspension permit almost complete removal of endoplasmic reticulum from the nuclei. The homogenate is layered on a cushion of 5 ml of buffer D and centrifuged at 45,000 g for 60 minutes in an SW-25.1 rotor (Beckman-Spinco). After centrifugation, the pellet con-

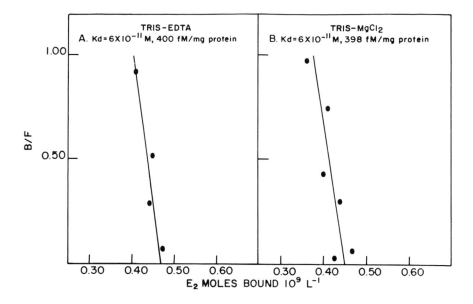

FIG. 1. Scatchard analysis of the binding of estradiol-³H to the cytosol fractions of C₈11RAP cells, according to the techniques described in Section II. (A) Data for cytosol prepared in buffer B [1.5 mM EDTA, 10 mM Tris-HCl (pH 7.4) at 4°C]. (B) Properties of a similar fraction extracted in buffer A [2 mM MgCl₂, 10 mM Tris-HCl (pH 7.4) at 4°C].

taining the purified nuclei is resuspended in an isotonic medium. With DNA being assayed by the Burton (1956) method, nuclear recovery is expressed as the ratio of DNA recovered in the nuclear fraction to the DNA present in the total homogenate. Following this purification, a recovery of 30–40% is observed.

Electron micrographs (Fig. 2) show that our preparations are composed of essentially intact nuclei free from contaminating organelles. Lactic dehydrogenase assays (Kornberg, 1955) of our nuclei indicate that the latter contain less than 0.5% of the enzymatic activity normally found in the cytosol.

E.   Preparation of Nuclear Extract

The nuclear pellet is resuspended in 1–2 vol of buffer E [500 mM KCl, 1.5 mM EDTA, 10 mM Tris–HCl (pH 7.4) at 4°C] and sonicated by a Model W185D Sonifier (Heat Systems-Ultransonics, Inc., Plainview, New York) using 50-W, 5-second pulses at 30-second intervals. In our laboratory, five pulses are sufficient to yield maximal extractions. After centrifugation at 129,000 g for 20 minutes in a T-65 rotor (Beckman-Spinco), the supernatant contains about 85% of the total KCl-extractable nuclear ER.

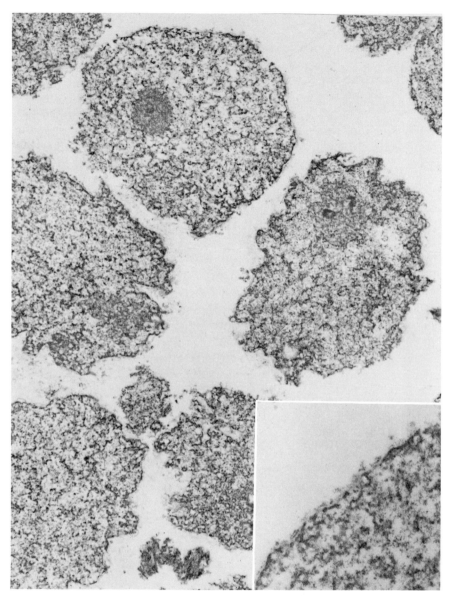

FIG. 2.   Electron micrograph of nuclear fraction of C_5U15 cells. 3000 ×. Inset: 10,000 × magnification of same preparation showing loss of outer nuclear membrane after treatment by Triton X-100, as described in Section II.

F.   Protein Assays

The amount of protein in the subcellular fractions was estimated by the microbiuret method (Goa, 1953).

G. ER Assays

Our determination of the properties and molar concentration of E_2 receptor sites in a subcellular extract is based on the competition for these specific binding sites, which are limited in number, by both labeled and unlabeled estradiol. In a total volume of 0.3 ml, cytoplasmic or nuclear fractions are incubated 12–18 hours with 0.5–5.0 nM $E_2[^3H]$ (New England Nuclear, Boston, Massachusetts, 90–110 Ci/mmole) and increasing amounts of unlabeled E_2 (0–12 nM) at 4°C. To estimate binding of E_2 to nonspecific sites (which are not saturable), the subcellular extract is incubated with $E_2[^3H]$ and a 100-fold excess of unlabeled steroid hormone. After equilibration is attained, bound E_2 is separated from the remaining free steroid by any one of four techniques: (1) adsorption of free E_2 by charcoal (Nugent and Mayes, 1966; Korenman, 1968); (2) adsorption of bound E_2 by hydroxylapatite (Erdos *et al.*, 1971); (3) adsorption of bound E_2 by DEAE filters (Santi *et al.*, 1973); (4) recovery of all bound E_2 by protamine sulfate precipitation (Steggles and King, 1970).

Our assays of the diminishing amounts of labeled E_2 that can be recovered with ER (as increasing amounts of unlabeled steroid hormone are included with the incubations) are carried out by the principles developed by Scatchard (1949). Implicit in this analysis are the assumptions that: (1) the separation of bound and free forms of E_2 is complete and quantitative; (2) the affinities of ER for labeled and unlabled E_2 are the same; (3) the binding reaction of ER to steroid is reversible (i.e., a measurable equilibrium constant exists); and (4) the receptor is homogeneous and monovalent.

Estimations of the amount of steroid bound and the dissociation constant (K_d) of the E_2—receptor complex can also be obtained by the Lineweaver-Burk method. This analysis, however, assumes an uniformity of variance of $1/B$ (the reciprocal of moles of steroid bound), as opposed to the more independent and uniformly variant B/F (moles of steroid bound/moles of free steroid) used in Scatchard analysis. In the Lineweaver-Burk treatment, therefore, the variance is greatly magnified at the larger values of bound steroid leading to estimations of the equilibrium constant of hormone–receptor interaction that are spuriously high. A detailed analysis of the various methods used for the evaluation of binding data is available (Robard., 1973).

It must be emphasized that the concentration of ER and extracted protein are critical for these data to be accurate. If the protein level is too high, nonspecific binding of steroid hormone will elevate the apparent equilibrium constant of steroid and ER; if ER is too dilute, however, the number of available binding sites will immediately become saturated with $E_2[^3H]$ and only the dilution of labeled steroid by increasing amounts of unlabeled steroid will be observed. In addition, dissociation of the receptor may occur. A portion of subcellular extract must therefore be chosen within the range of extract concentrations in which the extent of steroid hormone bound varies linearly with the amount of subcellular fraction included.

1. ADSORPTION OF FREE E_2 BY CHARCOAL

a. Preparation of Charcoal. Acid-washed charcoal (Norit Amend Drug and Chemical Co., Irvington, New Jersey) is exhaustively washed and titrated to neutrality, dried, and resuspended in buffer of the same ionic strength as the experimental samples. The concentration of the stock suspension of treated charcoal is 2.5 gm%; a Dextran-70 supplement (Pharmacia Fine Chemicals, Uppsala, Sweden) may be added to a final concentration of 0.025 gm%.

b. Method. After equilibration of the subcellular fraction with labeled and unlabeled E_2 is completed, 0.2 ml of charcoal suspension is added, and incubation at 4°C is carried out for 30 minutes. All samples are subsequently centrifuged at 800 *g* for 10 minutes at 4°C before 0.1 ml of each supernatant is transferred to a scintillation vial. Ten milliliters of toluene–POPOP–PPO solution is then added to each vial, and counting is executed in a Beckman LC-100S scintillation counter with a tritium efficiency of 41–46%.

c. Results. Figure 3 indicates that the dextran supplement in the charcoal stock suspension increases the efficiency of the removal of free

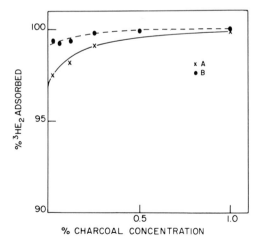

FIG. 3. Adsorption of free estradiol-³H as a function of the concentration of charcoal. Varying amounts of charcoal lacking (×——×) or containing (●---●) dextran (0.025 gm%) are added to samples of estradiol-³H (final concentration 3×10^{-9} *M*), as described in Section II.

hormone at virtually all concentrations of charcoal tested. The addition of bovine serum albumin, a common practice in the assay of steroid-binding proteins by the charcoal adsorption technique (Jungblut *et al.*, 1972), shows no perturbation of the adsorption of free E_2 at charcoal concentrations exceeding 0.1% (Fig. 4). The amount of bound E_2 that can be measured by this method varies in a linear manner with the concentration of subcellular fraction if the latter exceeds 0.065 mg of protein per adsorption tube.

2. Adsorption of Bound E_2 by Hydroxylapatite[2]

a. Preparation of Hydroxylapatite. A 90-gm portion of hydroxylapatite (Bio-Gel HTP, Bio-Rad Laboratories, Richmond, California) is resuspended in 1500 ml of buffer F [50 mM Tris–HCl, 5 mM KH$_2$PO$_4$ (pH 7.4) at 4°C], and settling is allowed for 10 minutes. Fines are then decanted, and the operation is repeated again before the remaining resin is resuspended in a final volume of 300 ml.

b. Method. At the conclusion of the incubations of subcellular fractions with labeled and unlabeled E_2, 0.3 ml of the hydroxylapatite slurry

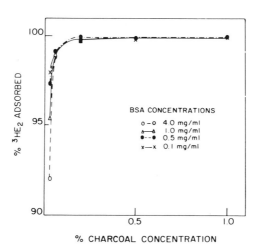

FIG. 4. The effect of bovine serum albumin (BSA) on the ability of charcoal to adsorb free estradiol. Sample of BSA at final concentrations of 0.1 mg/ml, 0.5 mg/ml 1.0 mg/ml, and 4.0 mg/ml are equilibrated with estradiol-^{3}H (final concentration 3×10^{-9} M) and are subjected to treatment by varying amounts of charcoal in order to separate bound and free estradiol, as described in the text.

[2] We wish to thank Dr. E. J. Peck, Jr., for his helpful suggestions regarding the use of the hydroxylapatite method.

is added, and incubation at 4°C is carried out for at least 15 minutes. Resin-treated samples are subsequently filtered at 4°C with suction through Whatman No. 1 filter papers supported on a multiple filter apparatus (Yeda Scientific Instruments, Rehovot, Israel) and washed twice with 5 ml of buffer F to remove free E_2. The dried filters are transferred into scintillation vials, 10 ml of toluene–POPOP–PPO is added, and counting is performed as described above.

   c. Adsorption Capacity of the Resin.   Portions of $E_2[^3H]$ labeled cytosol (0.5 ml) containing 6 mg of protein are incubated with increasing amounts of slurry. After 15 minutes of incubation at 4°C, samples are centrifuged in an Eppendorf Model 3200 centrifuge at 12,000 rpm for 10 minutes. Super-natants are used for protein determinations, while the pellets are washed six times in buffer F and counted. The capacity of the resin can thus be evaluated to be 12 mg of protein per milliliter of slurry.

   d. Results.   If samples are filtered and washed with increments of buffer from 0 to 40 ml after equilibration of a trial solution of free $E_2[^3H]$ and hydroxylapatite is completed, a constant background of 2.3% of the total radioactivity added to the resin is attained after a minimum of 10 ml of buffer is used. If a subcellular extract containing ER is included in the equilibration with hydroxylapatite, the amount of bound E_2 that can be thus recovered varies in a linear manner with the concentration of subcellular fraction if the latter exceeds 0.026 mg of protein per adsorption tube. As shown in Fig. 5, at least 0.3 ml of hydroxylapatite slurry is required to achieve the maximal adsorption of both protein and radioactivity to the resin in the presence of a subcellular fraction. The coincidence of these two adsorption maxima suggests that the radioactive E_2 recovered on the resin is associated with the protein thus retrieved.

3. ADSORPTION OF BOUND E_2 BY DEAE FILTRATION

   a. Method.   Samples equilibrated with labeled E_2 are spotted in 0.05 ml portions onto DEAE-cellulose filter disks (DE81, 2.4 cm Whatman) supported on a multiple-filter apparatus (Yeda Scientific Instruments, Rehovot, Israel) at 4°C. The samples are equilibrated for at least 1 minute before suction is applied, and 5 × 1 ml washes of buffer G [1.5 mM EDTA, 20 mM Tris–HCl (pH 7.8) at 4°C] are passed through each filter. The disks are subsequently transferred to scintillation vials; toluene–POPOP–PPO is added, and counting is performed as described above.

   b. Results.   The adsorption of ER by the filter is quantitative when the volume of the sample does not exceed 50–75 μl and the amount of protein per filter is below 2 mg. If ER is filtered in the presence of 0.2 M KCl, the recovery of bound E_2 is reduced by 40%.

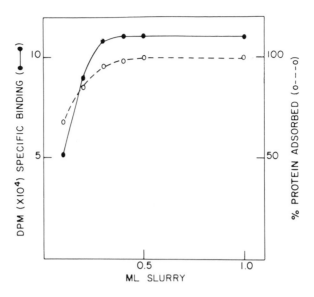

FIG. 5. The recovery of ER as a function of the concentration of hydroxylapatite. Samples of C₅U15 cytosol are equilibrated with estradiol-³H (final concentration $3 \times 10^{-9} M$) in a final volume of 0.5 ml to which increasing volumes of hydroxylapatite solution are added, as described in Section II. After hydroxylapatite is recovered by filtration, the residue is assayed for radioactivity by scintillation counting (●···●) Protein not adsorbed by resin is assayed in the filtrate, and the difference between the latter quantity and the concentration of protein in the cytosol before treatment by hydroxylapatite is assumed to represent the amount of protein actually adsorbed by the resin (○---○).

4. RECOVERY OF BOUND E₂ BY PROTAMINE SULFATE PRECIPITATION

a. Method. After equilibration of subcellular fractions with $E_2[^3H]$ is completed, samples are mixed with increasing amounts of protamine sulfate (Sigma Chemical Co., St. Louis, Missouri) in a final concentration of 0.5–50 mg of protamine and incubated at 4°C for 15 minutes in a total volume of 2.0 ml. The samples are subsequently passed by suction through glass-fiber filter papers (GF/C, Whatman) supported on a multiple-filter apparatus (Yeda Scientific Instruments, Rehovot, Israel) at 4°C and washed with 10 ml of buffer H [1.5 mM EDTA, 10 mM Tris–HCl (pH 7.8) at 25°C]. The disks are subsequently recovered for scintillation counting as described above.

b. Results. Figure 6 shows the specific binding of $E_2[^3H]$ after treatment with increasing amounts of protamine sulfate. Under our experimental conditions, background (nonspecific) adsorption represents 0.2% of the total radioactivity, and recovery of ER by protamine sulfate at a final concentration of protamine of 1–10 mg is as efficient as that obtained by the adsorption of bound steroid by hydroxylapatite.

III. Discussion

A comparison of the relative efficiencies of each of the techniques discussed above in determining the amount and properties of ER might best be summarized by the data presented in Table I. It is apparent that, in the presence of 0.5 M KCl used to extract nuclear ER (receptor II), a slight decrease in the number of moles of recovered receptor II is noted if ER is retrieved by its adsorption to DEAE, and a more significant (2-fold) decrease is observed if the charcoal adsorption method is followed. The effect of high ionic strength in reducing the amount of receptor that can be determined by the charcoal method, an observation that has recently been confirmed with the ER extracted from immature rat uteri (Peck and Clark, 1974), is particularly distressing in the study of receptor II, since high KCl concentrations are often used to extract the latter ER (Best-Belpomme et al., 1975; Jensen et al., 1967). The reduction we observed of the amount of receptor that can be recovered by DEAE filtration at high ionic strengths is in accord with the results of Santi et al. (1973).

TABLE I

PROPERTIES OF ER OF C$_s$UI5 SUBCELLULAR FRACTIONS, AS DETERMINED BY VARIOUS METHODS OF SEPARATION OF BOUND AND FREE ESTRADIOL[^{3}H][a]

Method	Cytosol receptor I		Nuclear extract receptor II	
	$K_d \times 10^{-10}\,M$	MB $\times 10^{-10}\,M$	$K_d \times 10^{-10}\,M$	MB $\times 10^{-10}\,M$
Charcoal	3.0	9.00	1.3	0.78
DEAE	3.0	9.40	1.0	1.30
Hydroxylapatite	2.0	10.00	0.8	1.70
Protamine sulfate	2.5	13.00	1.6	1.60

[a] The values shown represent ER assays of a single source of nuclear and cytosol fractions from which aliquots were taken to perform a uniform Scatchard analysis of the different methods tested, as outlined in the text. Subcellular fraction were prepared by the techniques described in Section II, from a tumor of the C$_s$UI5 series growing in a Wistar/Furth female rat that had been castrated and adrenalectomized 10 days before sacrifice. After equilibration of estradiol-^{3}H and subcellular fractions, separation of bound and free hormone was carried out by each of the methods indicated.

A further limitation of the charcoal method is the apparent ability of charcoal to adsorb ER as well as free E_2, an effect that is more dramatic at low concentrations of subcellular fraction. As indicated in Table II, the amount of bound ER that can be measured by the charcoal technique is the same as that observed with hydroxylapatite adsorption only at concentrations of cellular material approaching 1.3 mg of protein per adsorption tube. As the dose of subcellular fraction is reduced, the amount of ER that can be measured by the charcoal technique decreases to as little as 1% of that which can be recovered by hydroxylapatite at the same concentration of extract. That charcoal is capable of adsorbing significant amounts of protein in a nonspecific manner is demonstrated by our observation that the apparent concentrations of bovine serum albumin (BSA) (as determined by the microbiuret assay) are significantly reduced if the standard BSA solutions are exposed to the charcoal–dextran slurry routinely used in the separation of bound and free hormone. Over half the BSA is lost if the concentration of protein in each incubation with charcoal is less than 0.3 mg. The fraction of nonspecific binding included in the measurement of bound estradiol has been observed in other laboratories to decrease by a factor of 5 if hydroxylapatite or protamine sulfate is used instead of charcoal to separate bound and free hormone (Blondeau and Robel, 1975).

It appears from our data (Table I) that the measurement of ER by its precipitation by protamine sulfate is as desirable a method as the one involving adsorption of receptor to hydroxylapatite. Very little protamine is

TABLE II

SEPARATION OF BOUND AND FREE ESTRADIOL-$[^3\mathrm{H}]$ AS A FUNCTION OF
CONCENTRATION OF PROTEIN IN THE CYTOSOL OF C_5UI5 CELLS[a]

Protein (mg/ml)	Method of separation of bound and free hormone	
	Charcoal (dpm)	Hydroxylapatite (dpm)
1.3	15,619	17,100
0.13	414	1,970
0.09	131	1,375
0.043	7	578

[a]Cytosol fractions of C_5UI5 cells extracted in buffer A were equilibrated with saturating amounts of estradiol-^{3}H (final concentration of $1 \times 10^{-9}\ M$) and were subsequently diluted to various concentrations of protein in the presence of $1 \times 10^{-9}\ M$ estradiol. Hydroxylapatite or charcoal was then used to separate bound and free estradiol, according to the methods described in the text.

required to precipitate all the binding activity (Fig. 6), and the technique therefore seems to be an excellent means for separating ER from the bulk of the intracellular proteins. However, the precipitate formed by protamine sulfate is extremely fine and becomes highly gelatinous after 15 minutes at 4°C. Unless specific precautions are taken, loss of material on the walls of laboratory glassware thus becomes a likely possibility.

Because the measurement of ER by its adsorption to hydroxylapatite remains the only technique relatively independent of (1) variations in ionic strength, (2) different concentrations of protein in the subcellular fraction,

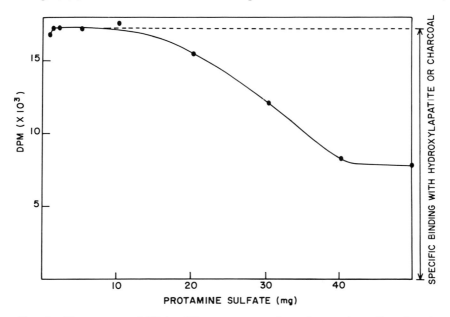

FIG. 6. The recovery of ER by different concentrations of protamine sulfate. Samples of C_5U15 cytosol at 4°C are equilibrated with estradiol-^{3}H (final concentration $3 \times 10^{-9} M$), mixed with increasing concentrations of protamine sulfate, and drawn through glass-fiber filters according to the techniques described in the text. Residues are subjected to scintillation counting to determine the amount of ER precipitated.

(3) time of equilibration with the adsorbant, and (4) possible loss of virtually invisible material on the walls of laboratory glassware, we choose this method as the one that is the most definitive in the study of the properties of ER in sex steroid target cells.

ACKNOWLEDGMENTS

The authors thank Heather Duram, Gary Murphy, and Daniel Kiracofe for their technical assistance, and Linda Wong for her typing of the manuscript. In addition, they are indebted to Ernest J. Peck, Jr., and H. I. Jacobson for their critical review of this communication. The research was supported by grants from the Public Health Service and from the American Cancer Society.

REFERENCES

Best-Belpomme, M., Mester, J., Weintraub, H., and Baulieu, E. E. (1975). *Europ. J. Biochem.* **57**, 537–547.

Blondeau, J. P., and Robel, P. (1975). *Europ. J. Biochem.* **55**, 375–384.

Burton, K. (1956). *Biochem. J.* **62**, 315–323.

Chauveau, J., Moule, Y., and Rouillier, Ch. (1956). *Exp. Cell Res.* **11**, 317–321.

Erdos, T., Bessada, R., Best-Belpomme, M., Fries, J., Gospodarowicz, D., Menahem, M., Reti, E., and Veron, A. (1971). *Advan. Biosci.* **7**, 119–135.

Furth, J., Clifton, K. H., Gadsden, E. L., and Buffett, R. F. (1956). *Cancer Res.* **16**, 608–616.

Gey, G. O., Coffman, W. D., and Kubicek, M. T. (1952). *Cancer Res.* **12**, 264–265.

Goa, J. (1953). *Scand. J. Clin. Invest.* **5**, 218–222.

Jensen, E. V., and Jacobson, H. I. (1962). *Recent Progr. Horm. Res.* **18**, 387–414.

Jensen, E. V., DeSombre, E. R., Horit, D. J., Kawashima, T., and Jungblut, P. W. (1967). *Arch. Anat. Microsc. Morphol. Exp. Suppl.*, **56**, 547–560.

Jungblut, P. W., Hughes, S., Hughes, A., and Wagner, R. K. (1972). *Acta Endocrinol. (Copenhagen)* **70**, 185–195.

King, R. J. B., and Mainwaring, W. I. P. (1974). "Steroid-Cell Interactions." Univ. Park Press, Baltimore, Maryland.

Korenman, S. G. (1968). *J. Clin. Endocrinol. Metab.* **28**, 127–130.

Kornberg, A. (1955). *In* "Methods in Enzymology" (S. P. Colowick and N. O. Kaplan, eds.), Vol. 1, pp. 441–443. Academic Press, New York.

Laskov, R., and Scharff, M. D. (1970). *J. Exp. Med.* **131**, 515–541.

Lebeau, M. C., Massol, N., and Baulieu, E. E. (1973). *Eur. J. Biochem.* **36**, 294–300.

Martin, L., and Stone, G. M. (1965). *Steroids* **6**, 473–483.

Mešter, J., Brunelle, R., Jung, I., and Sonnenschein, C. (1973). *Exp. Cell Res.* **81**, 447–452.

Nugent, C. A., and Mayes, D. M. (1966). *J. Clin. Endocrinol. Metab.* **26**, 116–1122.

O'Malley, B. W., and Means, A. R., eds. (1973). "Receptors for Reproductive Hormones." Plenum, New York.

Peck, E. J., Jr., and Clark, J. H. (1974). *J. Steroid Biochem.* **5**, 327–328.

Robard, D. (1973). *In* "Receptors for Reproductive Hormones" (B. W. O'Malley and A. R. Means, eds.), pp. 289–326, Plenum, New York.

Raspé, G., ed. (1971). "Advances in the Biosciences," Vol. 7. Pergamon, Oxford.

Samuels, H., Tsaj, J. S., and Casanova, J. (1974). *Science* **184**, 1188–1191.

Santi, D. V., Sibley, C. H., Perriard, E. R., Tomkins, G. M. and Baxter, J. D. (1973). *Biochemistry* **12**, 2412–2416.

Scatchard, G. (1949). *Ann. N. Y. Acad. Sci.* **51**, 660–672.

Shao, T. C., Castaneda, E., Rosenfield, R. L., and Liao, S. (1975). *J. Biol. Chem.* **250**, 3095–3100.

Sibley, C., and Tomkins, G. (1974). *Cell* **2**, 213–220 and 221–227.

Sonnenschein, C., Richardson, U. I., and Tashjian, A. H., Jr. (1970). *Exp. Cell Res.* **61**, 121–128.

Sonnenschein, C., Weiller, S., Farookhi, R., and Soto, A. M. (1974a). *Cancer Res.* **34**, 3147–3154.

Sonnenschein, C., Posner, M., Sahr, K., Farookhi, R., and Brunelle, R. (1974b). *Exp. Cell Res.* **84**, 399–411.

Sonnenschein, C., Soto, A. M., Colofiore, J., Farookhi, R., and Duram, H. (1975). Submitted for publication.

Steggles, A. W., and King, R. J. B. (1970). *Biochem. J.* **118**, 695–701.

Talwar, G. P., Segal, S. J., Evans, A., and Davidson, O. W. (1964). *Proc. Nat. Acad. Sci. U. S.* **52**, 1059–1066.

Tashjian, A. H., Jr., and Hoyt, R. F., Jr. (1971). *In* "Molecular Genetics and Developmental Biology" (M. Sussman, ed.), pp. 353–387. Prentice-Hall, Englewood Cliffs, New Jersey.

Tashjian, A. H., Jr., Yasumura, Y., Levine, L., Sato, G., and Parker, M. L. (1968). *Endocrinology* **82**, 342–352.

Toft, D., and Gorski, J. (1966). *Proc. Nat. Acad. Sci. U. S.* **55**, 1574–1581.

Tseng, L., and Gurpide, E. (1972). *Amer. J. Obstet. Gynecol.* **114**, 1002–1008.

Yasumura, Y., Tashjian, A. H., Jr., and Sato, G. (1966). *Science* **154**, 1186–1189.

Yokoro, K., Furth, J. and Haran-Ghera, H. (1961). *Cancer Res.* **21**, 178–186.

Chapter 11

Long-Term Amphibian Organ Culture[1]

MICHAEL BALLS

Department of Human Morphology, The Medical School,
University of Nottingham, Nottingham, England

DENNIS BROWN[2] AND NORMAN FLEMING[3]

School of Biological Sciences, University of East Anglia,
Norwich, England

I. Introduction	. . . .	214
II. Animals	. . . .	214
III. Methods	. . . .	215
A. Preparation of Donor Animal	. . . .	215
B. Dissection	. . . .	217
C. Setting up Cultures	. . . .	217
D. Whole-Organ Culture	. . . .	219
E. Culture Media	. . . .	219
F. Addition of Hormones, Drugs, or Toxic Substances	. . . . .	220
G. Sampling Frequency	. . . .	221
H. Histology and Metaphase Arrest	. . . . . .	221
I. Use of Radioactive Isotopes	. . . . . .	224
J. Organs Cultured	. . . .	224
IV. Results	. . . .	224
A. Retention of Normal Structure and Function	. . . .	224
B. Control of Cell Proliferation	. . . . . .	226
C. Hormones and Substrates	. . . .	230
D. Hormone Production	. . . .	231
E. Enzyme Production	. . . .	231
F. Drug Effects	. . . .	232

[1] This work was partly supported by grants from the Science Research Council (D.B.) and The Wellcome Trust (N.F.)

[2] *Present address:* Institut d'Histologie et d'Embryologie, École de Médecine, Université de Genève, Geneva, Switzerland.

[3] *Present address:* Anatomisches Institut, Universität Zürich, Zürich, Switzerland.

    G. Toxicological Studies 233
    H. Chemical Carcinogens *in Vitro* 233
    I. The Role of Serum Factors 234
  V. Concluding Remarks 235
    References 236

I. Introduction

Organ culture involves the maintenance of whole organs or parts of organs *in vitro* under conditions in which tissue structure, cell–cell relationships and functions, which are almost invariably lost when cell monolayer cultures are set up, are retained. Organ cultures of embryonic and neonatal tissues have proved useful in studies on morphogenesis and endocrinology, but the poor survivial of most mammalian and avian adult tissues *in vitro* (including liver, kidney, pancreas, and muscle) has limited their use to tissue slice and perfusion studies for comparatively short periods (Trowell, 1959; Jones, 1967; MacDougall and Coupland, 1967; Campbell and Hales, 1971; Balls and Monnickendam, 1976). By contrast, fragments of a variety of organs from adults of several amphibian species survive in organ culture for comparatively long periods (Monnickendam and Balls, 1973a). Apart from certain major differences, such as nitrogenous excretion in pre- and postmetamorphic stages and responses to thyroxine and prolactin, the basic morphology, physiology, and biochemistry of amphibian organs is very similar to that of homoiothermic vertebrates. Therefore long-term amphibian organ culture has many potential uses in biomedical research. In this chapter, we outline the methods used and summarize the results of our work to date [see Monnickendam and Balls (1973a) for a review of the methods and results of other research groups].

II. Animals

Animals of six amphibian species have been used in our main experiments. Apart from the *Rana*, which are kept at 4°C and are not fed after receipt, these animals are easy to maintain for long periods at room temperature and, after some initial coaxing in the case of *Amphiuma* and *Necturus*, readily feed themselves.

Amphiuma means (Congo eel) may be obtained from W. F. Prince, Silver

Springs, Florida, The Snake Farm, La Place, Louisiana, and Carolina Biological Supply Company, Burlington, North Carolina. They are kept singly in tanks of tap water, and thrive on chopped liver and heart.

Necturus maculosus (mud puppy), obtained from Carolina Biological Supply Company, are kept in aerated, deionized water, and are fed chopped liver or *Tubifex* worms. We have kept *Necturus* for more than 2 years without experiencing any of the feeding and disease problems listed by Kaplan and Glaczenski (1965).

Ambystoma mexicanum (Mexican axolotl), obtained from Gerrard and Haig Ltd., Worthing Road, East Preston, Nr. Littlehampton, Sussex, U.K., are kept in aerated tap water and fed mealworms or chopped liver.

Triturus cristatus carnifex (Italian great crested newt), obtained from Gerrard and Haig Ltd., and kept in a tank with a secure lid, containing a little water and some rocks. Newts are fed on *Tubifex* worms.

Rana temporaria (European common frog) are supplied by The Frog Farm, Rockfield Road, Kells, County Meath, Eire.

Xenopus laevis laevis (South African clawed toad) are obtained from Gerrard and Haig Ltd., kept in tap water, and fed chopped liver or heart.

Local anuran or urodele species would be expected to give results similar to those obtained with *R. temporaria* and *T. c. carnifex. Xenopus* and *Ambystoma* are obtainable from local dealers in many countries, but *Xenopus* may also be obtained direct from South Africa Snake Farms, P.O. Box 6, Fish Hoek, Cape Province, South Africa.

There are great differences between the sizes of mature individuals of the six species, and this affects the number of cultures that can be obtained from each (Table I). There is also variation in the time for which tissues survive in good condition. This appears to be related to respiration rate and cell size (Table I; Monnickendam and Balls , 1973b; Brown *et al.*, 1975c). The most suitable species for a particular experiment depends both on the number of replicate cultures required and on the planned culture period. For example, many experiments can be carried out with fifty 10-day cultures of *Rana* or *Xenopus* liver, whereas experiments for longer periods requiring more replicates are best carried out with *Amphiuma* liver cultures.

III. Methods

A. Preparation of Donor Animal

Many amphibians can live for several weeks or months without feeding and, depending on the experiment to be carried out, it may be necessary to

TABLE I

DETAILS OF ANIMALS USED

Species	Approximate weight of medium to large specimen (gm)	Order of hepatocyte size (1 = largest)	Order of respiration rate (6 = lowest)	Survival of liver pieces in good condition (days)	Approximate number of replicate cultures from one specimen	
					Liver flask cultures	Pancreas roller cultures
A. means	800–1200	1	6	70 or more	200–300	15–30
N. maculosus	80–150	2	5	42–56	50–100	10–15
A. mexicanum	50–120	3	4	21–28	30–60	5–10
T. c. carnifex	5–10	4	3	14–21	10–15	3–5
R. temporaria	25–45	5	1	10–16	10–15	3–5
X. l. laevis	80–150 ♀ 40–70 ♂	6	1	7–12	30–50	5–10

feed or starve an animal before use (e.g., *A. means* liver still has a 1% liver glycogen content after 6 weeks without feeding). After 1 hour in dilute $KMnO_4$ solution, the experimental animal is killed by a sharp blow on the head in the case of *Amphiuma* or by immersion in sterile 1% MS222 (Sandoz) for the other five species. Animals required for skin culture are kept in sterile water containing antibiotics and killed by decapitation and pithing.

B.   Dissection

The dissection, and all other stages in setting up organ cultures, are carried out under sterile conditions in a laminar flow cabinet [Pathfinder (experimental) 1974 Ltd., Solent Road, Havant, Hants., Hampshire, U.K.]. The skin is swabbed with absolute alcohol, and an incision is made along the length of the ventral surface of the animal which is fixed to a dissection board. The skin and ventral body wall muscle are pulled aside or removed to allow easy access to the viscera. The required organs are then removed and placed separately in petri dishes containing culture medium or amphibian phosphate-buffered saline (APBS). A $10 \times$ concentrate of APBS contains, per liter of double-distilled water, 60 gm NaCl, 2 gm KCl, 1 gm Na_2HPO_4, 0.2 gm KH_2PO_4, and 20 gm glucose, plus 10 ml 0.2% phenol red, and is sterilized by Millipore filtration. Immediately before use, dilute with sterile double-distilled water and add antibiotics at the concentrations used in culture media.

The liver is the largest organ (reaching 20 cm in length in larger *Amphiuma*), so it is taken out as three or four pieces which are placed in separate petri dishes. Thus contamination of one piece does not lead to loss of all the liver cultures. Particular care is necessary in removing the pancreas, which is attached to the intestinal wall. Since gut contents would contaminate other organs, the stomach and intestine are removed after all the other organs.

C.   Setting Up Cultures

Since it takes a long time to set up organ cultures from a large animal, such as *Necturus* or *Amphiuma*, organs such as the liver are first cut into 0.5- to 0.75-cm cubes. All organs can be stored in culture medium at 4°C until the investigator is able to deal with them. Glycogen assays performed on liver from an animal with a particularly high initial glycogen level (8.9% of wet weight) showed that after 24 hours at 4°C the glycogen content of a 0.75-cm cube was still 8.2%.

Liver, pancreas, kidney, and muscle are cut with fine scissors into 1- to 2-mm cubes, but cylindrical organs such as lung and intestine are cultured as 2- to 4-mm rings. Fragment size is important, since central necrosis occurs in larger pieces (3-mm cubes) of *A. means* liver, whereas no necrosis is detectable in smaller fragments, even after 9–10 weeks. Fragments smaller than 1 mm³ are avoided, because of the high ratio of damaged to undamaged cells. Organ fragments are transferred to culture vessels with fine forceps, and great care is taken not to damage them.

Organ fragments may be cultured in several ways (Fig. 1). Liver, kidney, and muscle are cultured in 25-ml Erlenmeyer flasks (10 pieces per flask = 40–50 mg) containing 10 ml medium. For lung, 8 to 10 rings are placed in each flask. Because of its nonsterile lumen, several changes of medium are passed through the intestine before it is cut into rings and cultured in flasks (6–8 pieces per flask). Flasks are loosely capped (Cap-O-Test, Switzerland) to allow free gaseous exchange with the atmosphere, and are shaken at 2 cps in a linear harmonic shaker at 25°C. Roller tubes are now routinely used for pancreas cultures—five fragments are incubated in plastic tubes contain-

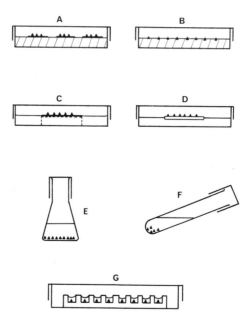

FIG. 1.   Organ culture methods. Culture on a solid substrate (medium–agar gel) in a 5-cm petri dish: (A) cultures supported by lens paper or a Millipore filter; (B) cultures on the medium–agar gel surface. Culture at the air–medium interface in a 5-cm petri dish: (C) cultures on a Millipore filter or on lens paper, supported by a titanium or stainless-steel mesh platform; (D) cultures supported by a floating raft made of rayon, a gelatin sponge, or a Millipore filter pad. Submerged or suspension culture: (E) cultures in a loosely capped 25-ml Erlenmeyer flask; (F) cultures in a test tube rotated or a roller tube apparatus; (G) microwell culture of individual organ fragments.

ing 2 ml medium when insulin is to be assayed (insulin tends to adhere to glass) or 7–8 pieces are incubated in 5 ml medium in glass tubes for other studies. Liver (6 fragments in 5 ml medium) and lung (4 fragments) also culture well in roller tubes, which allows the number of replicate cultures from one animal to be almost doubled. The disadvantage of roller tubes is that the addition of substances and sampling of medium or tissue is more difficult than with larger flasks. Microwell culture is particularly suitable for renin-release studies (one kidney fragment in 0.25 ml medium in each well) or spleen antibody production assays.

For skin culture, the ventral skin is removed, placed dermis downward on a Millipore filter (type HA, 47 mm) in a petri dish of APBS, and then cut into smaller pieces of the size required for culture. Skin cultures are maintained at the medium–air interface—either on titanium mesh platforms or on Millipore pad rafts (Fig. 1).

D. Whole-Organ Culture

Whole hearts, particularly those of smaller urodeles, will survive and beat *in vitro* for several months (Millhouse *et al.*, 1971). It is also possible to culture the intact urodele alimentary tract by slowly passing medium down the lumen.

E. Culture Media

The standard culture medium used consists of: 50% minimum essential medium [Auto-Pow-MEM (modified), Flow Laboratories, Irvine, Scotland], 40–45% double-distilled, deionized water, 5–10% fetal bovine serum (Flow Laboratories), with 15 mM HEPES (Hopkin and Williams, Chadwell Heath, Essex, U.K.), 1 mM L-glutamine (Flow Laboratories), 100 IU/ml benzylpenicillin (Glaxo, Greenford, Middlesex, U.K.), 100 μg/ml streptomycin sulfate (Glaxo) and 2 μg/ml Fungizone (Squibb, New York, N.Y.). We have also used media containing 50% basal medium Eagle [Auto-Pow-BME (modified), Flow Laboratories] or 50% Leibovitz L-15 (modified) (Flow Laboratories) in place of minimum essential medium. The abbreviations MEM, BME, and L15 will be used to denote the different complete media used.

Fifty percent MEM or BME (10% serum) is easily made up from autoclavable powder media as follows. (1) To a clean 1-liter bottle add 4.8 gm Auto-Pow-MEM (modified) or 4.6 gm Auto-Pow-BME (modified); add 3.6 gm HEPES and 880 ml double-distilled water, and then autoclave for 15 minutes at 15 lb/in²; (2) allow to cool (can be stored at 4°C for up to 1 month); then, immediately before use, add 5 ml 200 mM L-glutamine (bought as a sterile solution or made up and sterilized by Millipore filtration), 100 ml fetal bovine serum (bought sterilized), and the antibiotics; (3) adjust to pH 7.4 by adding approximately 9 ml 1 N NaOH (sterilized by autoclaving). L15 is also bought as a powder, which is dissolved in double-distilled water and sterilized by Millipore filtration. MEM and BME can also be made from the basic salts, amino acids, and vitamins. This is useful, since individual constituents, such as glucose or a particular amino acid, can be omitted to give glucose-free MEM or, say, alanine-free MEM.

Our early work was based on culture in L15, and viability was assessed on histological criteria. This medium was formulated by Leibovitz (1963) and contains large quantities of free-base amino acids which act as an effective buffer system. L15 thus allows cultures to be maintained in free gaseous exchange with the atmosphere, obviating the need for special culture vessels or the 5% CO_2–95% air mixture required for the bicarbonate–CO_2 buffering system. A range of zwitterionic buffers, developed by Good *et al.* (1966), was found to be satisfactory for cell culture by Williamson and Cox (1968) and for organ culture by Fisk and Pathak (1969). MEM or BME buffered with HEPES have been found to be more suitable than L15 for long-term amphibian organ culture (Fleming *et al.*, 1975), which confirms an earlier view (Monnickendam and Balls, 1973a) that specially formulated media, such as Wolf and Quimby's medium, are not necessary for amphibian cells and tissues.

In long-term culture, media are normally changed once per week. During this time, *A. means* liver cultures use less than 25% of the available glucose, but weekly medium changes prevent the buildup of nitrogenous excretory products to potentially toxic levels. Replenishment of the antibiotics also reduces the likelihood of contamination.

The precise osmolality and serum content of amphibian organ culture media are not important. Monnickendam and Balls (1972) cultured *A. means* liver and spleen in 30–90% L15 with 8% serum, and *A. means* kidney in 15–45% L15 with 0.8 or 8% serum. No major differences in the cultured tissues were detected. More recently, we have found that *A. means* liver fragments survive for at least 28 days in serum-free medium. The pH of the medium is, initially, 7.4 although *A. means* tissues may be able to tolerate a wider range than mammalian tissues. Liver glycogen is maintained at a constant level for at least 2 weeks in the range pH 6.9–7.5 in MEM with 5 or 10% serum.

F. Addition of Hormones, Drugs, or Toxic Substances

Additions to culture media are made in the smallest possible volume, to avoid altering the overall balance of the medium. The effective concentrations of hormones such as insulin, adrenaline, and glucagon, were found by means of dose–response curves, and then the lowest effective doses were used in subsequent experiments. In principle, it is very important that concentrations applied *in vitro* are close to the concentrations used *in vivo*. However, it is very difficult to decide whether to assume that a drug administered intravenously in man is dispersed in the total body weight (70 kg), total body fluid (40 liters), total extracellular fluid (15 liters), accessible extracellular fluid (plasma, lymph, and interstitial fluid—12 liters), plasma (3 liters), or some smaller volume (should the injected material circulate, at least initially, as a bolus). The problem is even greater with orally administered drugs. Thus, while recognizing that any general rule is open to criticism and is unlikely to apply equally well to all substances, we have based the concentrations applied *in vitro* on the assumption that, whether applied intravenously or orally, drugs are dispersed in 10 liters of fluid.

A further problem is the effective life of biodegradable hormones and drugs *in vitro*. Most hormones are rapidly degraded *in vivo* (Goodman and Gilman, 1970), but their half-lives *in vitro* are not known. Thus, how long will a dose added at time zero be effective, and how often should the dose be repeated?

G. Sampling Frequency

Related to the problems of dose level and frequency of application is that of sampling frequency. In short-term fragment or whole-organ perfusion cultures of mammalian tissues, the problem is one of collecting data before the preparation deteriorates. A 70-day culture period for *A. means* liver presents a different problem—that of estimating the most meaningful intervals between sampling medium or tissue. Studies on drug or hormone effects on specific processes require sampling at intervals of 24 hours or less, but studies on the long-term effects of added agents or on the general maintenance of tissue function require sampling at intervals of 7 or 14 days. The danger is that important changes occurring between sampling times will not be detected. Thus the strength of amphibian organ culture, the comparatively long culture periods available, presents difficulties as well as opportunities for the experimenter.

H. Histology and Metaphase Arrest

For routine histology, cultures are fixed in Carnoy's fluid, dehydrated in an ethanol–butanol series, embedded in paraffin wax, sectioned at 5 μm, and stained in Mayer's acid hemalum and eosin. For estimation of mitotic incidence (the number of mitotic figures per 10^5 cells), cultures are fixed

TABLE II

ANALYSIS OF FUNCTION IN CULTURED TISSUES

Organ	Culture methods	Parameters measured	Assay methods
Liver	10 fragments (45 mg) in 25-ml conical flasks containing 10 ml medium	Glycogen content	Hydrolysis of glycogen with amyloglucosidase; determination of glucose with hexokinase/glucose-6-phosphate dehydrogenase (Slein, 1965).
		Glucose uptake or release	GOD-PERID (Boehringer Corp.)
		Urea and ammonia production	Urea test combination (Boehringer Corp.)
		Lactate and pyruvate uptake or release	UV test combination (Boehringer Corp.)
		Arginase activity	Coulombe and Favreau (1963)
		OTC activity	Reichard (1957)
		Protein synthesis	Fleming (1974)
		Albumin production	Albumin determination (Sigma)
		Other enzyme activities (GOT, GPT, LDH, and so on)	UV test combinations (Boehringer corp.)
		Fructose-1, 6-diphosphatase	Pogell and McGilvery (1952)
		Renin substrase release	Angiotensin I radioimmunoassay
Kidney	10 fragments (45 mg) in conical flasks containing 10 ml medium or microwell culture (1 fragment per well in 0.25 ml medium)	Renin production	Angiotensin I radioimmunoassay
		Glucose uptake or release	As for liver
		Urea and ammonia production	
		Glycogen content	
Pancreas	5 fragments (20 mg) in 2 ml medium in plastic roller tubes, or 8 fragments (30 mg) in 5 ml medium in glass tubes	Insulin production	Insulin radioimmunoassay
		Amylase production	Amylase determination (Roche Diagnostica)

Tissue	Culture method	Parameter measured	Assay method
Skeletal muscle	8 fragments (80 mg) in 25 ml conical flasks containing 10 ml medium	Glucose uptake Lactate production Glycogen content	As for liver
Adipose tissue	10 pieces (50 mg) floating in 10 ml medium in a conical flask	Triglyceride storage and glycerol and nonesterified fatty acid release	UV method (Boehringer Corp.) Colorimetric method (Boehringer Corp.)
Skin	Raft or platform culture at medium-air interface in a petri dish	DNA synthesis and cell proliferation Keratinization	Autoradiography and liquid scintillation counting Protein determination (Lowry et al., 1951), histochemistry, cystine-^{3}H labeling and autoradiography
Intestine	Rings of tissue in 10 ml medium in a conical flask	Absorption of lipid droplets Mucus production	Histochemistry Histochemistry
Lung	Rings of tissue in 10 ml medium in a conical flask or 5 ml medium in a roller tube	Lactate production Urea and ammonia production Mucus production	As for liver Histochemistry
Heart	Whole-organ culture in a petri dish	Heartbeat rate; response to hormone and drug stimulation or inhibition	Direct observation

4–16 hours after the addition of 0.1–4 μg/ml Colcemid (Ciba, Basel) or 0.1–1 μg/ml vinblastine sulfate (Eli Lilly, New York). The time of exposure depends on the species concerned (the length of the M phase of the cell cycle is much longer in the larger-celled species) — 4–8 hours for *Xenopus* or 8–20 hours for *Amphiuma*. Godsell (1972) found that 0.25 μg/ml vinblastine sulfate gave the maximum number of arrested metaphases in *Xenopus* cell cultures.

I.   Use of Radioactive Isotopes

Protein, DNA, and RNA synthesis are easily studied in amphibian organ cultures by the addition of labeled precursors to the culture media. Tissue fragments are removed and prepared for liquid scintillation counting or autoradiography by standard techniques.

J.   Organs Cultured

A variety of organs have been cultured, and the retention of normal structure and function has been studied in several ways (Table II) in addition to ordinary light microscopy and electron microscopy (see also Brown *et al.*, 1975c; Fleming *et al.*, 1975).

IV.   Results

A.   Retention of Normal Structure and Function

Early studies were concentrated on establishing that normal structure and function were retained in organ culture. On this basis, later work was undertaken on the control of metabolic processes and on responses to substances added via the culture medium. As stated earlier, survival time *in vitro* varies with the species, appearing to depend on cell size and tissue metabolic rate. In general, *A. means* provides the longest organ culture periods, and *A. means* liver fragments retain their histological integrity and ultrastructure for at least 70 days (Fig. 2). *Necturus* liver and kidney also culture well (Figs. 3 and 4), and preliminary studies on long-term skin culture have been promising. Evidence for the retention of normal functions is summarized in Table III, to which could be added examples of the work of other groups, namely, the induction of urea cycle enzymes in adult *Rana catesbeiana* liver by thyroxine (Bennett *et al.*, 1969); vitellogenin synthesis by adult *Xenopus* liver (Wallace and Jared, 1969); and antibody production and release by *Xenopus* spleen fragments (Auerbach and Ruben, 1970).

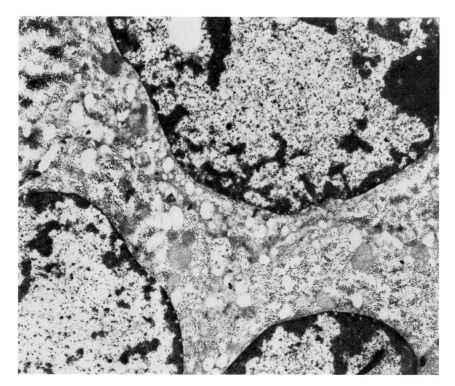

FIG. 2.   *A. means* liver after 70 days in culture (see also Fleming *et al.*, 1975). 9000 ×.

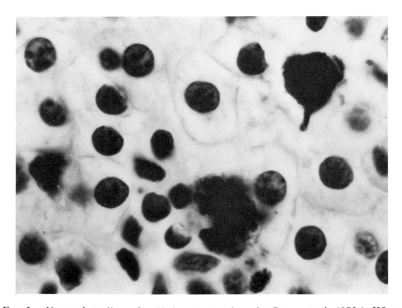

FIG. 3.   *N. maculosus* liver after 22 days *in vitro* (see also Brown *et al.*, 1975c). 500 ×.

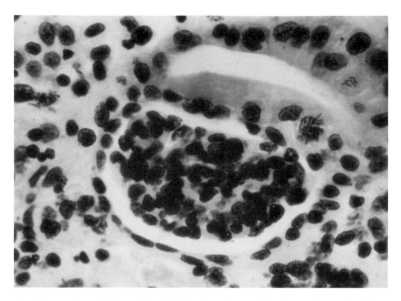

FIG. 4. *N. maculosus* kidney after 43 days in culture, showing a glomerulus, Bowman's capsule, and a cell in mitosis. 250×.

The rate of incorporation of L-leucine-^{14}C into protein by *A. means* liver cultures remained relatively constant throughout a 56-day culture period (Fig. 5). The retention of glycogen is also considered a valuable sign of normality in long-term liver or kidney cultures. It has been found that initial glycogen levels in *A. means* liver vary from 0.1 to 8.9% of wet weight of tissue. In culture, glycogen content tended to increase in liver fragments with low initial levels and to decrease in pieces from animals with a high initial level. Glycogen levels remained relatively constant for 10 weeks or more in fragments with initial levels of 2–4% (Fig. 6).

B.   Control of Cell Proliferation

Mitotic incidence is generally low in organ cultures (Trowell, 1959), and this is also the case in amphibian organ cultures (Table IV), although some increases over generally very low *in vivo* levels have been observed. This suggests that cell proliferation is controlled *in vitro*, and that further studies of the processes involved might be very rewarding, particularly in relation to the relative roles of systematic and local factors, the occurrence of tissue-specific mitotic inhibitors and the effects of drugs, toxins, and carcinogens. In two cases, namely, *A. means* kidney cultures

TABLE III

RETENTION OF NORMAL FUNCTION IN ORGAN CULTURE

Organ	Species	Parameters investigated	References
Liver	*A. means, N. maculosus*	Glycogen storage, glucose uptake, glyco-genesis, glycogenolysis	Monnickendam *et al.*, 1974; Brown *et al.*, 1975c
	A. means	Gluconeogenesis, transamination	Fleming *et al.*, 1975; Brown *et al.*, 1975d
	A. means, X. l. laevis, T. c. carnifex	Retention of specific enzyme patterns	Fleming *et al.*, 1975
Kidney	*A. means*	Urea and ammonia production	Monnickendam and Balls, 1975
	A. means	Renin release	R. T. S. Worley (unpublished results)
	N. maculosus	Glycogen storage, glycogenesis, glyco-genolysis	Brown *et al.*, 1975c
Pancreas	*A. means, N. maculosus*	Amylase release	Monnickendam and Balls, 1972; Gater *et al.*, 1975
Skin	*A. means*	Insulin release	Gater *et al.*, 1975
	A. means, X. l. laevis, R. temporaria	Epidermal basal cell proliferation and keratinization	Simnett and Balls, 1969; P. S. Moondi, unpublished results

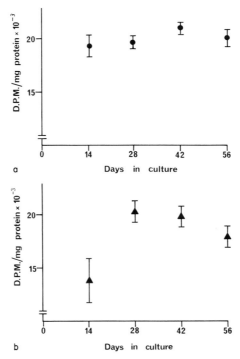

FIG. 5. (a) L-Leucine-¹⁴C incorporation into tissue protein during a 24-hour pulse (b) ¹⁴C activity in protein released into the medium over a 24-hour labeling period. At 14-day intervals, liver fragments from three or four replicate flasks were incubated at 25°C for 24 hours in the presence of 2.5 μCi/ml L-leucine-¹⁴C. The fragments were then blotted, weighed, and homogenized in a 0.1% sodium cholate solution (1:50 w/v ratio). The homogenate was centrifuged at 10,000 g for 10 minutes, and two 0.5-ml aliquots were taken from the clear supernatant fluid. One of the aliquots was diluted with distilled water to a final w/v ratio of 1:200, and its protein content determined by the method of Lowry *et al.* (1951). The second aliquot was mixed with an equal volume of 10% trichloracetic acid (TCA), and a precipitate was formed which included protein, RNA, and DNA. The mixture was centrifuged at 3000 g for 5 minutes, and the supernatant fluid discarded. The precipitate was rewashed and recentrifuged in two changes of 2 ml 5% TCA, and then boiled for 30 minutes on a water bath in a third change of 2 ml 5% TCA. At this temperature, RNA and DNA were solubilized in TCA, and protein remained precipitated. The mixture was centrifuged at 3000 g for 5 minutes, and the supernatant fluid discarded. The protein precipitate was washed in two changes of cold 5% TCA and centrifuged at 3000 g. The supernatant fluid was again discarded, and the protein pellet dissolved in 1 ml 1 M hyamine hydroxide (10- $\times$ in methanol). The solution was transferred to a glass scintillation vial with 15 ml scintillation fluid (toluene, 1 liter; Triton X-100, 500 ml; PPO, 7 gm), and ¹⁴C activity was measured in a Packard Tri-Carb liquid scintillation spectrometer. Results were corrected for counting efficiency, quenching, and chemiluminescence, and converted to disintegrations per minute (dpm) by the addition of a L-leucine-¹⁴C internal standard of known activity. Labeled protein in the culture medium was measured by the same technique, substituting 0.5 ml medium for the 0.5 ml supernatant fluid.

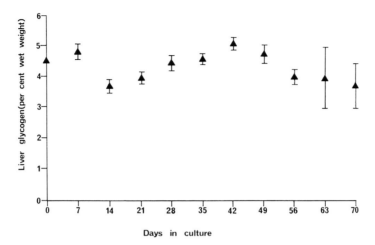

FIG. 6. Glycogen content of *A. means* liver fragments over a 10-week culture period (each point represents the mean ± S.E.M. of three samples).

TABLE IV

MITOTIC INCIDENCE IN CULTURED ORGAN FRAGMENTS

Species	Organs	Mitotic incidence *in vitro* compared with *in vivo* level	Reference
X. l. laevis	Liver, kidney, lung	Increased	Simnett and Balls, 1969
	Epidermis, spleen	Decreased	Simnett and Balls, 1969
	Spleen, pancreas	Increased	Balls *et al.*, 1969
T. c. carnifex	Liver, pancreas	Increased	Monnickendam *et al.*, 1970
	Intestine	Decreased	Monnickendam *et al.*, 1970
R. temporaria	Epidermis	Increased	P. S. Moondi, unpublished results
A. means	Liver, spleen, lung	No change	Monnickendam and Balls, 1972
	Pancreas, kidney	Increased	Monnickendam and Balls, 1972
	Kidney	Increased (greater increase with added urea)	Monnickendam and Balls, 1975
N. maculosus	Liver, spleen	No change	Brown *et al.*, 1975c
	Kidney	Increase	Brown *et al.*, 1975c

in medium with added urea (Monnickendam and Balls, 1975) and epidermal basal cells of *Rana* skin following *in vitro* damage to the keratinized layers (P. S. Moondi, unpublished results), there is evidence of a proliferative response to an increase in functional load and demand for fully differentiated cells. Extensive studies on cell proliferation in amphibian tissues *in vitro* have also been carried out by other groups (see Monnickendam and Balls, 1973a).

C. Hormones and Substrates

The effects of insulin, glucagon, and adrenaline on glycogenolysis, glycogenesis, and gluconeogenesis in *A. means* and *N. maculosus* liver cultures have been studied (Table V), and the results are compatible with the known effects of these agents in mammals. The hormone doses were very much lower than those used in most other *in vitro* systems, in which the application of high doses has affected the acceptability of the results obtained (Frieden and Lipner, 1971).

Cultured *A. means* liver retains the capacity to utilize pyruvate, lactate, and alanine as gluconeogenic substrates under the influence of glucagon (Brown *et al.*, 1975d). Alanine and glycine addition resulted in increased ammonia production by *A. means* liver cultures, whereas added glutamic acid lowered ammonia production (Fleming and Balls, 1976).

The problems of hormone dosage and longevity in culture were mentioned in Section III,F. Although hormones may be rapidly degraded, their influence may be effective over long periods. For example, glucagon added at weekly intervals exerts a continuing and substantial effect on gluconeogenesis, and insulin, which induces glycogen synthesis within minutes of addition, also causes net glycogenesis for several days. Similarly, cultures treated with adrenaline to lower their glycogen level do not synthesize more

TABLE V

EFFECTS OF HORMONES ON *A. means* AND *N. maculosus* LIVER CULTURES

Hormone	Dose	Responses	Reference
Adrenaline	$>5.5\times10^{-7}\,M$	Rapid glycogenolysis	Monnickendam *et al.*, 1974
Glucagon	$>1.4\times10^{-10}\,M$	Slower glycogenolysis; increased gluconeogenesis from pyruvate and alanine; increased GOT, GPT, and F1, 6DPase activity after 40 hours	Brown *et al.*, 1975d
Insulin	>0.1 mU/ml 10 mU/ml	*Necturus:* glycogen synthesis *Amphiuma:* glycogen synthesis; decreased GOT and GPT activity after 72 hours	Brown *et al.*, 1975c,d

glycogen when placed in fresh medium (without adrenaline) unless insulin is added or the glucose concentration in the medium is greatly increased. *In vivo*, of course, the situation would be more complex, and other nervous and hormonal actions would, for example, reverse the adrenaline-induced reduction in liver glycogen and lower the blood sugar level.

D. Hormone Production

Renin release into the medium by kidney fragments is assayed by the radioimmunoassay of angiotensin I, which is produced by the incubation of culture medium samples with homologous renin substrate. The renin substrate is produced by *A. means* liver fragments cultured in MEM in microwell trays. The radioimmunoassay is modified from the method of Haber *et al.* (1969), and uses labeled peptide and antiserum supplied by Gruppo Radiochimica (13040 Saluggia, Varcelli, Italy). The effects of selected antihypertensive drugs, such as pindolol (Sandoz), acebutolol (Sectral, May and Baker), and propranolol (ICI), on renin production are currently being investigated.

Insulin production and release by *Amphiuma* and *Rana* pancreas cultures is being followed by means of a direct radioimmunoassay designed for detecting human insulin levels (supplied by Eurotope Services Ltd., 104 East Barnet Road, New Barnet, Herts., Hertfordshire U.K.). The effects of drugs, hormones, and metabolites (e.g., propranolol, tolbutamide, glibenclamide, glucagon, glucose) on insulin production and release is currently being investigated. Although bovine insulin stimulates amphibian liver to synthesize glycogen, it is doubtful that the efficiency of the radioimmunoassay is as high for amphibian insulin as it is for mammalian insulin, although the relative amounts produced by treated and untreated amphibian pancreas cultures can be compared. Insulin production can also be detected by a bioassay method. *Rana* or *A. means* pancreas is cultured in glucose-free, normal glucose or high-glucose MEM, and then liver cultures (pretreated with adrenaline) are added. The glycogen level of the liver fragments after a further 24-hour insulation gives a semiquantitative indication of insulin release (Gater *et al.*, 1975).

E. Enzyme Production

Pancreas fragments from *A. means* and *N. maculosus* released amylase into the culture medium for 42 and 56 days, respectively (Fig. 7). Thus *Necturus* pancreas appears to survive for longer than *Amphiuma* pancreas *in vitro*, and further studies showed that MEM was a more suitable medium than L15 for pancreas cultures from both species.

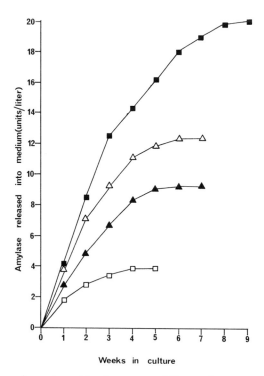

FIG. 7. Total amylase released by *A. means* and *N. maculosus* pancreas cultures (30 mg pancreas per flask). *A. means* fragments were cultured in MEM (▲) or BME (△) whereas *N. maculosus* fragments were cultured in MEM (■) or L15 (□).

One major problem in pancreas organ culture is that endocrine and exocrine products, which are normally separated *in vivo*, become mixed *in vitro*. This means that proteolytic enzymes, as well as insulin and glucagon, are released into the culture medium. Thus subsequent estimates of hormone or enzyme production may be artificially low because of partial enzymatic degradation. Attempts are being made to overcome this problem by the addition of the kallikrein inhibitor Trasylol (Bayer, Germany) to the medium (500–1000 kIU/ml).

F.   Drug Effects

Our preliminary experiments suggest that long-term amphibian organ cultures should prove very useful for investigations on the specific and general effects of drugs (Table VI; see also Brown *et al.*, 1975a,b; 1976).

TABLE VI

RESPONSES OF *A. means* LIVER CULTURES TO ADDED DRUGS[a]

Drug	Pharmacological effects	Effects on *A. means* liver
Diazoxide	Antihypertensive, hyperglycemic agent	Glycogenolysis induced; reduction in glucose uptake
Frusemide	Diuretic, hyperglycemic side effects	Glycogenolysis induced; reduction in glucose uptake
Propranolol	β-Adrenergic receptor blocking agent	Inhibition of glycogenolysis induction by adrenaline and isoprenaline
Pindolol	β-Adrenergic receptor blocking agent	Inhibition of glycogenolysis induction by adrenaline and isoprenaline
Isoprenaline	β-Adrenergic receptor stimulator	Glycogenolysis induced
Phentolamine	α-Adrenergic receptor blocking agent	Did not inhibit induction of glycogenolysis by adrenaline or isoprenaline
Phenformin	Oral hypoglycemic agent	Increase in glucose utilization and in lactate production; induction of glycogenolysis

[a] Data from Brown *et al.*, 1975a,b, 1976.

G.   Toxicological Studies

Preliminary studies (Gater *et al.*, 1976) indicate that the effects of paracetamol and ethanol on *A. means* liver cultures may be meaningful in terms of the known *in vivo* toxic effects of these agents in mammals. *Amphiuma means* liver fragments rapidly lose pigment when exposed to paracetamol at concentrations of 250 μg/ml or more (Fig. 8). The amount of pigment released appears to be related to the concentration of paracetamol applied, so pigment release may prove useful as an indicator of liver damage. Paracetamol also increased medium lactate and transaminase (GOT, GPT) levels, but ethanol (1–2 mg/ml), which itself inhibits lactate uptake by cultured liver fragments, reduced paracetamol-induced transaminase release when the two agents were added together. This reduction may have resulted from the inhibition of protein synthesis by the ethanol.

H.   Chemical Carcinogenesis *in Vitro*

Almost all the quantitative studies on chemical carcinogenesis *in vitro* have involved mammalian fibroblast monolayer cultures, but the low rate of

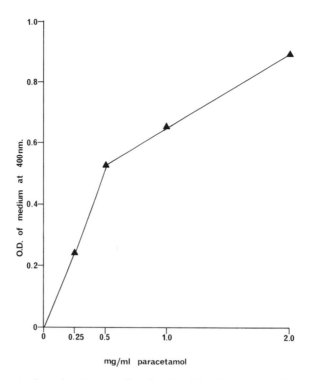

mg/ml paracetamol

FIG. 8.   Pigment release by *A. means* liver incubated in the presence of a range of paracetamol concentrations. In man, plasma concentrations of more than 300 μg/ml 4 hours after
ingestion are always associated with severe hepatic lesions (Prescott *et al.*, 1971).

chemically induced transformation and the disturbingly high rate of "spontaneous" transformation in untreated cultures complicates interpretation
of the results obtained. In addition, transformed fibroblasts induce sarcomas
when inoculated into suitable hosts, but 85% of mammalian tumors are car-
cinomas. Thus, as Heidelberger (1973) recently pointed out, there is a great
need for quantitative systems for the malignant transformation of epithelial
cells *in vitro*. The main problem is that normal, differentiated epithelial cells
cannot be maintained as dividing cell lines in monolayer culture. However,
we hope that long-term amphibian organ culture may provide such a system,
and we have begun experiments with nitroso compounds on lung, liver, and
spleen cultures.

I.   The Role of Serum Factors

Cells and tissues require serum for growth and survival in culture, so
strictly defined culture conditions are not usually available. This could be

a serious drawback for investigations of the effects on cell metabolism of various factors, such as metabolites and hormones, which may already be present in the serum. Various attempts have been made at replacing the serum requirement by adding hormones (e.g., insulin) or polypeptides (spermine), but it seems likely that a combination of many serum factors is necessary for the survival of almost all mammalian cells and tissues (Griffiths, 1972; Hosik and Nandi, 1974). We have recently found that *A. means* liver survives in serum-free MEM for at least 28 days, and that the glycogen level fell only slowly during this period (Fig. 9). Thus glycogen loss from *A. means* liver in serum-free medium is much slower than in rat liver cultures in serum-free medium, where 95% of the initial glycogen was lost within 96 hours (Campbell and Hales, 1971). Urea and ammonia excretion also continued at a relatively constant rate during a 28-day culture period in serum-free medium, although more ammonia was excreted than in culture medium containing 5% fetal bovine serum (Balls *et al.*, 1975). Adrenaline induced glycogenolysis by *A. means* liver fragments in serum-free medium, but insulin did not induce glycogenesis, and albumin production was greatly reduced. We are currently investigating recovery following serum-free culture.

V.   Concluding Remarks

The techniques described for amphibian organ culture, unlike those commonly used for maintaining mammalian tissues *in vitro*, are technically

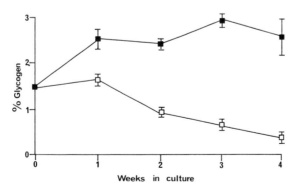

FIG. 9.   Glycogen content (mean ± S.E.M. of four samples) of *A. means* liver cultured for 28 days in 5% serum MEM (■) or serum-free MEM (□).

very simple. The use of zwitterionic buffers (L15, HEPES-buffered MEM or BME) is both the key to the simplicity of the culture methods (for no complex culture chambers or gassing systems are necessary) and an important contributor to the long-term survival of cultured amphibian tissues. More conventional media, buffered by the $NaHCO_3$–CO_2 method, require a high CO_2 gas phase, but this might be expected to be harmful to amphibian tissues, since the pCO_2 of amphibian blood is much lower than that of mammals (Table VII). We have summarized the results of some of our own studies, from which it is clear that amphibian tissues respond *in vitro* to metabolites, hormones, drugs, and toxins in ways that are compatible with their known effects on mammalian tissues, both *in vivo* and *in vitro*. The methods described offer to the experimenter a choice of animal and culture procedure to suit the requirements of a variety of experiments. Most important of all, they provide viable cultures of a range of tissues for long culture periods which, in the absence of comparable long-term methods for adult mammalian tissues, have many potential uses in biomedical research.

TABLE VII

PARTIAL PRESSURE OF CO_2 IN AMPHIBIAN AND HUMAN BLOOD

Species	pCO_2 in arterial blood (mm Hg)	References
Necturus maculosus	4.4	Lenfant and Johansen, 1967
Amphiuma tridactylum	6.1	Lenfant and Johansen, 1967
Rana ridibunda	6–9	Emilio, 1974
Rana catesbeiana	8.2	Lenfant and Johansen, 1967
Homo sapiens	36–44	Campbell *et al.*, 1968

ACKNOWLEDGMENTS

We gratefully acknowledge the help given by our colleagues, Elizabeth Arthur, R. H. Clothier, S. Gater, Marjorie Monnickendam, P. S. Moondi, J. S. Pryor, Ranjini Rao, and R. T. S. Worley, both in the development of these techniques and in the preparation of this chapter.

REFERENCES

Auerbach, R., and Ruben, L. N. (1970). *J. Immunol.* **104**, 1242.
Balls, M., and Monnickendam, M. A., eds. (1976). "Organ Culture in Biomedical Research." Cambridge Univ. Press, London and New York (in press).

Balls, M., Simnet, J. D., and Arthur, E. (1969). *In* "Biology of Amphibian Tumors" (M. Mizell, ed.) p. 385. Springer-Verlag, Berlin and New York.

Balls, M., Brown, D., Clothier, R. H., and Rao, R. (1975). In preparation.

Bennett, T. P., Kriegsten, H., and Glenn, J. S. (1969). *Biochem. Biophys. Res. Commun.* **34**, 412.

Brown, D., Pryor, J. S., and Balls, M. (1975a). *Comp. Biochem. Physiol.* (in press).

Brown, D., Pryor, J. S., and Balls, M. (1975b). *Comp. Biochem. Physiol.* (in press).

Brown, D., Fleming, N., Monnickendam, M. A., and Balls, M. (1975c). *J. Morphol.* **145**, 319.

Brown, D., Fleming, N. and Balls, M. (1975d). *Gen. Comp. Endocrinol.* **27** (in press).

Brown, D., Pryor, J. S., and Balls, M. (1976). *In* "Organ Culture in Biomedical Research" (M. Balls and M. A. Monnickendam, eds.). Cambridge Univ. Press, London and New York (in press).

Campbell, A. K., and Hales, C. N. (1971). *Exp. Cell Res.* **68**, 33.

Campbell, E. J. M., Dickinson, C. J., and Slater, J. D. H. (1968). "Clinical Physiology," 3rd ed. Blackwell, London.

Coulombe, J. J., and Favreau, L. (1963). *Clin. Chem.* **9**, 102.

Emilio, M. G. (1974). *J. Exp. Biol.* **60**, 901.

Fisk, A., and Pathak, S. (1969). *Nature (London)* **224**, 1030.

Fleming, N. (1974). Ph.D. Thesis. University of East Anglia, Norwich.

Fleming, N., and Balls, M. (1976). In preparation.

Fleming, N., Brown, D., and Balls, M. (1975). *J. Cell Sci.* **18**, 533.

Frieden, E., and Lipner, H. (1971). "Biochemical Endocrinology of the Vertebrates." Prentice-Hall, Englewood Cliffs, New Jersey.

Gater, S., Brown, D., and Balls, M. (1975). In preparation.

Gater, S., Brown, D., and Balls, M. (1976). *In* "Organ Culture in Biomedical Research" (M. Balls and M. A. Monnickendam, eds.), Cambridge Univ. Press, London and New York (in press).

Godsell, P. M. (1972). Ph.D. Thesis, University of East Anglia, Norwich.

Good, N. E., Wingett, G. D., Winter, W., Connolly, T. N., Izawa, S., and Singh, R. M. M. (1966). *Biochemistry* **5**, 467.

Goodman, L. S., and Gilman, A., eds. (1970). "The Pharmacological Basis of Therapeutics," 4th ed. Macmillan, New York.

Griffiths, J. B. (1972). *Exp. Cell Res.* **75**, 47.

Haber, E., Koerner, T., Page, L. B., Kliman, B., and Purnode, A. (1969). *J. Clin. Endocrinol. Metab.* **23**, 1349.

Heidelberger, C. (1973). *Advan. Cancer Res.* **18**, 317.

Hosik, H. L., and Nandi, S. (1974). *Exp. Cell Res.* **84**, 419.

Jones, R. O. (1967). *Exp. Cell Res.* **47**, 403.

Kaplan, H. M., and Glaczenski, S. S. (1965). *Lab. Anim. Care* **15**, 151.

Leibovitz, A. (1963). *Amer. J. Hyg.* **78**, 173.

Lenfant, C., and Johansen, K. (1967). *Resp. Physiol.* **2**, 247.

Lowry, O. H., Rosenbrough, N. J., Farr, A. L., and Randall, R. (1951). *J. Biol. Chem.* **193**, 265.

MacDougall, J. B. D., and Coupland, R. E. (1967). *Exp. Cell Res.* **45**, 385.

Millhouse, E. W., Chiakulas, J. J., and Scheving, L. E. (1971). *J. Cell Biol.* **48**, 1.

Monnickendam, M. A., and Balls, M. (1972). *J. Cell Sci.* **11**, 799.

Monnickendam, M. A., and Balls, M. (1973a). *Experientia* **17**, 1.

Monnickendam, M. A., and Balls, M. (1973b). *Comp. Biochem. Physiol. A* **44**, 871.

Monnickendam, M. A., and Balls, M. (1975). *Comp. Biochem. Physiol. A* **50**, 359.

Monnickendam, M. A., Millar, J. L., and Balls, M. (1970). *J. Morphol.* **243**, 676.

Monnickendam, M. A., Brown, D., and Balls, M. (1974). *Comp. Biochem. Physiol. A* **47**, 567.

Pogell, B. M., and McGilvery, R. W. (1952). *J. Biol. Chem.* **177**, 293.

Prescott, L. F., Wright, N., Roscoe, P., and Brown, S. S. (1971). *Lancet* **1**, 519.

Reichard, H. (1957). *Scand. J. Clin. Lab. Invest.* **9**, 311.

Simnett, J. D., and Balls, M. (1969). *J. Morphol.* **127**, 363.

Slein, M. W. (1965). *In* "Methods of Enzymatic Analysis" (H. U. Bergmeyer, ed.), p. 117. Academic Press, New York.

Trowell, O. A. (1959). *Exp. Cell Res.* **16**, 118.

Wallace, R. A., and Jared, D. W. (1969). *Develop. Biol.* **19**, 498.

Williamson, J. D., and Cox, P. (1968). *J. Gen. Virol.* **2**, 309.

Chapter 12

A Method for the Mass Culturing of Large Free-Living Amebas

LESTER GOLDSTEIN AND CHRISTINE KO

Department of Molecular, Cellular and Developmental Biology
University of Colorado
Boulder, Colorado

I. Introduction	239
II. The Basic Ameba Culture Procedure	240
III. The Culturing and Harvesting of the Food Organism *Tetrahymena pyriformis*	241
IV. The Cleaning and Harvesting of Ameba Cultures	243
V. Concluding Remarks	245
References	246

I. Introduction

The increasing use of free-living amebas in biochemical studies (Jeon, 1973) creates a greater need for practical methods with which mass amounts of the amebas can be cultured. Several methods have been employed in the past (see, e.g., Griffin, 1973) with some success. We describe here a method that has been used in our laboratory for several years and which seems to be a significant improvement over older methods. The latter method is not, however, a radical departure from those used by others, and there remains room for improvement.

Our procedure routinely produces for "consumption" about $20–25 \times 10^6$ *Amoeba proteus* per week under conditions that result in an ameba doubling time of 7–10 days. The amebas are used primarily for the extraction of RNAs and proteins as part of a variety of biochemical experiments.

II. The Basic Ameba Culture Procedure

The amebas used in our work come from a strain of *A. proteus* (Bk strain) originally isolated in Berkeley, California in 1954 and since maintained in continuous culture. The cultures are grown in Pyrex glass baking dishes with inside dimensions of 33 × 21 × 4.5 cm; on the bottom of each dish is a plastic platform about 30 × 20 cm on small plastic feet 2–3 mm high. This platform provides an added surface area on which the amebas can grow. Each dish normally contains ameba medium to a depth of 1–2 cm. The ameba medium is made up of 50 mg $CaHPO_4$, 60 mg KCl, 40 mg $MgSO_4$, and 1 liter glass-distilled H_2O. Each culture dish normally contains 0.5–1.5 × 10^6 amebas. All estimations of ameba number are based on our finding that 4–5 × 10^5 living amebas pack into a volume of 1 ml when centrifuged for 1 minute at 200 g.

We routinely maintain 25 to 30 baking-dish cultures as shown in Fig. 1. The dishes are stacked so that the one above covers the one below, and an empty dish serves as the topmost cover. Every day (7 days a week) each culture dish receives 150 ml harvested *Tetrahymena* (see Section III) as the sole administered food source. The culture dishes are kept at room temperature (21°–22°C).

Every third or fourth day the cultures are decanted and fresh ameba medium is added. About 10 minutes prior to this decantation, 25 ml of the

FIG. 1. Five stacks of baking-dish cultures on a laboratory bench. The top dishes are inverted to serve as covers of the uppermost dish containing amebas. On the shelf above the stacks are three Fernbach flasks containing harvested tetrahymenas.

usual *Tetrahymena* suspension is given to each culture dish, which ensures that most amebas will attach securely to the substratum. They do this as they prepare to capture food. The ameba decantates from three or four dishes are pooled to form new cultures in fresh dishes, and this represents, in effect, the growth increment since the previous decantations 3 or 4 days earlier.

III.   The Culturing and Harvesting of the Food Organism *Tetrahymena pyriformis*

The only food deliberately administered to the cultures is *T. pyriformis*. However, since no attempt is made to maintain the ameba cultures aseptic or axenic, all sorts of airborne contaminants may serve as minor nutrient sources (Griffin, 1973).

The *Tetrahymenas* are cultured in a rich, complex medium in 14-liter culture bottles, which are rotated horizontally at 60 rpm (Fig. 2). Each culture bottle contains 2 liters medium prepared as follows:

1. To a large culture bottle is added 400 ml of the following salt and vitamin stock mixtures (which stocks are frozen for storage). (a) A salt solution of the following contents is prepared: K_2HPO_4, 2.00 gm; $MgSO_4 \cdot 7$

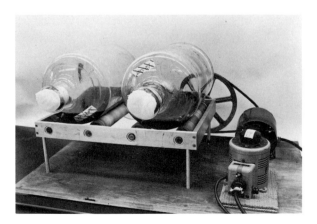

FIG. 2.   Two *Tetrahymena* culture bottles on a roller apparatus. The two bottles are inoculated approximately 3 days apart. The variable transformer serves to regulate the speed of rotation.

H_2O, 0.20 gm; $Zn(NO_3)_2 \cdot 6\,H_2O$, 0.10 gm; thiamine HCl, 0.02 gm; pyridoxyl HCl, 0.04 gm; sodium acetate, 20.00 gm; $FeSO_4 \cdot 7\,H_2O$, 0.01 gm; $CuCl \cdot 2\,H_2O$, 0.01 gm; cholesterol, 0.02 gm; and glass-distilled H_2O, 4 liters. (b) To the above is added 10 ml of the following vitamin solution: riboflavin, 0.10 gm; calcium pantothenate, 0.10 gm; niacin, 0.10 gm; folic acid, 0.01 gm; thioctic acid, 0.01 gm; glass-distilled H_2O, 500 ml. (c) After mixing both solutions well, they are divided into 10 ca. 400-ml quantities, which are separately frozen in plastic bottles.

2. After adding 400 ml of the above thawed solution to the large culture bottle, a solution prepared as follows is also added to the bottle: Eighty grams proteose-peptone (Difco Laboratories) and 8 gm liver fraction L (Nutritional Biochemical Corp.) are dissolved, with constant stirring, in 1 liter warm, glass-distilled H_2O.

3. Approximately 600 ml glass-distilled H_2O is added to the large culture bottle to make a final volume of 2 liters.

4. The bottle is cotton-stoppered and autoclaved. It is ready for inoculation when the solution has cooled to room temperature.

After the bottle is cool, about 100 ml axenic 2% proteose-peptone culture of *Tetrahymenas* inoculated 3–4 days earlier with 10 ml of a rich *Tetrahymena* culture is inoculated sterilely into the roller culture bottle. The bottle is rotated to provide the aeration necessary for high growth rates and yields. (We have found that aeration of a stationary culture by bubbling with compressed air is unsatisfactory, because the tetrahymenas become so fragile that they cannot withstand the centrifugation involved in harvesting.)

After about 3 days of rotation, the tetrahymenas grown in the bottle culture are harvested with a cream separator centrifuge (Figs. 3 and 4). The centrifuge is started with the flow valve closed while the rheostat is brought to 80 V. When this voltage is reached, 500 ml *Tetrahymena* culture is added to the bowl, and the rest of the bottle culture is returned to the roller apparatus for continued rotation. (If rotation of the full-grown culture is stopped for more than a few minutes, the cells begin to deteriorate.) Following the addition of 500 ml *Tetrahymena* culture to the cream separator, the bowl is filled with 3–4 liters distilled water. When the centrifuge motor has reached maximum speed, the flow valve is opened and the "supernatant" fluid begins to flow out of the spouts. As soon as the bowl is empty, the flow valve is closed and the centrifuge motor is shut off. The centrifuge is disassembled as quickly as possible, and the tetrahymenas that have pelleted in the cone assembly are rinsed off with distilled water into a beaker. The centrifuge is reassembled, the procedure is repeated three more times with the same tetrahymenas to ensure that they are washed thoroughly free of all *Tetrahymena* growth medium (since traces of this medium are toxic to amebas), and then they are finally suspended in ameba medium. The remaining 1500 ml of the *Tetrahymena* culture in the growth medium

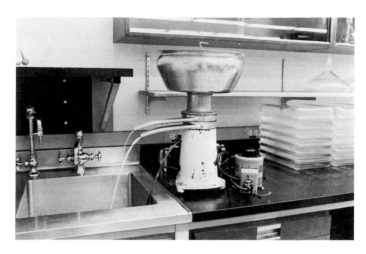

FIG. 3. In the center of the photograph is the cream separator centrifuge in operation showing the "supernatant" flowing out of the spout assembly.

is similarly treated, and finally all the harvested cells are pooled. This final *Tetrahymena* suspension is diluted with ameba medium until the suspension yields a turbidity reading of 50 units in a Klett-Summerson photoelectric colorimeter with a blue (no. 42) filter. The usual 13- to 14-liter yield of diluted tetrahymenas is then placed, 1 liter to a flask, in Fernbach flasks and stored at room temperature as shown in Fig. 1. These cell suspensions serve as the food source for the ameba cultures. For the 20 to 30 ameba cultures usually maintained, such a *Tetrahymena* preparation is made every 3 or 4 days.

The yield of tetrahymenas in the roller culture is about 6×10^5 cells/ml (which is 5 to 10 times that obtained in a nonagitated culture) and the final, diluted *Tetrahymena* suspension (Klett-Summerson reading of 50) is about 8×10^4 cells/ml. From this we estimate that the average ameba is fed 10 to 25 tetrahymenas per day. Amebas certainly can eat much more than this, and presumably under another feeding regimen they would grow more quickly. We know that in smaller, more carefully tended cultures amebas of the same strain can grow with a doubling time of 2–3 days.

IV. The Cleaning and Harvesting of Ameba Cultures

At the same time that the ameba cultures are decanted to increase the number of baking-dish cultures, all the cultures are cleaned. Decanting

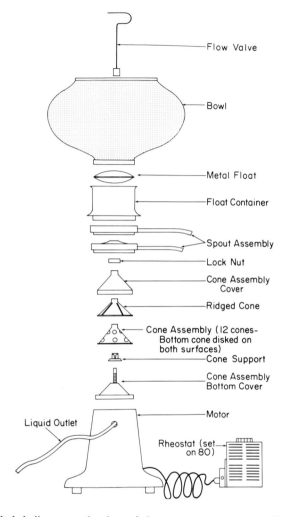

FIG. 4. Exploded diagrammatic view of the cream separator centrifuge. See text for details of operation.

itself obviously is part of the cleaning process for the older cultures, since it serves to remove floating debris and contaminants. The cells that remain attached to the substratum after decanting are dislodged, if necessary, with a gentle stream of ameba medium, and the suspension of amebas is placed in a large plastic dish in a relatively large volume of ameba medium. The amebas are allowed to settle out of suspension; this settling—because the amebas are so large—occurs more rapidly than the settling of almost everything else in the suspension. This rinse medium is then decanted, and the washed amebas are placed in fresh, clean baking dishes with fresh ameba medium. If further cleaning seems necessary, about 25 ml of tetrahymenas are added to a culture dish, and 10 minutes later the dish is decanted and fresh medium once again added.

The amebas are harvested by dislodging them from the substratum of their dishes and placing them in relatively tall beakers of ameba medium, in which they are allowed to settle to the bottom; this provides more of a rinse than the procedure described above. [The amebas are usually fasted for 1 or 2 days before harvesting in order to ensure that tetrahymenas (or their remains) in ameba food vacuoles will not contribute to the material to be extracted from the harvested amebas.] After most of the amebas have settled to the bottom of the tall beakers, the supernatant medium is siphoned off and the remaining suspensions are placed in 125-ml, pear-shaped, oil-testing centrifuge tubes, which have a capillary-like extension at the bottom, and centrifuged at 200 g for 1 minute. Upon completion of centrifugation the supernatant fluid is removed, and the pelleted cells are ready for lyophilization, extraction, or some other processing.

V.   Concluding Remarks

The method described here currently yields $20-25 \times 10^6$ amebas (or 40–50 ml packed cells) per week. Given sufficient resources, the system obviously can be scaled up to provide considerably greater yields if desired. As it is now used, the entire process requires about 20 hours per week of relatively unskilled college student help for all phases of the procedure. Most students are suitably trained in a few hours.

Considerable care in maintaining the cleanliness of all vessels with which the amebas and tetrahymenas come in contact is essential. This includes being certain that all detergents and other cleaning materials are thoroughly soaked away; wherever practical and reasonable, the utensils are extensively rinsed, soaked, and autoclaved. All the ameba cultures must be carefully scrutinized daily to be certain that each culture is healthy. Should a culture appear abnormal in any way, it must be isolated immediately to ensure that other cultures do not become "infected" by the abnormality. A loss of most of the cultures means that *at least 2 months* may be required to return to the basic ameba population of 25 to 30 backing-dish cultures.

The procedure currently in use probably could be automated to a fair extent with no great engineering achievement required—but we have not yet attempted that. However, new procedures may be developed in the not-too-distant future, and interested readers should contact us in the event that more up-to-date information may be available.

REFERENCES

Griffin, J. L. (1973). *In* "The Biology of Amoeba" (K. W. Jeon, ed.), pp. 83–98. Academic Press, New York.
Jeon, K. W., ed. (1973). "The Biology of Amoeba." Academic Press, New York.

Chapter 13

Induction and Isolation of Mutants in Tetrahymena

EDUARDO ORIAS

Section of Biochemistry and Molecular Biology,
University of California at Santa Barbara,
Santa Barbara, California

AND

PETER J. BRUNS

Section of Genetics, Development and Physiology,
Cornell University,
Ithaca, New York

I. Introduction	248
II. Elements of *Tetrahymena* Genetics	249
A. Three Genetic Compartments	249
B. Conjugation and Its Genetic Consequences	250
C. Random Distribution of Macronuclear Gene Copies	251
III. Strains	253
IV. Media	256
A. Growth Media	256
B. Mating Media	257
V. Routine Methods	257
A. Useful Tools	258
B. Cloning and Mass Subculturing	260
C. Routine Procedures	264
VI. Calibration and Suggested Doses of Mutagens	269
VII. Strategies and Protocols for Mutant Isolation	271
A. Dominant Micronuclear Mutations	271
B. Recessive Micronuclear Mutations	272
C. Macronuclear Mutations	273
D. Cytoplasmic Mutations	274
VIII. Genetic Analysis of the Mutants	275
A. Relevance of Genetic Analysis to Utility of the Mutants	275
B. Minimal Program of Genetic Analysis of Mutants	275

IX. Additional Information for Nongeneticists   278
   A. Designing Conditions for Mutant Isolation   278
   B. Some Powers and Limitations of the Use of Mutants   280
   References   281

I. Introduction

Tetrahymena pyriformis is a unicellular animal well suited to studies of cell and molecular biology. It can be grown in pure culture to high concentrations, in practically unlimited volumes, with a remarkably short vegetative doubling time (down to 2 hours under favorable conditions), and is thus extremely useful for the production and purification of cellular organelles and macromolecules. It is also well suited for genetic studies, having a diploid germ line with only five pairs of chromosomes, which are transmitted during conjugation according to the standard rules of Mendelian genetics. The cells can be cloned efficiently, and conjugation is under precise experimental control. Clones can be readily frozen and indefinitely stored under liquid nitrogen.

This ability to combine the methodology of genetics with that of cell and molecular biology makes *Tetrahymena* a promising animal cell for genetic dissection. Since essential specifications for elements of any biological machinery are contained in a collection of genes (specifying structural, catalytic, and regulatory components), mutation in *any one* of them can readily result in the loss of activity of one component and cause failure of the whole mechanism. Comparison of the properties of the mutant to its wild-type parent allows delineation of components and inferences about their role. Because each mutation affects only a single or, at most, a few genes, *one or very few* components are *primarily* affected in a given mutant; this gives unusual strength to the inferences drawn. Because *in vitro* biochemistry can be related to an *in vivo* phenotype, this approach provides automatic guidance toward uncovering the most *biologically relevant* biochemistry.

To exploit genetic dissection, one must first be able to isolate mutants. This requires three steps: (1) induction of the mutation, (2) expression of the mutant phenotype, and (3) isolation of the mutant from among an overwhelming majority of wild-type organisms. In *Tetrahymena*, steps 1 and 3 present no theoretical problems. A battery of mutagens is available for mutant induction, although we show in Section VI that biological considerations may influence the specific choice. The efficiency of isolation of the mutant presents practical problems, often unique to the kind of mutant sought, and which in general can be circumvented with a degree of success that depends on the ingenuity of the investigator. The expression of recessive mutations is a problem in *Tetrahymena* since the germ line is diploid. *Tetrahymena* is not alone in this difficulty; every animal with a diploid germ line shares this problem. Fortunately, *Tetrahymena* possess specialized genetic features which have been exploited to facilitate the isolation of recessive mutants.

This chapter covers the methods used in the induction and isolation of *Tetrahymena* mutants, including (1) recent technical advances in handling the large numbers of cells required to isolate rare mutants, and (2) recently developed strategies designed to increase the frequency of mutants. Since the chapter is addressed to biologists interested in using genetic dissection but not necessarily familiar with *Tetrahymena* genetics, we have included short sections not strictly on methods, e.g., sections describing elements of *Tetrahymena* genetics, and genetic analysis and uses of the mutants. We have also included sections on the design of conditions for isolating desired mutants, and on some powers and limitations of the use of mutants.

II. Elements of *Tetrahymena* Genetics

A. Three Genetic Compartments

Genes have three known locations in *Tetrahymena* cells: one micronucleus, one macronucleus, and many (500 or more) mitochondria. The macronucleus and micronucleus of a cell both arise from a single diploid zygote nucleus during conjugation.

The *micronucleus* has a diploid complement of five pairs of chromosomes, which are clearly resolved only at meiosis (Ray, 1956). No gene expression is known to occur in the micronucleus. It serves as the repository of the nuclear genetic information (micro- and macronuclear) for the subsequent sexual generation, obtained by conjugation. This genetic information is faithfully duplicated and distributed at binary fission. Amicronucleate strains (i.e., lacking a micronucleus) are frequently isolated in nature (Elliott and Hayes, 1955), and are represented among the longest kept laboratory strains. Such clones have lost (or never had) the capacity to conjugate, and are thus not amenable to standard Mendelian genetic analysis.

The *macronucleus* is the site of gene expression by transcription, and therefore determines the phenotype of the cell. Experimentally obtained cells, whose macro- and micronuclei differ genetically, express only the macronuclear determined phenotype; such cells are called functional heterokaryons (Bruns and Brussard, 1974b). The macronucleus is approximately 45-ploid (Woodward et al., 1972; Allen and Gibson, 1972; Doerder and DeBault, 1975). At least 80% of micronuclear DNA sequences are represented in the macronucleus (Yao and Gorovsky, 1974). The mode of distribution of macronuclear genetic information is discussed in Section II,C.

It must be emphasized that this nuclear compartmentalization in *Tetrahymena* (*nuclear dimorphism*) is analogous with the same phenomenon in higher organisms, leading to the distinction between germ line and soma. The micronucleus corresponds to the nucleus of those gonad cells that will undergo oogenesis or spermatogenesis; these cells play no role in the vegetative biology of the organism. The macronucleus is analogous with the aggregate of somatic nuclei that specify the vegetative biochemistry.

As in other eukaryotes, mitochondrial DNA specifies the ribosomal RNA of mitochondrial ribosomes (Chi and Suyama, 1970), and probably some mitochondrial transfer RNA and some mitochondrial membrane and regulatory proteins. If all mitochondrial DNA molecules are identical, they contain a very small fraction (less than 0.1%) of the basic genetic complement of the cell (Suyama and Miura, 1968; Flavel and Jones, 1970). The mode of distribution of daughter mitochondria at binary fission is unknown.

B. Conjugation and Its Genetic Consequences

Conjugation involves the pairing of two cells of different mating type, followed by a reciprocal exchange of nuclear genetic information and cytoplasmic macromolecules. Mitochondria are probably not usually exchanged at conjugation (Roberts and Orias, 1973a).

The following nuclear events take place during normal conjugation (Ray, 1956; Nanney, 1953). The micronucleus undergoes meiosis in each cell. Only one product of meiosis survives, and two identical copies of it are made. One copy in each cell is transferred to the other cell, where it fuses with the nontransferred copy. This reciprocal exchange leads to the formation of a single diploid fertilization nucleus in each cell. This nucleus divides, and the products differentiate, giving rise to the micronucleus and macronucleus of the sexual progeny. As the new macronucleus is developed, the old one is destroyed. Figure 1 gives a more detailed account of conjugation.

The reciprocal exchange and subsequent fusion of haploid nuclear copies identical to those that remain behind has an important genetic consequence: Both exconjugants of a pair receive an identical nuclear genome. Thus when phenotypic differences between the two exconjugant clones are regularly observed, differences in parental nuclear genes can be ruled out. Different cases of this have been ascribed to (1) cytoplasmic (probably mitochondrial) genes [e.g., chloramphenicol resistance (Roberts and Orias, 1973a)]; (2) complex mechanisms of gene expression involving macronuclear dif-

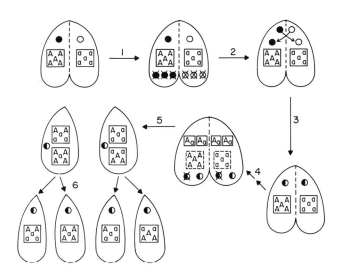

FIG. 1. Normal conjugation. Circles, Germinal nuclei; squares, somatic nuclei. 1, The micronuclei undergo meiosis; three of the four haploid products are eliminated. 2, the retained haploid nuclei divide mitotically, producing migratory and stationary pronuclei; migratory pronuclei are reciprocally exchanged. 3, The two pronuclei in each cell fuse to produce diploid zygote nuclei. 4, The zygote nuclei divide twice, producing two macronuclear anlagen and two micronuclei in each conjugant. 5, The macronuclear anlagen grow by repeated rounds of DNA synthesis; the old macronuclei are eliminated; one of the two new micronuclei is eliminated. 6, At the first cellular division the new micronucleus divides, but the macronuclei do not; the two macronuclei in each conjugant segregate to the two daughter cells.

ferentiation [e.g., mating-type determination (Nanney, 1956)], or (3) self-perpetuating differences in structural organization of the cell cortex which most likely do not depend on genetic differences [e.g., cortical inheritance of the number of ciliary rows (Nanney, 1966)]

Meiosis, followed by exchange of gametic pronuclei and fertilization, has another fortunate consequence: The genotypic and phenotypic ratios expected among the progeny follow the standard rules of diploid genetics. Thus, if two *Aa* heterozygotes are crossed, the two exconjugant clones from one-fourth of the pairs will be *AA*, one-half will be *Aa*, and one-fourth will be *aa*. If one allele is dominant, the standard 3:1 phenotypic ratio will be found among the exconjugant pairs.

C.   Random Distribution of Macronuclear Gene Copies

The macronucleus contains approximately 45 functional copies of most genes studied (Orias and Flacks, 1975). During each cell cycle all these copies are apparently replicated (Andersen, 1972; Andersen *et al.*, 1970).

The total number of copies is *randomly distributed* to each daughter macro-nucleus at cytokinesis (Allen and Nanney, 1958; Schensted, 1958; Orias and Flacks, 1975); and each macronucleus receives approximately one-half of the total macronuclear DNA (Cleffmann, 1968; Doerder *et al.*, 1975). Random distribution means that the two daughters of each of the 45 gene copies have approximately equal chances of going to the same or opposite daughter macronuclei (see Fig. 2). (Note the difference from mitosis in higher animals or plants, in which the two gene copies normally go to the opposite daughter nuclei).

Random distribution has no genetic consequences in a cell whose macro-nucleus arose from a homozygous fertilization nucleus; all gene copies in the

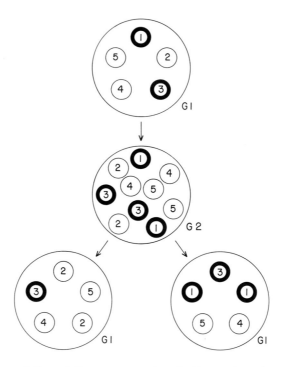

FIG. 2. Representation of the sequence of replication and random distribution of macronuclear gene copies during binary fission in syngen 1 of *T. pyriformis*. This diagram summarizes much genetic, cytochemical, and biochemical evidence. The large circles represent the macronuclear membrane, and the small circles represent functional copies of *the same gene* in an *Aa* hetero zygote. The open circles represent *A* alleles, while the partly solid circles represent *a* alleles. Only 5 numbered copies are shown for simplicity's sake, instead of 45, the probable average number. G_1 and G_2 refer to stages in the cell cycle. Nonfunctional (repressed) gene copies, if they exist at all, are not depicted.

macronucleus are identical (except for rare somatic mutations), and all the vegetative descendants of that cell will have identical phenotypes. However, in heterozygotes, random distribution leads to vegetative descendants expressing different phenotypes, a phenomenon called *phenotypic assortment* (Sonneborn, 1974) (also called allelic repression, allelic exclusion, macronuclear drift, or allelic assortment). Consider a cell whose macronucleus arose from a fertilization nucleus *Aa*. When the macronucleus is formed, it will contain a certain number of *A* and *a* functional copies. When random distribution begins, the ratio of functional *A* to *a* copies will drift among the descendant macronuclei. Descendants are eventually produced that have lost all functional copies of one type. Such cells have reached a stable macronuclear condition; all their descendants will have identical genotypes and phenotypes. Thus a clone started with an *Aa* mixed macronucleus comes to contain a mixture of descendants with macronuclei pure for functional *A* copies, pure for functional *a* copies, and mixed (both *A* and *a* copies). Heterozygotes with macronuclei pure for functional *a* copies are phenotypically identical to *aa* homozygotes, while those with macronuclei pure for the functional *A* copies are phenotypically identical to *AA* homozygotes. It must be emphasized that phenotypic assortment involves the *macronucleus only*. The germ line remains heterozygous, and behaves as such at conjugation. The nature of the assortment mechanism is still unclear (Sonneborn, 1974); the resolution of the problem should not affect the methods discussed in this chapter.

Phenotypic assortment has been exploited for the isolation of mutants, because it permits the expression of the recessive phenotype in vegetative descendants of heterozygotes (See section VII,B). It also makes possible the isolation of mutants in cells without a germ line (i.e., amicronucleate strains; B. C. Byrne, personal communication).

III. Strains

A great potential for confusion and wasted effort exists with *T. pyriformis* because of the abundance of unrelated strains. Much biochemical work has been performed on strains not readily susceptible to standard mutagenesis and genetic analysis, since they contain no germinal nucleus (the so-called amicronucleate strains). Furthermore, many of these strains are mislabeled (Borden *et al.*, 1973). Among the micronucleate strains, 12 genetically isolated subgroups, syngens 1 through 12, have been identified (Elliott, 1970). Each syngen has its own system of mating types; no fruitful matings between syngens have been reported. The syngens have been shown to be extremely diverse and should be regarded as separate species rather distantly related to one another (Allen and Li, 1974; Borden *et al.*, 1974).

Of all these groups, syngen 1 has been most used genetically and is the system used with all the methods described in this chapter. This syngen has the following advantages. (1) It has a short but readily detectable immaturity period (a growth of 40 to 80 fissions following conjugation, necessary before the cells are able to again form pairs) (Nanney and Caughey, 1953). (2) Its mechanism of mating-type determination results in the appearance of many different mating types among genetically identical progeny, facilitating inbreeding (Nanney, 1956). (3) Well-characterized, inbred strains are available which generally yield a high frequency of viable progeny in crosses (Allen and Gibson, 1973; E. M. Simon, unpublished). (4) Genomic exclusion, a process accompanying mating with certain strains (see Section VI), is well characterized and can be controlled in this syngen (Allen, 1967). This phenomenon is useful for isolating mutants, inbreeding, and performing genetic analyses. (5) All but one of the loci studied in syngen 1 undergo phenotypic assortment, another phenomenon useful in mutant isolation and genetic analysis (Sonneborn, 1974). (6) Detailed information on the specific conditions necessary for successful conjugation has been collected for this syngen (Bruns and Brussard, 1974a). (7) Syngen 1 is the best characterized genetically; many technically useful markers are available and linkage maps are being constructed (Sonneborn, 1974; J. W. McCoy, personal communication).

A set of inbred strains of syngen 1 has been constructed by crosses of wild strains followed by inbreeding (Nanney, 1959). Some of these strains are no longer fertile, while others continue to grow and mate successfully (E. M. Simon, personal communication). Table I lists lines of the inbred strains known to perform well in crosses; we include at least two mating types of each strain except C3. The notation used to designate these strains is explained in the footnote. A more complete list of all syngen 1 strains, together with their genealogy can be found in the review by Allen and Gibson (1973). Samples of the strains in Table I are stored in liquid nitrogen in the laboratory of D. L. Nanney, Provisional Department of Genetics and Development, University of Illinois, Urbana, Illinois, and are available to interested investigators.

J. W. McCoy (personal communication) has found that different inbred strains may generate differences in the amount of recombination (map distance) between the same linked genes. The basis for these differences is still unknown. To avoid such problems, *we now isolate all our mutants in one strain*. We have chosen strain B1868, since it has an immaturity period

TABLE I
USEFUL INBRED STRAINS AVAILABLE FROM FROZEN STORES[a]

A-1768IIa	B3-368II
A-1667Ib	B3-368VIa
Al-1667Id	C3-368V
Al-1667IIId	
	D-1968IIIa
B-1868III	D-1968-V
B-1868IV	
B-1868VIa	D1-1768III
B-1868VII	D1-1768Va
B2-868V	
B2-767VII	

[a]All strains listed are available from D. L. Nanney, Provisional Department Genetics and Development, University of Illinois, Urbana, Illinois. Notation includes strain designation, number generation of inbreeding, year of inbreeding (in arabic numbers), mating type (in roman numerals), and designation of line if more than one was isolated in a particular inbreeding. Thus, for example, strain A-1768IIa is the seventeenth inbred generation, produced in 1968, mating-type II, subclone a in that particular inbreeding cross, of strain A.

shorter than other syngen-1 strains, and this generation has shown no loss of progeny viability accompanying old age (6 years of continuous growth). Different phenotypes for some traits may make other listed strains more useful for certain studies. Should any of the currently used lines begin to show signs of senility, new samples can be taken from the liquid-nitrogen frozen stores.

Finally, a set of strains derived from strain B1868, which contain a germinal (micro-) nucleus heterozygous, or even homozygous, for a dominant allele for drug resistance, but which have a macronucleus expressing sensitivity, have been isolated and are available. These strains are functional heterokaryons (Bruns and Brussard, 1974b) and are extremely useful for mutant isolation. We return to them in Section VII. Table II lists the strains currently available from the author (P.B.) and explains the notation used for their designation. It should be noted that this collection will expand rapidly as mutants resistant to other drugs are isolated; updated lists can be obtained from P.B.

A list of syngen-1 mutants and their phenotypes is given in Table II of Sonneborn's (1974) review.

TABLE II

FUNCTIONAL HETEROKARYONS[a]

Chx-2/Chx-2$^+$ (cy. sens. II)
Chx-2/Chx-2$^+$ (cy. sens. IV)
Chx-2/Chx-2 (cy. sens. II)
Mpr/Mpr$^+$ (6-mp. sens. IV)
Mpr/Mpr$^+$ (6-mp. sens. VI)
Mpr/Mpr (6-mp. sens. IV)

[a]In keeping with the system proposed by Bruns and Brussard (1974b), the micronuclear genotype is listed first; *Chx-2* and *Mpr* are two separate dominant mutations conferring resistance to either 25 μg/ml cycloheximide or 15 μg/ml 6-methylpurine. The symbols in parentheses indicate phenotype: sensitivity to either cycloheximide or 6-methylpurine, and the mating type in roman numerals.

IV. Media

Although *Tetrahymena* can be grown using one of a variety of smaller organisms as food (monoxenic culture), it is unique among the ciliates in regard to the ease and long history of growth exclusively on nutrient broth or completely defined media (axenic culture). Most *genetic* studies reported in the literature have been performed with cells in monoxenic media, but methods have recently been developed so that all the manipulations necessary for mutagenesis and genetic analysis can be done in axenic culture. We describe only axenic media here, because we feel that, in comparison with monoxenic conditions, axenic conditions (1) are at least as effective as monoxenic conditions, (2) are generally simpler and can be more readily standardized, controlled, and automated, and (3) yield higher concentrations of cells.

A. Growth Media

1. NUTRIENT BROTH

The medium common to almost all work is proteose peptone (Difco). Although not necessarily true for some of the amicronucleate strains, the inbred strains of syngen 1 can be grown quite well in simple 1% peptone. In fact, this medium has proven best for growth prior to starvation for mass matings (P. W. Bruns and T. B. Brussard, unpublished). However, some lots of Difco peptone (identified by "control" number) have been associated with varied syngen-1 responses to metabolic inhibitors (Roberts and Orias, 1974; S. Moore and C. T. Roberts, Jr., personal communication; C. J. Husa, personal communication) and mating conditions (T. B. Brussard, unpublished). Various workers have supplemented this medium with a variety of broths, carbon sources, and salts; the following are commonly used variants.

1. PPY: 20 gm proteose peptone, 1 gm Bacto yeast extract (Difco), 1 ml salts solution in 1000 ml H_2O. Salts solution: 10 gm $MgSO_4 \cdot 7\ H_2O$, 5 gm $ZnSO_4 \cdot 7\ H_2O$, 0.5 gm $FeSO_4 \cdot 7\ H_2O$, 0.5 gm $CaCl_2 \cdot 2\ H_2O$, 1000 ml H_2O.

2. PPY + PS: PPY medium supplemented with streptomycin sulfate and penicillin G, each at a final concentration of 250 μg/ml. (Made fresh, added as powder to the sterile medium.)

2. DEFINED MEDIUM

A nearly minimal defined medium that syngen-1 cells grow well in was reported by Elliott *et al.* (1954). A recipe for this and other media, together with a discussion of preservatives, can be found in another article in this series (Everhart, 1972).

B. Mating Media

Although it has been known for some time that starvation is necessary for mating, it is now known that too high a tonicity will inhibit the initial reaction (Bruns and Brussard, 1974a). The following are two salt solutions that allow mating.

1. 10 mM Tris–HCl (pH 7.5): A 1:100 dilution of a 1 M stock solution is autoclaved and stored sterile. Stock solution: 242.2 gm Tris (Sigma Trizma base), 138 ml concentrated HCl, 1620 ml H_2O. Note: (a) Different pH electrodes may indicate different pH values with Tris. (b) There has been some indication of poorer results with Tris other than Sigma Trizma.

2. Dryl's salt solution (Dryl, 1959): 0.5 gm sodium citrate, 0.14 gm NaH_2PO_4, 0.14 gm Na_2HPO_4, 985 ml H_2O. After autoclaving, add 15 ml of a separately autoclaved solution of $CaCl_2$ (0.735 gm in 50 ml H_2O).

V. Routine Methods

In this section we first describe several tools that find useful application in the isolation and characterization of mutants. We then discuss various methods for cloning cells and other simple routine operations. For the reasons described in Section IV, we consider only axenic culture methods. Our experience has shown that axenic genetic methods are suprisingly trouble-free and do not require any kind of special facilities to maintain sterility; only a germicidal UV lamp, an autoclave and a dry-heat oven are needed. With only moderate care, all the operations described in this chapter can be performed on an open, standard laboratory bench, without fear of contamination.

A.   Useful Tools

1.   MICROPIPETTE

A useful method of cloning, manual single-cell isolation, requires drawing cells into a micropipette (tip 60–100 μm in diameter) and releasing them one at a time into individual drops of medium or medium-containing depressions. The micropipette is fitted with a rubber bulb or tubing and mouthpiece. The life-span of the pipette is increased if the tip is flexible (thin-walled) and about 3 cm long (see Fig. 3). To make such a pipette, the tapered end of a piece of soft-glass tubing or a Pasteur pipette is heated in the flame of a bunsen burner. The red-hot tip is then engaged with a piece of cooler, hard-glass tubing and pulled, still over the flame, to give a bridge about 5 mm long. The bridge is kept over the flame for a second or two until it is about to break, and then the two pieces of tubing are quickly pulled *simultaneously sideways* (away from one another) and *forward* (away from the flame). If the attempt fails, or if a good pipette breaks, the narrow tip is broken at the base, and the process repeated. A good preliminary test of a pipette is (1) to check to see if its tip is flexible and (2) to bubble air through it into a beaker of water and look at the bubble size. Experience quickly teaches if the pipette shows promise by this test. The ultimate test is single-cell isolation of course, and here individual preference determines the final range of acceptability.

2.   REPLICATOR

In screening for mutants, often the desired phenotype can be ascertained only under conditions in which the cells are killed; to save the presumptive mutants one must test only a sample of each of the clones being screened. At other times, either when isolating mutants or when testing progeny phenotypes, one may want to subject the same collection of mutants to a variety of phenotypic tests, not all of which can be performed on the same sample. A replicator speeds up all this work, by enabling the simultaneous subculturing of many clones in one operation. It consists of a device containing a large number of metal prongs in a regular array, matching a corresponding physical arrangement of the cultures to be replicated, and the media containers to replicate into. Given the high cell densities obtained in axenic

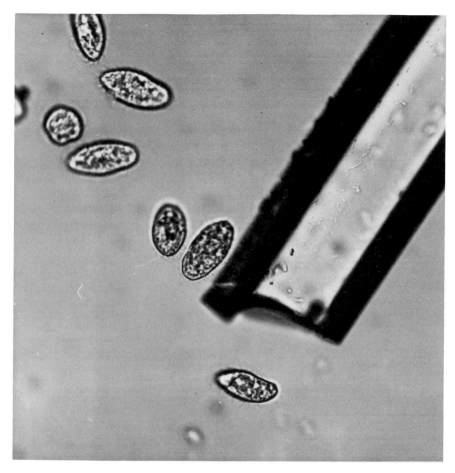

FIG. 3. Micrograph showing the size relationship between *Tetrahymena* cells and the micropipette used for manual single-cell isolation. The cells are 50–55 μm in length.

culture ($>$ 5×10^5 cells/ml), 20 to 100 cells are transferred in the droplet of medium that adheres to the tip of each replicator prong when the latter is dipped into the culture. The prong is sterilized between different replications by quickly dipping it in 90% alcohol (to kill *Tetrahymena* and possible contaminant cells, such as bacteria or mold) and quickly passing it through a flame (*not* to further sterilize, but only to eliminate the leftover alcohol by burning it). The replicator should be allowed to cool down to a safe temperature before it is used again (in the order of 30 seconds, but variable depending on the thickness of the prongs). It is efficient to work with two replicators, alternatively allowing one to cool down while the other is used for replicating. A 32-prong replicator, suitable for petri plate cultures (Section V,B,1,a) is available from the Melrose Machine Shop, Woodlyn, Pennsylvania. Figure 1 in Roberts and Orias (1973a) shows a 48-prong replicator which matches microtiter plate wells (Section V,B,1) and can be used with petri plates.

3. CAPILLARY TUBES

Capillary tubes, originally designed for freezing-point depression measurements (1 mm ID × 100 mm), are extremely useful for a variety of operations, substituting for bulkier serological pipettes when exact volume quantitation is unnecessary, or for Pasteur pipettes. The approximate volumes taken in and dispensed can be regulated, either by exploiting capillary action and varying the vertical angle of the capillary tube, or by fitting at one end a rubber bulb with a hole at the top which can be closed with a finger. Convenient bulbs ready to fit capillary tubes are available commercially for use with Drummond measuring micropipettes. Hundreds of capillary tubes can be easily washed and rinsed at once in a large beaker; clean capillaries are then distributed into 25 × 115 mm shell vials, capped with aluminum caps or foil, and dry-heat-sterilized. Immediately after use they are discarded into a beaker with soapy water. Uses of capillary tubes include the following:

1. Making approximate dilutions; a capillary tube holds approximately 0.1 ml, and fractions can be estimated by eye.

2. Subculturing; particularly useful when 10–50 μl have to be transferred to fresh medium from each of a large number of cultures. A single-bulb-fitted capillary is used, and is "sterilized" between successive uses by dipping in a beaker of nearly boiling water.

3. Lifting a drop culture on a petri plate (see Section V,B,1) and transferring to a test tube or depression.

4. Sampling cultures, as for staining, microscopic examination, or testing for bacterial contamination.

4. INOCULATING LOOPS

Inoculating loops, sterilized by holding in a flame until white hot, are useful for transfers. Preformed loops which deliver 0.01 and 0.001 ml are commercially available.

B. Cloning and Mass Subculturing

1. TYPES OF CULTURE

Tetrahymena can be grown in pure culture to high concentrations [routinely 10^6 cells/ml in PPY media, 10^7 cells/ml in specialized media (L. Rasmussen and E. Orias, unpublished)] with remarkable facility.

Tetrahymena can be handled by the liquid culture methods used with non-fastidious bacteria, like *Escherichia coli*, the only difference being the biological time constants (doubling times of 150 versus 20 minutes, respectively). Any kind of glassware or plasticware routinely used to grow bacteria, mammalian cells, or other microbes is suitable for use with *Tetrahymena* cells. Higher yields are obtained when aeration is provided in excess, although excessive shaking slows down growth. Culture containers include fermenters, carboys (forcefully aerated or on rollers), test tubes, Erlenmeyer or Fernbach flasks, standard petri plates containing up to 20 ml of culture, and so on.

More specialized methods of culturing *Tetrahymena* cells include the following. (1) Drops of PPY medium (containing up to 50 μl) on sterile plastic petri plates (the drops cling to the plate with remarkable tenacity, and with only moderate care the plate can even be turned upside down without the drops merging): This method of culture is the easiest way to do single-cell or pair isolations. By placing the drops in a regular array, many cultures can be replicated at once for test purposes. Alternatively, P.B. uses 0.1 ml drops of 1% peptone arranged in the petri plates to match the pattern of the commercially available 32-place aluminum replicator (see Section V,A,2); the eight peripheral prongs and one other are not used, providing an asymmetric, easily oriented pattern. Drop cultures inoculated with single cells are fully grown in 3 days at 30°C. Extremely dense cultures are achieved if 0.003% Sequestrene (Geigy, Ardsley, New York) is added to the medium (B. C. Byrne, unpublished). (2) Microtiter plates: These are rectangular plastic plates (85 × 125 mm), having 96 wells in an 8 × 12 regular array (see Fig. 1 in Roberts and Orias, 1973a). Each well can hold 200 μl. These plates can be filled semiautomatically by the use of a manifold which delivers medium or culture through 12 channels simultaneously and is fitted with an automatic pipette. A multichannel dispenser, including the manifold, the automatic pipette, and a moving platform with a ratchet mechanism, is available from Cooke Laboratory Products (Alexandria, Virginia). The plates come sterile initially, and can be reused, by soaking in water with detergent, rinsing with cold water, drying, and sterilizing under a germicidal lamp. (They are not autoclavable.) Some care to ensure thorough rinsing should be taken (see Section V,C,3). The automatic pipette and dispenser are cleaned and sterilized by rinsing with 95% alcohol and near-boiling water; this is much simpler than autoclaving, which is also possible. The assembly is rinsed with water and then with alcohol after each use. Microtiter plates are available either with U-shaped or flat bottoms. The former are most useful for simple growth tests, while the latter are most useful for mating-type tests or tests in which examination with a compound microscope is necessary, in which case an inverted microscope with long focal length is very useful. Microtiter plates can also be used for single-cell isolation. (3) Three-spot depression slides: These are autoclavable glass slides, about $84 \times 34 \times 13$ mm, having three depressions, each of which can hold 0.9 ml of culture. Although depression slides have been used extensively in the past, in our laboratory they have been replaced by plastic microtiter and petri plates. All these vessels are incubated in covered chambers containing a layer of water (plastic refrigerator crispers or sweater boxes work well).

In addition, two new promising techniques have been described (Gardonio *et al.*, 1973, 1974) which use semisolid media.

2. Cloning

Cloning refers to any procedure that generates a group of cells from a single cell by binary fission. In a mutant hunt, cloning plays a crucial role in the physical isolation of a mutant from the majority of wild-type cells that usually accompany the mutant.

a. Individual Cell Isolation. This is the traditional method of cloning *Tetrahymena*, using a micropipette to deposit a single cell into a drop or well with medium, under direct observation with a dissecting microscope. The micropipette is "sterilized" before each use by dipping in a beaker of near-boiling water. Individual cell isolation is an essential technique for certain forms of genetic analysis, but *not* for mutant isolation. Facility with this technique can be achieved readily, and an experienced person can do up to 600 isolations per hour (though not for many hours in a row).

b. Poisson Lottery. This method consists in distributing a thoroughly mixed cell suspension to containers in such a ratio of cells to containers that many containers receive a single cell. If the distribution is assumed to be described by Poisson statistics, the highest achievable frequency of containers with a single cell is 37% (e^{-1}), achieved when the ratio of cells to containers is 1. This ratio can be fixed experimentally by varying the cell concentration and/or the volume to be dispensed to each container. When this ratio equals, 1, 37% of the containers will remain empty, while 18, 6, and 1.5% of the containers will have two, three, and four cells respectively. The general formula is:

$$P_k = \frac{e^{-m} m^k}{k!}$$

where P_k is the fraction of containers having k cells, m is the mean number of cells per container, and e is the base of the natural system of logarithms. Since mutants are rare, it can be safely assumed that a mutant will be accompanied by wild-type cells in those containers with more than one cell. For certain traits (e.g., heat sensitivity of growth), the phenotype of the wild-type will mask that of the mutant, and only the single-cell containers are usable. For other traits, it may still be possible to identify the mutant cells among the mixture; if so, the percentage of usable containers goes up to about 60%. In spite of the waste represented by containers with no cells or multiple cells, this method is extremely useful because the cells are deposited in regular arrays which promote automation. For mutant isolations the Poisson lottery method is more efficient (in time) than individual cell isolation by hand.

c. Limiting Dilution. Dilutions of a well-mixed culture are made, and many aliquots of each dilution are distributed to individual containers. At high concentrations, most of the containers receive more than one cell, while at sufficiently low concentrations most of the containers have no cells, and of those that have at least one cell most have only one. Again this distribution is assumed to be determined by Poisson statistics. How far to dilute and how many empty containers to "waste" is determined by the margin of safety placed on the requirement that a culture that grows up is indeed a single clone. For cloning, this method is much more expensive and time-consuming than manual individual cell isolation, but it is useful for determining the frequency of rare mutants with selective advantage (Roberts and Orias, 1973a) or of survivors of any treatment that kills most of the cells.

d. Spray Cloning. Cells can be cloned using liquid medium in a manner analogous to growing discrete colonies on nutrient agar plates, by spraying a fine mist of microdrops onto a petri plate. Since several thousand drops of a size sufficient to support several rounds of cell division can be deposited on a single plate, dilutions ensuring no more than one inoculating cell per drop are easily established. Plates are sprayed under a sterile hood with a chromatography sprayer (A. E. Thomas, Philadelphia, Pennsylvania), previously sterilized by autoclaving or by simply spraying 95% ethanol before use. Since high-surface tension prevents spreading of the microdrops, the growth medium must not contain a high concentration of protein. The defined medium (Section IV) supplemented with 0.04% peptone provides good drops and supports growth well. After spraying with a dilution of 2000 cells/ml, plates are incubated upside down in humid chambers for 1–2 days. Cells are transferred from the microdrops to a drop of peptone on a master plate containing drops in the replicator pattern. The transfer is easily made under a dissecting microscope by using a microloop made of 0.008-inch-diameter platinum-palladium wire (Ted Pella Co., Tustin, California) imbedded in the end of a sealed Pasteur pipette. Placing the plate on a petri plate top inscribed with about 10 lines, 1 cm apart, provides landmarks to identify those colonies that remain to be transferred. The loop is sterilized between each transfer by flaming; the fine wire cools down so fast that no waiting after flaming is necessary. This method of cloning can serve as a replacement for the standard micropipette approach for those who cannot perform the necessary micropipette manipulations easily.

e. Cloning on Agar Plates. Two promising methods of cloning on agar plates have been developed (Gardonio *et al.*, 1973, 1974). These methods may also be extremely useful for the screening of certain specialized mutants.

3. Mass Subculturing

By "mass subculturing" we refer to any procedure by which a container with culture medium is inoculated with more than one cell, such that no purification of a cell type is expected. It is useful in the routine maintenance and propagation of strains, in raising cells to maturity, and in testing clone samples. A flame-sterilized inoculating loop, as well as sterile serological or Pasteur pipettes, are frequently used. Sterile capillary tubes (with or without a rubber bulb, see Section V,A,3) are useful for the transfer of approximate volumes between 10 and 50 μl. Simultaneous transfer of minute samples (of the order of 1 μl or less) of many cultures is accomplished with a replicator.

C. Routine Procedures

1. Maintenance of Strains

Two different objectives must be kept in mind: (1) short-term maintenance of strains, to have healthy cells available every day for experiments, and (2) long-term preservation of wild-type and important laboratory-derived strains.

There is a great deal of variety in and individual preference for short-term maintenance. E.O. uses 25-ml cultures in 125-ml Erlenmeyer flasks, transferred once a week by a 1-ml inoculum and kept at 18°C. Cultures propagated in this way are discarded after 6 months. Alternatively, standard laboratory stocks suitable for subculturing are maintained in P.B.'s laboratory by loop transfer to 10 ml 1% peptone culture in 18 × 150 mm culture tubes. If stored at room temperature, the tubes can be used 4 days after inoculation and remain suitable as a source of cells for at least 4 weeks. For use, a subculture at a ratio of 1:100 in a flask of 1% peptone at 30°C provides cells in midlog phase in 14–18 hours.

Long-term maintenance in strains containing induced mutations is often influenced by a progressive loss of the ability to give rise to sexual progeny, as a function of growth (number of vegetative fissions) after conjugation. The problem can be circumvented in two ways. (1) By making several serial backcrosses to the parental inbred, wild-type strain and selecting the most fertile progeny of each backcross until fertility approaches that of the inbred parental strain. (2) By freezing young mutant clones and storing them under liquid nitrogen; this is by far the most convenient and safest method for a large strain collection. Methods have been described (Simon, 1972, 1973; Simon and Flacks, 1974). Special equipment, commercially available, is desirable to control the rate of freezing of the cells. Detailed procedures and equipment specifications for the freezing of axenic cultures are available by writing to E.O. David L. Nanney maintains a center of frozen *Tetrahymena* strains at the University of Illinois; investigators who do not wish to provide their own liquid nitrogen storage may wish to send their valuable mutant strains to this center for safe-keeping.

2. CELL COUNTS

A Coulter-type counter provides the easiest and quickest way to count cells. An orifice of 180–200 μm is used. Electrolytes successfully used include 0.89% w/v NaCl and 10, 35, and 70 mM Tris (pH 7.4). When this type of counter is not available, a convenient method is to deposit a known volume of cells on the surface of an agar plate, wait until the liquid is absorbed, and count the cells on the agar surface. Under these conditions, the apparent size of the cells is about twice their real size. The agar near each cell is punctured with a colored pencil or a needle to avoid double-counting. Other methods of counting have been described (Everhart, 1972; Simon and Flacks, 1975). Viable counts can be obtained by the method of limiting dilution (Section V,B,2,c) (Heaf and Lee, 1971; Roberts and Orias, 1973a). The precision of limiting dilution is determined by the number of aliquots plated at each dilution of the culture. As previously stated, it is most useful for determining the number of rare mutants with selective advantage, or the number of survivors of any treatment that kills most of the cells. Colony formation has also been obtained on semisolid media by Gardonio *et al.* (1973, 1975). Precision here is limited by the number of colonies counted. Accuracy depends on prior calibration of plating efficiency, which can be relatively high.

3. MAKING CROSSES

Mass matings with a high degree of synchrony can be easily performed on cultures of a few milliliters to several liters in volume. Synchrony is important for efficient use of short-circuited genomic exclusion (see Section VII,B) and is useful in many other applications. The method, based on a study of prepairing events (Bruns and Brussard, 1974a), utilizes fast swirling of a mixture of two mating types in 10 mM Tris buffer to allow starvation but block pair formation. Equal numbers of mid-log-phase cells of different mating types are mixed, concentrated by a 1-minute centrifugation at the intermediate speed of a tabletop International clinical centrifuge, and gently washed twice by centrifugation with sterile 10 mM Tris–HCl (pH 7.5). The mixture is centrifuged a final time and resuspended in fresh, sterile 10 mM Tris buffer at a concentration of 10^5 to 10^6 cells/ml. Mating mixtures of 10–25 ml are put into 250-ml Erlenmeyer flasks. Mixtures of larger volumes are kept at a volume ratio of at least 10:1 flask capacity/mating mixture volume. The flasks are immediately put on a culture shaker, such as a New Brunswick Model G76 gyrotary water bath shaker, and swirled at 200 rpm for at least 3 hours at 30°C, but can be held on the shaker for over 24 hours. Since pair formation begins within an hour after the cessation of swirling, and reaches a maximum by 4 hours, mating mixtures with over 90% of the cells in pairs can be produced at any convenient time by attaching a time-operated switch to turn off the shaker. Since proteose peptone can reverse 1- to 2-hour-old pairs (P. J. Bruns and T. B. Brussard, unpublished), firm pairs, suitable for isolation in nutrient broth predominate 6–8 hours after the shaker has been turned off. If no shaker is available, mixing cells of two mating types, prestarved for at least 3 hours, provides the same pair-formation kinetics.

A word of caution on two elements of this approach is necessary. First, swirling apparently prevents the correct cellular contacts necessary for pair formation (co-stimulation as well as pair formation, see Bruns and Brussard, 1974a; Bruns and Palestine, 1975), and so the swirling must be fast enough to prevent this. Since angular momentum is a function of both speed of rotation and flask radius, we do not use a flask smaller than 250 ml on the shaker at this swirling rate; pair formation occurs in smaller flasks during shaking. The second point is that the glassware must be kept extremely clean. It has recently become apparent that glassware with even a trace amount of chemical contaminants can reduce the mating response to less than 50% cells in pairs (R. F. Palestine and M. A. Gorovsky, personal communication). In P. B.'s laboratory glassware is routinely washed in low-phosphate soap (Teepol G10, Shell Chemical Co.), and any glassware that has been exposed to drugs or other harmful reagents is soaked in a chromic acid bath.

4. Design of Matings to Eliminate Parental and Nonconjugant Cells

Mating success in *Tetrahymena* is seldom complete; the best mating mixtures approximate but seldom achieve 100% pairs, and usually not all pairs complete conjugation. In addition, cultures exposed to mutagens form fewer pairs, and not all pairs yield immature, viable exconjugants. Thus some method is necessary to eliminate both those cells that never paired, and those that paired but never developed new macronuclei.

Two approaches have recently been used successfully:

1. Recessive conditional lethals (Roberts and Orias, 1973b). Crosses can be arranged such that each strain is homozygous for a different conditional lethal recessive marker. Progeny are obtained by a selection for cells expressing both dominant phenotypes. For example, a cross between two strains, each homozygous for different recessive heat-sensitive mutations, can be made. Following mating, incubation of the cells at a temperature restrictive to both markers will allow growth of only the progeny.

2. Functional heterokaryons (Bruns and Brussard, 1974b). As described in Section III, functional heterokaryons are cells with different macro- and macronuclei. Several strains have been constructed containing a micro- nucleus heterozygous for a dominant allele conferring resistance to specific drugs, but a macronucleus expressing the phenotype of the recessive allele (drug sensitivity). A cross involving any one of these strains results in progeny containing the dominant allele in both the new micronucleus and the new macronucleus. These progeny will be resistant, and all other cells in the mating mixture will be sensitive. Thus addition of the drug after con- jugation is completed will eliminate all but the resistant class of successful exconjugants.

5. Testing for Successful Conjugation: Maturity Tests

If no strains are available to select for successful exconjugants (see Section V,C,4), or if there is a reason to avoid selective pressure, advantage can be taken of the immaturity phase found in syngen 1. Exconjugant cells with a newly formed macronucleus are immature; vegetative growth of 40 to 80 fissions following conjugation is necessary before the cells will form pairs (Nanney and Caughey, 1953; Bleyman, 1971). A test that indicates whether or not immediate vegetative progeny of an isolated pair can mate with known mature cells reveals whether or not the pair really completed conjugation. The following test, using drop cultures in petri plates (see Section V,B,1) was perfected by B. C. Byrne (Bruns and Brussard, 1974b). The unknowns to be tested are reared 2 days in 0.07-ml drops of 1% peptone and 0.003% Sequestrene (Geigy, Ardsley, New York) in the replicator pat- tern. Petri plates are set up with drops of 0.07 ml of 10 mM Tris in the pattern of the replicator. Mature cells of a nonparental and the two parental mating types are separately washed in 10 mM Tris and concentrated to 5×10^5 cells/ml; 20 ml of each suspension is put into sterile petri plates. Separate mating mixtures of the unknowns and knowns are established by replicating each into the 10 mM Tris drops. The unknowns need not be washed, because the dilution of nutrients during replication into the Tris drops is sufficient for starvation and mating. The mixtures are incubated in humid boxes at 30°C and observed after 24 and 48 hours. Nearly all tests are complete at the first observation. The absence of any pair formation in all three mating-type combination is judged adequate proof of immaturity. Drop cultures of known mature cells should be included as controls. This test can be adapted for microtiter plates.

6.   Raising Cells to Maturity

Since progeny of a cross are immature, they must undergo 40 to 80 fissions before they can be crossed again, although some "early mature" strains have been described (Bleyman, 1971). The most convenient way to raise many clones to maturity is by serial replication every second day to microtiter plates containing 0.2 ml of medium per well or to drops in petri plates. Cells are generally mature after five such transfers in microtiter plates. The last transfer is usually done by single-cell isolation (rather than mass subculturing) to ensure that the mature clone is pure for the mating type.

Cells can also be raised to maturity by serial subculturing with capillary tubes, loop transfers, or by single-cell isolation. These methods become very time-consuming when large numbers of clones are involved.

7.   Mating-Type Tests

The operational test of mating type is performed by presenting the unknown with each of the five to seven possible mating types (depending on the inbred strain) in a mating mixture and observing which mixture has no pairs. For small tests, both unknowns and testers are subcultured (0.1 ml into a 10-ml 1% peptone tube) 24 hours before needed. The test is set up by concentrating enough of each tester (about 0.5 ml per unknown), washing each twice into 10 mM Tris, and resuspending in 10 mM Tris to the original volume. Similarly, 0.5 ml unknown per tester is concentrated (if all seven mating types are being checked, this will be 3.5 ml unknown culture), washed twice in 10 mM Tris, and resuspended to the original volume. The matings are made by mixing equal volumes of unknowns and testers in a suitable vessel (such as 2 drops of each into wells of a flat-bottom microtiter plate). The tests are incubated in a humid chamber at room temperature for 18–24 hours, and then observed for pairs.

Large numbers of unknowns can be tested in petri plates by the replica plating method used for maturity testing (see Section V,C,5), except that all the mating-type testers are included in the test.

Whenever mating-type tests are performed, the known testers must also be mixed in all possible paired combinations, to be certain that the testers are all sexually reactive and of different mating type.

VI. Calibration and Suggested Doses of Mutagens

Calibration of the effects of mutagens in *Tetrahymena* is complicated by the nuclear dimorphism of separate germinal and somatic nuclei. Determination of killing doses directly after mutagen treatment reveals only part of the picture; although effect of a mutagen on somatic viability is a factor that must be taken into account, production of mutations in the germinal nucleus is the desired end product and can be assayed only after conjugation and the production of a new macronucleus. Thus the need to interpose conjugation between induction and expression of mutations is common to both the calibration of mutagens and the actual search for desired mutations.

A solution to this problem is provided by genomic exclusion, an induced self-fertilization triggered by matings with certain strains containing defective micronuclei, notably C* (Allen, 1967). Figure 4 illustrates the events that occur when a normal cell pairs with C*. Two rounds of mating are involved; the first yields cells with new, homozygous micronuclei derived from the non-C* parent, but with parental macronuclei. These cells express parental phenotypes. Since round-2 mating proceeds normally, if round-2 pairs are formed from round-1 exconjugants of the same pair, the round-2 exconjugants will be homozygous, containing genes derived exclusively from the non-C* parent.

The effect of mutagens was measured using genomic exclusion in the following manner (P. J. Bruns and T. B. Brussard, unpublished). Mutagen was applied to cells of strain B1868VII. These cells were then mated with C*, and round-1 pairs were isolated into growth medium. After the isolated cells had cloned, they were allowed to mate again (for round 2), and two pairs from each mating mixture were isolated into fresh growth medium. Lethality was used as the measure of the effect of the various mutagen treatments. Since the round-2 progeny were the first cells with a functional macronucleus derived from the mutagenized micronucleus, the death of both round-2 isolated pairs was taken as an indication of the production of at least one lethal mutation in the original micronucleus. The frequency of viable cells was taken as the zero class in a Poisson distribution of cells containing lethal mutations, such that e^{-x} = frequency of viables, and x = average frequency of lethal mutations among the homozygous progeny. These germinal killing effects were compared with somatic killing effects obtained by counting the number of successful clones of isolated cells fol-

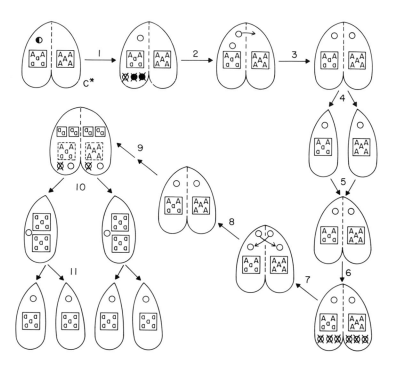

FIG. 4. Genomic exclusion. As in Fig. 1, somatic and germinal nuclei are represented by circular and square outlines, respectively. 1, The non-C* conjugant undergoes normal meiosis and elimination of three of the products; the C* conjugant loses its micronucleus. 2, The non-C* haploid nucleus divides mitotically and the migratory pronucleus is transferred; the C* conjugant does neither. 3, The new germinal nucleus in each conjugant becomes diploid, probably by endoreduplication; the two nuclei are thus homozygous at every locus. 4, Normal development of new macronuclei aborts, the old macronuclei are retained, and the conjugants separate; at this point they are heterokaryons with identical homozygous micronuclei, but parental macronuclei. 5–11, The cells remate and undergo a normal second-round conjugation (as in Fig. 1).

TABLE III

FREQUENCY OF GERMINAL LETHALS PER HOMOZYGOUS PROGENY OF
GENOMIC EXCLUSION AT 50% SOMATIC KILLING

Mutagen	Dose	Germinal lethals
Ethyl methanesulphonate	2%, 1 hour, 30°C	0
Ultraviolet light	2000 ergs/mm²	0
N-methyl-N'-nitro-N-
nitrosoguanidine	10 µg/ml, 3 hours, 30°C	0.8
X ray	100 kr, 22°C	0.5

lowing mutagen treatment and vegetative growth under nonmating conditions. Table III presents the frequency of germinal lethals in the progeny with four different mutagens at doses that cause about 50% somatic killing. As can be seen, ethyl methanesulfonate (EMS) and UV produced no measureable germinal lethals at these doses, while N-methyl-N'-nitro-N-nitrosoguanidine (NG) and x-irradiation produced measureable germinal lethals in several repeated experiments. It should be remembered that the induction of somatic lethality is a real problem, since the cells must be capable of completing the complex set of events of conjugation before any germinal mutations will be expressed. Increasing doses of "somatic-active" mutagens (i.e., mutagens with a high ratio of somatic lethality to production of germinal lethal mutations, such as EMS and UV) to increase germinal mutations may so increase the somatic damage that the probability of successful conjugation by a cell containing a desired germinal mutation may be extremely low. Thus for *Tetrahymena* we consider NG and x rays better mutagens than UV or EMS.

The following protocols have been used successfully for mutagenesis. For chemical mutagenesis, 10 μg/ml of NG is added to a culture actively growing in 1% proteose peptone. A 100× stock solution of the mutagen is prepared at 1 mg/ml in ethanol just prior to use. Following a 3- to 6-hour incubation in the mutagen at 30°C, the cells are washed twice by centrifugation into fresh 1% peptone. They are allowed to recover in growth medium for 3–6 hours before starvation for conjugation; no measurable growth occurs for the first 6 hours. For mutagenesis by irradiation, an actively growing culture is exposed to 100 kr in 50 × 12 mm plastic petri dishes with snap-on lids (Falcon Plastics). The treated culture is allowed to recover approximately 6 hours in standard growth conditions before mating procedures are started.

VII. Strategies and Protocols for Mutant Isolation

A. Dominant Micronuclear Mutations

As described earlier, in order to be expressed, a micronuclear mutation must be inserted into a macronucleus. This is achieved by interposing a mating between induction and selection. The use of two strains homozygous for different recessive selectable markers, or use of functional heterokaryons (see Section V,C,4) ensures the elimination of all "nonprogeny." Use of the second method has the added advantage that only one of the two strains need be a specialized strain.

Log-phase cells of strain B1868IV are treated with 10 μg/ml NG and

allowed to recover in peptone, as described in Section VI. Next, the mutagenized cells are mixed with an equal number of cells of a suitable functional
heterokaryon [such as Mpr/Mpr^+ (6-mp. sens., VI)] to give final concentration of 10^5 cells/ml in the desired volume. The mixture is concentrated and
gently washed twice in sterile 10 mM Tris by centrifugation. The mating
mixture is then allowed to stand undisturbed for 48 hours in a flask with a
capacity at least 10 times the volume of the mixture. Next, an equal volume
of 30 μg/ml 6-methylpurine in 2% peptone is added. The culture is allowed to
grow for 2 days at 30°C. The surviving cells (only successful progeny) are
washed out of the drug medium and are ready for appropriate selection or
screening procedures.

B.   Recessive Micronuclear Mutants

All the methods described in this section take into account both the
need for a new macronucleus following mutagenic treatment, and the requirement that this new macronucleus express recessive phenotypes.

1.   Phenotypic Assortment

Phenotypic assortment (see Section II,C) provides cells that stably
express only the recessive member of an allelic pair after conjugation
and a period of vegetative growth. A scheme employing mutagen treatment, followed by conjugation and growth (about 40 fissions) before
screening for mutant phenotypes, has been successfully employed (Carlson, 1971; Doerder, 1973; Doerder et al., 1975). Cells suitable for screening may be obtained by using the protocol for dominant micronuclear
mutations (Section VII,A), but modified by allowing vegetative growth
for at least 40 fissions before screening for mutant phenotypes.

We feel that this method is presently most appropriate for mutant phenotypes that confer selective advantage during vegetative growth. In general,
we recommend the use of short-circuited genomic exclusion (Section
VII,B,2).

2.   Short-Circuited Genomic Exclusion

Although the scheme of events in genomic exclusion outlined by Allen
(1967) occurs in most participants (Fig. 4), the use of functional heterokaryons has revealed that some round-1 conjugants (apparently about 1%)
develop new macronuclei, at least some of which are homozygous (Bruns
et al., 1975). This means that some of the products of round-1 genomic
exclusion are cells with both micro- and macronuclei derived from the same
homozygous zygote nucleus. Although the mechanism of production of such
cells is not yet known, they are extremely valuable material for finding recessive micronuclear mutations. The following scheme, including mutagen

treatment, mating to C*, prevention of round-2 mating, and selection of cells with new macronuclei, is called "short-circuited genomic exclusion" and has been used to isolate recessive micronuclear mutations (Burns *et al.*, 1975). The use of dilutions to clone successful progeny was devised by E. Fleck (personal communication).

Log-phase cells of an appropriate functional heterokaryon, for this example *Mpr/Mpr*$^+$ (6-mp. sens., IV), are treated with mutagen and washed into peptone as described in Section VI. Next, the mutagenized cells are crossed with C* by the method described in Section V,C,3. After at least 3 hours of starvation, the shaker is turned off (or the two mating types are mixed), and the mating mixture is incubated without shaking for 6–8 hours. Next an equal volume of 2% peptone is added (to prevent the second round of mating), and the culture is incubated another 6–8 hours (to allow most of the round-1 conjugants to finish conjugation). Then the culture is diluted 2×10^{-2}, 10^{-2}, and 5×10^{-3}-fold with 1% peptone (or PPY–P + S); 0.1 ml is inoculated into each well of sterile round-bottom microtiter plates. The plates are incubated at 30°C for 24 hours, and then an equal volume of 30 μg/ml sterile 6-methylpurine in 1% peptone is added to each well. The plates are incubated at 30°C for 4–5 days and then scored for living cultures in the wells. The plates containing about one-third wells with no survivors are retained, because the Poisson distribution predicts that these conditions provide the maximum number of wells inoculated with one cell (see Section V,B,2,b). Cells from the chosen plates can then be replicated into other media for selection or screening of the desired mutant phenotypes.

3. OTHER METHODS

Two other methods have been successful for the isolation of recessive mutants: "normal" genomic exclusion (Orias and Flacks, 1973) and selfing (McCoy, 1973). The first is more efficient and has been more extensively used than the second. Because we feel that both methods will be made obsolete by short-circuited genomic exclusion, we do not give protocols; the interested reader can find detailed procedures in the respective publications.

C. Macronuclear Mutations

In view of the high ploidy of the macronucleus, it may seem surprising that one can obtain expression of rare mutations induced in the macronucleus. Success is ensured, however, by the random distribution of macronuclear gene copies; the number of functional mutant copies in the macronucleus can increase from 1 to any number up to 45 in the course of vegetative growth of the clone, assuming that the mutation does not confer selective disadvantage. Even recessive macronuclear mutations can in principle be isolated, even though on the average more cell divisions than for a dominant mutation are required before the number of mutant copies accumulates to a threshhold value, called "phenotype threshold" (Orias and Flacks, 1975, in which the mutant phenotype is expressed. Macronuclear mutations have the disadvantage that they cannot yet be analyzed genetically (see Section VIII,A), nor can they yet be combined with other mutations to generate more useful strains (see Section IX,B). Nevertheless, they are the only kind of nuclear gene mutation that can be isolated in amicronucleate strains. B. C. Byrne (personal communication) has isolated 6-methylpurine-resistant mutations in an amicronucleate strain which, by analogy to similar mutations isolated by him in a syngen-1 strain (Byrne and Bruns, 1974), are probably macronuclear. Orias and Newby (1975) isolated mutants resistant to cycloheximide in syngen 1 which have been shown, by appropriate crosses and segregation kinetics, to have macronuclear mutations.

The strategy for isolating macronuclear mutations is simple: (1) mutagenize a functional heterokaryon (as in Section VI); (2) grow in proteose peptone to allow mutant-allele buildup (through random distribution) and phenotypic expression; and (3) select or screen for the desired mutant. The optimal amount of growth in peptone will in general be specific for each particular mutant. Three fissions (e.g., a 10-fold dilution in peptone) are sufficient to obtain dominant cycloheximide-resistant mutants (Orias and Newby, 1975). The most recessive mutation possible (i.e., one for which all the 45 gene copies in the macronucleus must have the mutant allele in order to express the mutant phenotype) is expected to require 65 and 140 fissions, respectively, from the time of mutation of a single allele copy until the time that 0.1% and 1% of the descendants express the mutant phenotype (based on calculations by Doerder, Lief and Doerder, 1975). Other mutations will clearly fall between these limits. When subculturing to achieve the desired number of fissions, it is important to keep in mind that mutants whose frequency is less than the reciprocal of the number of cells in the inoculum will likely be lost.

If a mutation confers selective advantage, mutant frequency can be measured by the method of limiting dilution (Orias and -Newby, 1975) (see Section V,B,2,c), although each aliquot must be allowed to grow for enough generations (empirically determined) to ensure that at least one member of each mutant clone has reached the "phenotypic threshhold."

D. Cytoplasmic Mutations

Cytoplasmic mutations to drug resistance have been isolated in *Tetrahymena* (Roberts and Orias, 1963a; K. -Y. Ling, personal communication) and are located in the mitochondria. The strategy used consists of mutagenizing (as in Section VI) and then exposing to the selective drug for several days at 30°C. Since the mechanism of distribution of daughter mitochondrial molecules to daughter mitochondria and cells is not known, the feasibility of isolating mitochondrial mutations that do not confer selective advantage under proper conditions cannot be ascertained.

VIII.   Genetic Analysis of the Mutants

A.   Relevance of Genetic Analysis to Utility of the Mutants

A variant can be very useful, even if nothing is known about the genetic properties of its mutation(s). Eventually, however, lack of this knowledge can seriously limit the depth of the insights that can be derived from a given mutant. For example, important conclusions can be drawn if one can be relatively confident that every feature of the mutant phenotype follows from the alteration of a single macromolecule, or if one can be confident that a mutation is cytoplasmic rather than nuclear. Moreover, phenotypic assortment can cause heterogeneity within a clone, if the mutation is heterozygous. This potential irreproducibility and seemingly uncontrolled variation can be avoided if the mutation's phenotype is studied only in homozygotes. Homozygotes can be obtained by the crosses outlined in Section VIII,B, or by isolating the appropriate progeny of genomic exclusion (see Section VI).

We provide below what we consider a minimal program of genetic analysis which will yield the information most valuable to the nongeneticist using mutants as a research tool. The use of functional heterokaryons in the isolation of mutants provides a built-in check for normal transmission of nuclear genes in the course of genetic analysis. We also anticipate that the use of functional heterokaryons with different markers will considerably simplify the genetic analysis of mutants. Fruitful collaboration with specialists in *Tetrahymena* genetics can undoubtedly be arranged for further genetic analysis of particularly interesting mutants.

B.   Minimal Program of Genetic Analysis of Mutants

1.   Location and Dominance or Recessiveness of the Mutation

Cross the mutant to the inbred parental strain in which it was induced; the two exconjugants should be isolated and separately tested for at least a sample of the pairs isolated. The simplest possible results and their most likely interpretation are as follows:

1. Result. All F_1 progeny are phenotypically wild type. Interpretation: The mutant (a) is homozygous or (b) is heterozygous for a recessive micronuclear mutation or (c) has a macronuclear mutation. To distinguish, backcross at least six different F_1 progeny to the *mutant* strain. If the mutant was homozygous for a micronuclear mutation, all the backcrosses will give a 1:1 ratio of mutant to wild type; if the mutant was heterozygous, half of the backcrosses will give a 3:1 ratio, and the remainder will give all wild-type progeny; finally, if the mutant has a macronuclear mutation, all F_2 progeny will express the wild-type phenotype.

2. Result. All F_1 progeny are mutant. Interpretation: the mutant is homozygous for a dominant nuclear mutation. Confirmation: Backcrosses of at least six F_1 progeny to the *wild-type* strain will give a 1:1 ratio of wild-type to mutant progeny.

3. Result. Both exconjugants from one-half of the pairs are mutant, and those from the other half of the pairs are wild type. Interpretation: the mutant strain is heterozygous for a dominant mutation. Confirmation: Backcross at least six phenotypically mutant F_1 progeny to the *wild-type* strain; a 1:1 ratio of wild-type to mutant progeny is predicted.

4. Result. One exconjugant from every pair is normal, and the other is mutant. Interpretation: the mutation is in a nonexchangeable extranuclear determinant. Confirmation: Backcross wild-type progeny to the mutant strain; results identical to F_1 results are expected.

Protocols for genetic analysis crosses are as in Section V,C,3 up until pair formation. Single pairs are then isolated into peptone drops, not earlier than 2 hours after the majority of the culture pair (to avoid pair dissolution) and not later than 6 hours after the first pairs are formed (to avoid a second round of conjugation). It is recommended that 48 to 60 pairs be isolated, because not all pairs wil give rise to viable, immature progeny. The drops are incubated 3 days and are then tested for maturity (Section V,C5) and for the desired phenotypes. If the progeny are to be crossed again, they must be raised to maturity (Section V,C,6), usually by serial replication. In crosses in which exconjugants are to be isolated, it is best to time mixing of the parental cells so that the isolation of pairs can be performed in the late afternoon or early evening; by early the next morning the exconjugants will have separated and can be isolated before they divide. It is convenient to isolate pairs into drops in every third "column," leaving columns 2 and 3 for the isolation of the two exconjugants of each pair. Regardless of the type of cross made, it is advisable also to inoculate separate drops in the plate with a sample of each unmixed parental strain. These provide built-in controls when maturity and phenotype tests are performed, since clones that are not true progeny will be phenotypically like either parent (or like a mixture of parents if exconjugants were not isolated).

2. SINGLE VERSUS SEVERAL MUTATIONS

The backcrosses indicated in the previous section will also comment on this question. Obtaining the predicted ratios implies that the mutant has either a single or several tightly clustered mutations. (These two alternatives can never be distinguished by breeding analysis exclusively.) If the results show distortions of the predicted ratios, it is possible that there are two (or more) mutations which may be epistatic and not closely linked to one another. Faced with such results, the nongeneticist should consult a standard genetics textbook or seek the advice of an expert in the field. It has been common experience that by far most laboratory-induced nuclear mutations in *Tetrahymena* have conformed to one of the simple expectations. It should be kept in mind that NG is capable of inducing tight clusters of multiple mutations in bacteria and yeasts (Drake, 1970).

By isolating revertants, a second line of evidence relevant to the question of one versus two or more mutations can be obtained for mutations that are conditionally lethal (see Section IX,A). If revertants that are indistinguishable from the wild-type strain can be isolated in reasonable frequency, a good (but not absolutely rigorous) argument can be made that a single mutation determines the phenotype of the mutant strain. This is a line of evidence especially appropriate for analysis of macronuclear or cytoplasmic (mitochondrial) mutations.

3. GROUPING MUTATIONS ACCORDING TO GENES (COMPLEMENTATION GROUPS)

If all works well, and a large collection of mutants expressing the same phenotype becomes available, the investigator might well want to determine the component affected by each mutation. The fact that many independent mutations can occur in the same gene, and therefore that all affect the same component, offers a possibility for much wasted effort. It is thus very important to group the mutations affecting the same gene, and then only one or very few representative members of each group need to be put through structural, physiological, or biochemical studies.

Two recessive mutations that give the same phenotype can be in the same or in each of two different genes. A double heterozygote is constructed, and its phenotype determined. If it has the wild-type phenotype, the mutations are said to "complement" one another, and the simplest interpretation is that the two mutations are in different genes. If the double heterozygote has the mutant phenotype (i.e., the mutations fail to complement), then the simplest interpretation is that the mutations are in the same gene. Phenomena that can complicate the results of complementation tests are known (described in any good genetics textbook), but if enough mutants are available and enough crosses are made, an accurate grouping can be arrived at, in spite of the complications. The most extensive complementation grouping in *Tetrahymena* has been carried out by Frankel *et al.* (1974) with reference to division-block mutants, and some complementation tests have been done with mutants whose growth is heat-sensitive (Orias and Flacks, 1973).

The most efficient and reliable way to set up complementation tests is to cross homozygotes for the two mutations in question, using the procedure described in Section V,C,3. The required homozygotes can be either the original mutant strains or those generated by the genetic program described in Section VIII,B,1. There is not enough experience with *Tetrahymena* to predict to what extent macronuclear recombination (if it occurs at all) could occur between two chromosomes having mutations in the same gene to create a wild-type allele (Orias and Flacks, 1973; Orias, 1973). If a wild-type phenotype resulted, the complementation test would give a false positive result. The limited experience available (Frankel *et al.*, 1974) encourages the belief that such false positive results will be rare, at least when conditions required for the phenotypic test do not give selective advantage to the hypothetical wild-type recombinant.

IX.   Additional Information for Nongeneticists

This section contains material familiar to most experienced geneticists. It is included because we are convinced that mutants will find increasing use as tools by cell biologists whose primary interest and experience is nongenetic. This section calls attention to important factors to be taken into account when designing conditions for isolating mutants, and to some powers and limitations of the use of mutants as an additional tool for studying a biological mechanism of interest.

A.   Designing Conditions for Mutant Isolation

The first challenge faced by the mutant hunter is to focus the search on that small fraction of mutants likely to comment incisively on the particular mechanism under study. Mutants are easy to find; although the probability of mutation of an individual gene is small, many different genes are available to mutate, although most mutations will be irrelevant or very indirectly related to the mechanism under study. Arriving at a screening procedure that eliminates unwanted mutants requires setting up conditions under which the failure of the mechanism in question leads to easily observable results. Here there is no substitute for the creativity and ingenuity of the investigator. Since the properties of mutants give insights into the mechanism, and this in turn improves mutant screening procedures, a productive positive feedback loop can be established.

A second challenge to the mutant hunter is to find a way of selecting rather than screening the mutant. By *screening* we mean that the phenotype of every clone must be examined to identify the mutant clones, whereas by *selecting* we mean that conditions are provided under which all or most nonmutant cells die or are physically separated from the mutants. Drug resistance (the wild types die, and the mutants live) has been used successfully with *Tetrahymena* (Roberts and Orias, 1973a,b; Orias and Newby, 1975; Byrne and Bruns, 1974). Unfortunately, the desired genetic failure will generally confer growth disadvantage under standard growth conditons, or at best be selectively neutral. Examples of elegant selection schemes addressed to this problem in other systems include 5-bromodeoxyuridine "suicide" for the selection of auxotrophs (Puck and Kao, 1967) in cultured mammalian cells, and penicillin "suicide" for similar purposes in bacteria (Davis, 1948). The use of tantalum particles and equilibrium density centrifugation to enrich for *Tetrahymena* mutants unable to phagocytize (Orias and Pollock, 1975), as well as the use of cell density in the absence of added particles (Bruns, 1973), are initial attempts in this direction.

A third problem results if the mechanism of interest is essential to the viability of the cell, and its failure is therefore lethal to the mutant. The isolation of a mutant in which that failure is compatible with survival provides dramatic evidence that a mechanism is not essential, but the investigator usually cannot afford to gamble on this possibility, or is certain that the contrary is true. This problem can be solved by searching for conditional lethal mutants, i.e., mutants that have a normal phenotype (and can be propagated) under a given set of conditions (permissive conditions) but have a lethal mutant phenotype under other conditions (restrictive conditions). Temperature (heat or cold) sensitive mutants (Horowitz and Leupold, 1951) are probably the most generally useful conditional lethal mutants. The generality of this approach depends on the fact that probably any polypeptide can be altered (through mutation of the gene that specifies its amino acid sequence) so that the specific three-dimensional conformation required for its function is normal (or nearly normal) at one temperature, but labile at another. Work done in bacteria and particularly bacteriophage T4 gives confidence that nearly every essential gene can mutate to an allele determining temperature sensitivity. Practical problems to be considered when searching for temperature-sensitive mutants include the possibility that the expression of the mutant phenotype is irreversible, i.e., the phenotype test kills the mutant. This problem is solved by keeping each clone always at a permissive temperature, and testing only a sample of each clone at the restrictive temperature; many clones can be easily transferred with the replicator. A second consideration is allowing enough time for the expression of the mutant phenotype at the restrictive temperature before attempting to score the phenotype. Even though the temperature-sensitive gene product may be inactivated within seconds or a few minutes, the gene product may be required only for a brief moment once every cell cycle, so that one doubling time may have to elapse before the mutant phenotype becomes evident. If the temperature-sensitive gene product is required for the development of a cell organelle that is conserved in one daughter at cell division (such as much of the oral apparatus assembly), this causes an additional delay before the vast majority of the cells express the mutant phenotype (Orias and Pollock, 1974).

B.   Some Powers and Limitations of the Use of Mutants

The isolation of mutants provides a powerful means of "dissecting" out the various structural and functional components of any biological mechanism. Perhaps the earliest clear use of this approach was the dissection of biochemical pathways (Horowitz et al., 1945), an approach still in use. Among numerous recent examples are the dissection of morphogenesis in the T4 bacteriophage (Wood et al., 1968) and membrane electrogenesis in Paramecium (Satow and Kung, 1974). A general method based on the construction of double mutants has recently been described (Jarvik and Botstein, 1973). It is designed to order steps in cases in which a mechanism functions by a linear sequence of steps, and has been used effectively in analyzing cell cycle mutants in yeasts (Hereford and Hartwell, 1974). A very useful general feature of temperature-sensitive mutants is the ability to turn a particular mechanism on and off by a simple temperature change, an experimental manipulation whose specificity is matched by the action of only very few drugs.

Among the most commonly encountered limitations are the residual uncertainty that the mutant phenotype is caused by a single mutation and what might be called "the cascade of indirect consequences." The first was discussed in Section VIII,B,2. The second limitation applies to the experimental method in general; it is the recognition that every malfunction in the cell (whether genetic or environmentally determined) sets off a chain of indirect effects. The challenge to the investigator is to dissect out the primary event from its consequences, a task that is seldom easy. In spite of these limitations, the isolation of mutants provides a wealth of experimental material capable of stimulating the imagination and generating material for studies in many diverse areas of biology.

ACKNOWLEDGMENTS

We dedicate this review to Dr. David L. Nanney, who, with remarkable insight into the potential utility of *Tetrahymena*, pursued, directed, and inspired the acquisition of our present knowledge of *Tetrahymena* genetics.

We are grateful to Drs. Sally L. Allen (University of Michigan), Bruce C. Byrne (Wells College, New York), F. Paul Doerder (University of Pittsburgh), Earl Fleck (Whitman College, Washington), Joseph Frankel (University of Iowa), Martin A. Gorovsky (University of Rochester), J. Wynne McCoy (University of Illinois), David L. Nanney (University of Illinois), Ronald E. Pearlman (York University, Ontario) Ellen Simon (University of Illinois) and Ms. Roberta Palestine (University of Rochester) for kindly communicating unpublished observations. In addition, we have been priviledged to have the following colleagues and friends read the manuscript and contribute valuable corrections and suggestions: Marina Adamich, Sally L. Allen, Lonnie C. Baugh, Trudy B. Brussard, Bruce C. Byrne, Marvin Cassman, Martin A. Gorovsky, Les Jenkins, David L. Nanney, Charles Samuel, Ellen M. Simon, and Charles T. Roberts, Jr. To all we express our sincere gratitude. We are grateful to Nina A. Pollock for Fig. 3. E.O.'s current research is supported by NIH grants GM-19290 and GM-21067; P.J.B.'s by NSF grant GB-40153.

REFERENCES

Allen, S. L. (1967). *Genetics* **55**, 797.
Allen, S. L., and Gibson, I. (1972). *Biochem. Genet.* **6**, 293.
Allen, S. L., and Gibson, I. (1973). *In* "The Biology of Tetrahymena" (A. M. Elliott, ed.), p. 307. Dowden, Hutchinson & Ross, Stroudsburg, Pennsylvania.
Allen, S. L., and Li, C. I. (1974). Biochem. Genet. **12**, 213.
Allen, S. L., and Nanney, D. L. (1958). *Amer. Natur.* **92**, 139.
Andersen, H. A. (1972). *Exp. Cell Res.* **75**, 89.
Andersen, H. A., Brunk, C. F., and Zeuthen, E. (1970). C. R. Trav. Lab. Carlsberg **38**, 123.
Bleyman, L. K. (1971). *In* "Developmental Aspects of the Cell Cycle" (I. L. Cameron, G. M. Padilla, and A. M. Zimmerman, eds.), p. 67. Academic Press, New York.
Borden, D., Whitt, G. S., and Nanney, D. L. (1973). *J. Protozool.* **20**, 693.
Borden, D., Miller, E. T., Whitt, G. S., and Nanney, D. L. (1974). *Evolution* (in press).
Bruns, P. J. (1973). *Exp. Cell Res.* **79**, 120.
Bruns, P. J., and Brussard, T. B. (1974a). *J. Exp. Zool.* **188**, 337.
Bruns, P. J., and Brussard, T. B. (1974b). *Genetics* **78**, 831.
Bruns, P. J., and Palestine, R. F. (1975). *Develop. Biol.* **42**, 75.
Bruns, P. J., Brussard, T. B., and Feinman, J. (1975), In preparation.
Byrne, B. C., and Bruns, P. J. (1974). *Genetics* **77**, Suppl., S7.
Carlson, P. S. (1971). *Genetics* **69**, 261.
Chi, J. C. H., and Suyama, Y. (1970). *J. Mol. Biol.* **53**, 531.
Cleffmann, G. (1968). *Exp. Cell Res.* **50**, 193.
Davis, B. D. (1948). *J. Amer. Chem. Soc.* **70**, 4267.
Doerder, F. P. (1973). *Genetics* **74**, 81.
Doerder, F. P., and DeBault, L. E. (1975). *J. Cell Sci.* **17**, 471.
Doerder, F. P., Frankel, J., Jenkins, L. M., and DeBault, L. E. (1975). *J. Exp. Zool.* **192**, 237.
Doerder, F. P., Lief, J. H., and Doerder, L. E. (1975). *Genetics* **80**, 263.
Drake, J. W. (1970). "The Molecular Basis of Mutation," p. 155. Holden, Day, San Francisco, California.

Dryl, S. (1959). *J. Protozool.* **6**, Suppl., 25.

Elliott, A. M. (1970). *J. Protozool.* **17**, 162.

Elliott, A. M., ed. (1973). "Biology of Tetrahymena." Dowden, Hutchinson & Ross, Stroudsburgh, Pennsylvania.

Elliott, A. M., and Hayes, R. E. (1955). *J. Protozool.* **2**, 75.

Elliott, A. M., Brownell, L. E., and Gross, J. A. (1954). *J. Protozool.* **1**, 193.

Everhart, L. P., Jr. (1972). *In* "Methods in Cell Physiology" (D. M. Prescott, ed.), Vol. **5**, pp. 220–288. Academic Press, New York.

Flavell, R. A., and Jones, I. G. (1970). *Biochem. J.* **116**, 811.

Frankel, J., Doerder, F. P., Jenkins, L. M. Nelsen, E. M., and DeBault, L. E. (1974). *In* "Molecular Biology of Nucleocytoplasmic Interactions in Unicellular Organisms" (S. Puiseux-Dao, ed.). Elsevier, Amsterdam (in press).

Gardonio, E., Crerar, M., and Pearlman, R. E. (1973). *J. Bacteriol.* **116**, 1170.

Gardonio, E., Crerar, M., and Pearlman, R. E. (1975). *In* "Methods in Cell Biology" (D. M. Prescott, ed.), Vol. **9**, pp. 329–348. Academic Press, New York.

Heaf, D. P., and Lee, D. (1971). *J. Gen. Microbiol.* **68**, 249.

Hereford, L. M., and Hartwell, L. H. (1974). *J. Mol. Biol.* **84**, 445.

Horowitz, N. H., and Leupold, U. (1951). *Cold Spring Harbor Symp. Quant. Biol.* **16**, 65.

Horowitz, N. H., Bonner, D., Mitchell, H. K., Tatum, E. L., and Beadle, G. W. (1954). *Amer. Natur.* **79**, 304.

Jarvik, J., and Botstein, D. (1973). *Proc. Nat. Acad. Sci. U.S.* **70**, 2046.

McCoy, J. W. (1973). *Genetics* **74**, 107.

Nanney, D. L. (1953). *Biol. Bull.* **105**, 133.

Nanney, D. L. (1956). *Amer. Natur.* **90**, 291.

Nanney, D. L. (1959). *Genetics* **44**, 1173.

Nanney, D. L. (1966). *Genetics* **54**, 955.

Nanney, D. L., and Caughey, P. A. (1953). *Proc. Nat. Acad. Sci. U.S.* **39**, 1057.

Orias, E. (1973). *Biochem. Genet.* **9**, 87.

Orias, E., and Flacks, M. (1973). *Genetics* **73**, 543.

Orias, E., and Flacks, M. (1975). *Genetics* **79**, 187.

Orias, E., and Newby, C. J. (1975). *Genetics* **80**, 251.

Orias, E., and Pollock, N. A. (1975). *Exp. Cell Res.* **90**, 345.

Puck, T. T., and Kao, F. T. (1967). *Proc. Nat. Acad. Sci. U.S.* **58**, 1227.

Ray, C., Jr. (1956). *J. Protozool.* **3**, 88.

Roberts, C. T., Jr., and Orias, E. (1973a). *Genetics* **73**, 259.

Roberts, C. T., Jr., and Orias, E. (1973b). *Exp. Cell Res.* **81**, 312.

Roberts, C. T., Jr., and Orias, E. (1974). *J. Cell Biol.* **62**, 707.

Satow, Y., and Kung, C. (1974). *Nature (London)* **247**, 69.

Schensted, I. V. (1958). *Amer. Natur.* **92**, 161.

Simon, E. M. (1972). *Cryobiology* **9**, 75.

Simon, E. M. (1973). *Cryobiology* **10**, 421.

Simon, E. M., and Flacks, M. (1975). *In* "Round Table Conference on the Cryogenic Preservation of Cell Cultures." (A. P. Rinfret and B. LaSalle, eds.), p. 37. *Nat. Acad. Sci.*, Washington, D.C.

Sonneborn, T. M., (1974). *In* "Handbook of Genetics" (R. C. King, ed.), Vol. **2**, pp. 433–467. Plenum, New York.

Suyama, Y., and Miura, K. (1968). *Proc. Nat. Acad. Sci. U. S.* **60**, 235.

Wood, W. B., Edgar, R. S., King, J., Lielausis, I., and Henninger, M. (1968). *Fed. Proc., Fed. Amer. Soc. Exp. Biol.* **27**, 1160.

Woodward, J., Kaneshiro, E., and Gorovsky, M. A. (1972). *Genetics* **70**, 251.

Yao, M., and Gorovsky, M. A. (1974). *Chromosoma* **48**, 1.

Chapter 14

Isolation of Nuclei from Protozoa and Algae

D. E. BUETOW

Department of Physiology and Biophysics,
University of Illinois,
Urbana, Illinois

I. Introduction	. . . .	284
II. Monitoring the Preparation	. . . .	284
III. *Tetrahymena*	. . . .	285
A. *Tetrahymena pyriformis* (Method of Ringertz *et al.*., 1967) .	. .	285
B. *Tetrahymena pyriformis* (Method of Lee and Scherbaum, 1965) .	. .	286
C. *Tetrahymena pyriformis* (Method of Mita *et al.*, 1966)	. . .	288
D. *Tetrahymena pyriformis* (Method of Byfield and Lee, 1970)	. .	289
E. *Tetrahymena pyriformis* (Method of Gorovsky, 1970a)	. . .	290
IV. *Paramecium*	. . . .	292
A. *Paramecium aurelia* (Method of Wolfe, 1967) .	. . . .	292
B. *Paramecium aurelia* (Method of Stevenson, 1967)	. . .	294
C. *Paramecium aurelia* (Method of Cummings, 1972)	. . . .	294
D. *Paramecium caudatum* (Method of Skoczylas *et al.*, 1963) .	. .	296
V. *Blepharisma*	. . . .	297
Blepharisma intermedium (Method of Saxena, 1966)	. . .	297
VI. *Didinium* .	. . . .	298
VII. *Spirostomum*	. . . .	298
VIII. *Amoeba* .	. . . .	299
Amoeba proteus (Method of Tautvydas, 1971)	. . . .	299
IX. *Euglena*	. . . .	301
Euglena gracilis (Method of Lynch and Buetow, 1975) .	. .	301
X. Algae	. . . .	303
A. *Griffithsia globulifera* (Method of Nasatir and Brooks, 1966)	. .	303
B. *Peridinium trochoideum* (Method of Rizzo and Noodén, 1973)	. .	304
C. *Peridinium cinctum* (Method of Rizzo and Noodén, 1973) .	. .	305
D. *Gymnodinium nelsoni* (Method of Mendiola *et al.*, 1966)	. .	306
E. *Gymnodinium nelsoni* (Method of Rizzo and Noodén, 1973)	. .	307
F. *Gyrodinium cohnii* (*Cryptothecodinium cohnii*) (Method of Rizzo and Noodén, 1973) .	. . . . . .	308
G. Diatoms (Method of Mendiola *et al.*, 1966)	. . . .	309
Note Added in Proof.	. . . .	310
References	. . . .	310

I. Introduction

An extensive literature is available on the isolation of nuclei from animal tissues (e.g., Roodyn, 1969; Busch and Smetana, 1970; Muramatsu, 1970). In contrast, techniques for the isolation of nuclei from protozoa and algae are more recent and fewer in number. This chapter is designed to bring together these more recently developed techniques for the mass isolation of nuclei from the latter cell types.

For each cell type discussed, information is given on the nuclear isolation technique as well as on the properties of the isolated nuclei. For each isolation method presented, the methods used for growth and harvesting of the cells involved are also included. Successful application of a particular isolation method may depend to some degree on the conditions used for cell growth. For example, in our laboratory, the ease and degree of breakage of the tough outer pellicle of *Euglena gracilis* was found to depend, in part at least, on the growth medium used. Also, in some cases, investigators have specified that certain makes of instruments were used in developing a particular isolation method. Actual instruments so specified are included in this chapter to serve as guides to successful use of the cited method. Methods presented in detail here were chosen from the literature on the basis that they were published with microscopic evaluations (light, phase-contrast, or electron microscope) of the purity and intactness of the final nuclear pellet.

II. Monitoring the Preparation

The degree of purity of nuclei at any stage in an isolation method usually can be monitored with the aid of a phase-contrast microscope. A light microscope usually can be used if the nuclei are stained. Some of the nuclear stains used include aqueous methyl green for *Paramecium caudatum* (Skoczylas *et al.*, 1963), methyl green in acetic acid for dinoflagellates (Mendiola *et al.*, 1966) and *E. gracilis* (Aprille and Buetow, 1973), methyl green–Pyronine for dinoflagellates (Jensen, 1962), azure B for *Blepharisma intermedium* (Seshachar and Saxena, 1968), azure C for *Tetrahymena pyriformis* (Muramatsu, 1970), toluidine blue for *T. pyriformis* (Stone and Cameron, 1964) and for chromatin from *E. gracilis* (Lynch *et al.*, 1975b), acetocarmine for paramecia (Moses, 1950), iron alum–acetocarmine for algae (Godward, 1948) and *E. gracilis* (Leedale, 1958), and Giemsa stain for *T. pyriformis* (Stone and Cameron, 1964). A useful compilation of nuclear stains and staining methods is given by Busch and Smetana (1970).

Recoveries of isolated nuclei can be estimated by staining them and counting them in a hemocytometer or by measuring the DNA content of the final nuclear pellet. Electron microscopy is useful in determining the ultrastructural integrity of isolated nuclei and identifying the nature of any contaminants in the final nuclear pellet.

III. *Tetrahymena*

Rosenbaum and Holz (1966) reported the isolation of intact macronuclei from *T. pyriformis* (strain WH14, mating type II, var. 1) lysed with digitonin, but did not present any microscopic studies on their macronuclear preparations. Once-washed macronuclei, however, yielded a pH 5 fraction active in amino acid activation. Twice-washed macronuclei were less active, and macronuclei washed three times were inactive (Rosenbaum and Holz, 1966). Prescott *et al.* (1966) and Muramatsu (1970) previously gave the details of two methods for isolating macronuclei and micronuclei from *Tetrahymena* in Volumes II and IV, respectively, of this series. Additional isolation methods are detailed below.

A. *Tetrahymena pyriformis* (Method of Ringertz *et al.*, 1967)

1. CULTURE AND COLLECTION OF CELLS

Tetrahymena pyriformis GL is grown at 22°C on PPYS medium containing 2% proteose peptone and 0.1% yeast extract and supplemented with salts according to Plesner *et al.* (1964). For quantitative cytochemical techniques, only small cultures of 10–120 ml need be grown. No aeration or shaking is necessary. Generation time is about 3 hours. Cultures grown to $5–30 \times 10^4$ cells/ml can be used without further concentration when lysed with ethanol. For lysis with indole, cells are concentrated by centrifugation at 5000 g for 15 minutes at 4°C.

2. ISOLATION OF MACRONUCLEI

a. Lysis with Ethanol.   Add 1 vol chilled 30% ethanol to 1 vol cell suspension in PPYS medium. Homogenize cells by hand in a loose-fitting glass homogenizer for 1 minute and check for cell breakage with a phase microscope. Continue homogenization until all cells are disrupted.

b. Lysis with Indole.   Suspend concentrated cells in pH 7.5 buffer containing 0.01 *M* Tris–HCl and 0.5 m*M* MgCl$_2$ and saturated with indole (Eadie and Oxford, 1954; Whitson *et al.*, 1966). Shake gently. All cells lyse within a few minutes. Buffer is at 4° C, and shaking is done at the same temperature.

With both methods of cell lysis, a concentrated suspension of macronuclei is obtained by centrifugation at 200 g for 10 minutes. The yield of macronuclei is very low, however, and this step is not done when macronuclei are to be used for cytochemical analysis. The average dry mass per macronucleus is 11.5×10^{-11} gm in exponentially growing cells, or about 5% of the dry mass of the whole cells (Ringertz et al., 1967). Composition of macronuclear dry mass is about 80% protein, 15–19% DNA, and about 3% RNA.

B. Tetrahymena pyriformis (Method of Lee and Scherbaum, 1965)

1. Culture and Collection of Cells

The growth medium (Scherbaum et al., 1959) for T. pyriformis GL contains 2% (w/v) proteose peptone (Difco), 0.1% (w/v) liver fraction L (Wilson Laboratories), 0.5% (w/v) Bacto dextrose (Difco), and sulfates and chlorides as in medium A of Kidder and Dewey (1951). Stock cultures are maintained in 3 ml of this medium in the dark at 29°C for optimal growth rates, or at lower temperatures for slower growth rates. Large cultures (7 liters) can be grown in carboys if aerated with sterile air. Cells are collected at low-speed centrifugations, i.e., 250–280 g for 5–7 minutes. Higher speeds cause many of the cells to rupture.

Another growth medium for T. pyriformis GL contains (Lee and Scherbaum, 1966) 2% (w/v) proteose peptone, 0.1% (w/v) Bacto dextrose, 0.1% sodium acetate, and 0.1% K_2HPO_4. This growth medium has also been used when supplemented with 0.1% yeast extract (Byfield and Lee, 1970). Large-scale cultures can be grown in constant-temperature stainless-steel tanks (Scherbaum and Jahn, 1964) containing 6–8 liters of vigorously shaking medium.

2. Isolation of Macronuclei

Macronuclei can be isolated by the following method from cells growing exponentially at 29°C, from cells growing exponentially at 29°C and then incubated for 2 hours at 34°C (conditions used to synchronize cell division, Scherbaum and Zeuthen, 1954), and from cells in the maximum stationary phase of growth.

1. Organisms are washed once by low-speed centrifugation (250–280 g) at room temperature in diluted Ringer's sodium phosphate (RSP) buffer $[0.047\ M\ \text{NaCl},\ 2\ \text{m}M\ \text{KCl},\ 1\ \text{m}M\ \text{MgSO}_4,\ 12.5\ \text{m}M\ \text{Na}_2\text{HPO}_4\ (\text{pH}\ 7.3)]$ and resuspended in RSP buffer not to exceed 4×10^6 cells/ml. Shake gently to avoid cell aggregation.

2. From the cell suspension, remove 100 ml with a plastic pipette in 10-ml portions and add slowly to 100 ml Medium A (0.5% v/v Triton X-100 and 0.5 M sucrose in RSP) in a 500-ml Erlenmeyer flask kept in crushed ice. For mixing, use an L-shaped glass rod (2 × 27 cm, blade is about 15 mm²) rotating at 200 rpm. Most cells lyse by 5 minutes after addition of the last aliquot of cells. At this time, with continued stirring, add 200 ml Medium B [2% w/v proteose peptone (Difco) and 0.1% w/v liver fraction L (Wilson Laboratories)].

3. Slowly pass the lysed cell suspension through a cotton filter at room temperature, but collect the suspension in an ice bath. Prepare a filter by covering sterilized absorbent cotton on both sides with cheesecloth and cutting 6 × 6 inch squares which are put into the bottom part of a polyethylene Büchner funnel (11 cm diameter).

4. Layer the filtrate (40-ml portions) on 30 ml 2% (v/v) Triton X-100 in 0.5 M sucrose in RSP layered over 10 ml 1 M sucrose in RSP in a 90-ml, round-bottom, plastic centrifuge tube. Centrifuge in the No. 240 head of an International PR-2 centrifuge for three 5-minute steps at 70, 250, and 800 g. Manually control deceleration to avoid mixing the three layers. Nuclei form a creamy, membranous pellet at the bottom of the tube.

5. Rapidly remove the supernatant by suction, except for the 5 ml covering the pellet; the latter 5 ml is poured out gently. Add 1 ml 1 M sucrose to the pellet. Make a homogeneous suspension of the pellet by repeatedly drawing it gently into a plastic pipette and letting the mix flow out.

6. Add an equal amount of 2 M sucrose in RSP to the suspension, mix carefully, layer a 2-ml portion over 2 ml 1.75 M sucrose in RSP in a 5-ml cellulose tube, and centrifuge for 20 minutes at 10,000 rpm in the SW rotor of a Spinco Model L ultracentrifuge.

7. Decant the supernatant. Add 1 ml 0.5 M sucrose to the gelatinous nuclear pellet and resuspend the nuclei by gentle pipetting as in step 5.

Isolated nuclei are pure and intact (Lee and Scherbaum, 1965). Lysis in Medium A (step 2) starts at the anterior of the cell and proceeds along the long axis to the posterior of the cell. Medium B (step 2) stabilizes nuclei and prevents their clumping and attachment to cytoplasmic fragments. Over 95% of the nuclei are released by lysis in step 2 from heat-treated and stationary-phase cultures, but only 50% of log-phase cells are lysed by this method.

Modifications of the above method are reported to improve nuclear stability during isolation (Lee and Scherbaum, 1966). These are the addition of an equal volume of 4% (w/v) polyvinylpyrrolidone (PVP) in RSP buffer instead of culture medium immediately after lysis of the cells (step 2), and the addition of 1% (w/v) PVP and the reduction of Triton X-100 to 1% concentration at step 4. With these modifications, the yield of nuclei is 30%.

Nuclei from stationary-phase cells are, in percent weight, 2.3% RNA, 22.4% DNA, 72% total protein, and 44% histone.

C. *Tetrahymena pyriformis* (Method of Mita *et al.*, 1966)

1. CULTURE AND COLLECTION OF CELLS

The amicronucleate strain GL is grown with gentle shaking at 28°C in proteose peptone. About 10 ml of packed log-phase cells is harvested at 180 g for 3 minutes.

2. ISOLATION OF MACRONUCLEI

All operations are done at 4°C.

1. Wash harvested cells twice with buffer containing 0.25 M sucrose, 0.01 M Tris–HCl (pH 7.5), 0.001 M MgCl$_2$, and 0.003 M CaCl$_2$ with centrifugation at 180 g for 3 minutes each time. Suspend washed cells in buffer to give a 10% cell suspension.

2. Make a 1% (v/v) solution of Nonidet P40 (surface-active agent, a noniongenic ethanoxy compound being an isooctyl phenol condensed with an average of six ethylene oxide groups, Shell Chemical Co.) in the buffer of step 1. Add 30–100 ml cell suspension (for more dilute cell suspensions, add proportionately less Nonidet). Shake the sample by hand for several seconds. Lysis of all the cells occurs quickly and can be monitored by microscopy.

3. Place 60 ml buffer as in step 1 (except here containing 0.33 M sucrose) in a 12-cm-long centrifuge tube of 100-ml capacity. Layer the lysate (30 ml) carefully over the buffer. Centrifuge at 1200 g for 5 minutes in a swinging-bucket centrifuge. Discard the supernatant.

4. Wash the nuclear pellet by centrifugation at 400 g with sucrose buffer as in step 1.

CaCl$_2$ is necessary to protect the macronuclei from karyolysis in the above procedure. Nonidet P40 is more effective in dispersing the cytoplasm than other surface-active agents such as sodium deoxycholate, indole, and digitonin. Also, Nonidet minimizes flocculation of cytoplasm, thus allowing isolation of purified nuclei.

A variation of the above procedure involves freezing of the harvested cells in a Dry Ice–trichloroethylene bath followed by thawing at room temperature before suspension in Nonidet P40 (Iwai *et al.*, 1965). This method produces fragmented macronuclei as well as whole macronuclei; however, this preparation is useful for the isolation of histones.

Isolated macronuclei have an RNA/DNA ratio of 0.17 and a histone/DNA ratio of 0.9 and show RNA polymerase activity (Mita *et al.*, 1966). The histone fraction is lysine-rich, with an amino acid composition like that obtained from freeze-thawed *Tetrahymena* (Iwai *et al.*, 1965). The multiple nucleoli of a macronucleus appear as an almost continuous layer in the periphery of the nucleus (Nilsson and Leick, 1970) and are attached to an unidentified structure (possibly the nuclear envelope). Nucleoli are better preserved in macronuclei isolated from exponentially growing cells than from stationary-phase cells (Nilsson and Leick, 1970). Chromatin granules with long interconnecting filaments as narrow as 10 nm are found in isolated macronuclei (Wolfe, 1967). Although chromatin bodies and nucleoli are intact in macronuclei isolated by the above Nonidet P40 procedure, parts of the nuclear membrane are lost (Leick, 1969). Studies on the processing of rRNA in *Tetrahymena* macronuclei indicate that 17 S RNA has a shorter half-life in the macronucleus than 25 S RNA, and that the rate of migration of rRNA lable from the nucleus to the cytoplasm proceeds about 10 times faster than in the case of HeLa cells (Leick, 1969; Leick and Engberg, 1970).

D. *Tetrahymena pyriformis* (Method of Byfield and Lee, 1970)

1. CULTURE AND COLLECTION OF CELLS

Tetrahymena pyriformis GL is grown on the medium of Byfield and Lee (Section III,B,1) to the late log phase of growth. Cells are harvested and washed once with buffer I [10 mM Tris–HCl (pH 7.4) containing 2 mM $CaCl_2$ and 1.5 mM $MgCl_2$] by centrifugation at 90 g for 4 minutes at room temperature.

2. ISOLATION OF MACRONUCLEI

Steps 1 and 2 are done at room temperatures; steps 3 to 5 are at $0°–2°C$ (Byfield and Lee, 1970; Lee and Byfield, 1970).

1. Resuspend washed cells in the above buffer I (2×10^6 cells/ml) at room temperature and slowly mix with 100 ml buffer I containing 0.2% (v/v) Triton X-100 and 0.5 M sucrose. All cells lyse within 5 minutes, but nuclei remain intact.

2. Slowly add 200 ml buffer I containing 4% PVP and 0.25 M sucrose.

3. Filter mixture through cotton (see Section III,B,2) previously stored in a freezer. Pour the mixture evenly on the cotton filter with a 10-ml pipette.

4. Layer each 40 ml of filtrate over 40 ml buffer I containing 0.5 M sucrose and 2% PVP in a 100-ml, round-bottom, plastic centrifuge tube. Centrifuge stepwise by 5-minute accelerations to 70, 250, and 800 g. Slowly decelerate. The nuclei pack loosely at the bottom of the tube.

5. Wash the nuclei twice in 50 mM Tris–HCl buffer (pH 8.0) containing 3 mM $MgCl_2$ and 0.5 M sucrose, by centrifugation at 500 g for 5 minutes.

Nuclei are free of cytoplasmic contamination and show RNA polymerase activity (Byfield and Lee, 1970; Lee and Byfield, 1970). The yield of nuclei is about 30%.

E. *Tetrahymena pyriformis* (Method of Gorovsky, 1970a)

1. CULTURE AND COLLECTION OF CELLS

Cells of mating type I, var. 1 are grown axenically in a medium containing 2.0% proteose peptone, 0.2% glucose, 0.1% yeast extract, and 0.003% sesquestrine (an iron–EDTA complex, Ciba-Geigy Corporation, Ardsley, New York). A 5-ml stock culture is added to 25 ml medium in a 250-ml Erlenmeyer flask and grown for 48 hours (cell density reached is $1–2 \times 10^6$ cells/ml). The 48-hour culture is added to 3 liters medium in a 5-liter diphtheria-toxin bottle. Cells are grown for 24–72 hours at $23°–28°C$ with vigorous aeration produced by bubbling air through a gas-dispersion tube. Sterile Antifoam B (Dow-Corning) is added if necessary to prevent foaming. A cell density up to 2×10^6 cells/ml is obtained, but cultures are used at densities of $0.5–1.2 \times 10^6$ cells/ml.

Six liters of culture are concentrated to 500 ml by passing the medium through a modified cream separator (Model 100, De Laval Separator Co., Chicago, Illinois). Cells are pelleted by centrifugation at 1000 g for 10 minutes.

2. ISOLATION OF MACRONUCLEI

All steps are done at $0°–5°C$. Centrifugation is done with an International PR-2 or PR-6 centrifuge equipped with a swinging-bucket head No. 253.

1. Suspend harvested cells in 500 ml Medium A $[0.1 M$ sucrose, 4.0% gum arabic, 1.5 mM MgCl$_2$ (pH 6.75) with NaOH] and wash them twice with Medium A by centrifugation at 1000 g for 5 minutes each time. Gum arabic is purified of large contaminants by centrifuging a 20% stock solution at 3000 g for 1 hour.

2. Resuspend 0.5- to 4-ml quantities of cells in 40 ml Medium B (0.63 ml *n*-octyl alcohol per 100 ml Medium A). Homogenize each 40-ml aliquot of cells in Medium B in the semimicro attachment of a Waring Blendor. Two 6-second periods at the higher speed is usually sufficient.

3. Centrifuge the homogenate at 1000 g for 10 minutes. The nuclei sediment. The supernatant and the "skin" of nonnuclear components at the top of the tube are collected, shaken vigorously, and centrifuged again to sediment more nuclei. Most nuclei are collected with three such centrifugations, but more than three centrifugations are needed at high cell density (4 ml cells/40 ml Medium B).

4. Crude nuclear fractions are pooled, resuspended in 80 ml Medium A, and washed twice by centrifugation at 1000 g for 10 minutes each time and then once at 250 g for 15 minutes.

The final pellet contains mainly macronuclei, being 90–95% macronuclear DNA and 5–10% micronuclear DNA. Recoveries of macronuclei average 56%, with a range of 43–71% reported in 10 measurements (Gorovsky, 1970a). Nuclear preparations are free of whole cells and cytoplasmic organelles and show an RNA/DNA mass ratio of 0.2–0.4. The occasional smaller cytoplasmic contaminants can be removed, if necessary, by resuspending nuclei in Medium A, layering the suspension over a denser sucrose solution (0.5–1.0 M), and centrifuging at 1000 g for 15–30 minutes. Losses of nuclei by this procedure are substantial, however. Gum arabic prevents macronuclei from swelling during isolation; however, after isolation, nuclei can be washed repeatedly in gum arabic–free solutions without damage. Gum arabic contaminates RNA and histone preparations from isolated nuclei, so it must be removed by washing before such preparations are made.

In the above method, macronuclei are isolated from well-aerated, growing cultures. Nuclei can be isolated from stationary-phase cells by adding 0.01% spermidine to Medium B and extending the time of homogenization, or by doubling the concentration of octanol in Medium B. Spermidine stabilizes nuclear structure when either growing or stationary-phase cultures are used. In growing cultures, the use of spermidine results in an increased yield (over 70%) of macronuclei. Spermidine cannot be used if micronuclei are to be isolated (see Section III,E,3), since spermidine-treated macronuclei are not broken by further homogenization and thus micronuclei cannot be isolated separately.

Techniques for using these isolated macronuclei for the isolation of histones (Gorovsky, 1970b, 1973; Gorovsky et al., 1974) and for determining the processing of rRNA (Kumar, 1970) have been published.

3. ISOLATION OF MICRONUCLEI

Micronuclei can be prepared from pooled, purified macronuclear fractions prepared in spermidine-free media (see Section III,E,2). Macronuclear fractions can be stored for 1 day to 4 weeks at −25°C before use (Gorovsky, 1970a).

1. Suspend macronuclear fractions in 40–80 ml Medium B (see Section III,E,2) and homogenize for 15 seconds at high speed in a Waring Blendor.

2. Centrifuge the homogenate at 250 g for 10 minutes. The resultant supernatant is collected, shaken vigorously, and centrifuged again at 250 g for 10 minutes. Repeat this procedure three times to give three pellets.

3. Resuspend all three pellets in Medium B, rehomogenize, and add back

to the 250 *g* supernatant remaining from step 2. This mixture, containing all the material of the original macronuclear fraction, is centrifuged at 1000 *g* for 10 minutes.

4. Suspend pellets from three to four collections as in step 3 in 40 ml Medium A (Section III,E,2) and centrifuge at 250 *g* for 10 minutes. Repeat this centrifugation three times. The final supernatant contains 20 to 40 micronuclei per large macronuclear fragment.

5. Centrifuge the supernatant at 1000 *g* for 30 minutes.

The final pellet contains 50–75% micronuclear DNA, 25–50% macronuclear DNA, and some particulates smaller than micronuclei. The pellet is not pure therefore, but is enriched in micronuclei compared to the macronuclear fraction (Section III,E,2). Histones can be isolated from the enriched micronuclear fraction (Gorovsky, 1970b, 1973).

IV. *Paramecium*

In 1950, Moses reported that both macronuclei and micronuclei could be freed intact from *Paramecium caudatum* if the cells were homogenized in 2% citric acid at 4°C for 5–15 seconds, but gave no information whether or not these nuclei could be isolated free of cytoplasmic contamination. Schwartz (1956) presented cytological studies on the RNA of individual macronuclei isolated from *Paramecium bursaria*. Kimball *et al.* (1960) isolated macronuclei from individual paramecia (*Paramecium aurelia*) and described methods for determining the dry weight of individual macronuclei. Prescott *et al.* (1966) previously gave the details of a method for isolating macronuclei and micronuclei from *P. caudatum* in Volume II of this series. Additional isolation methods are detailed below.

A.   *Paramecium aurelia* (Method of Wolfe, 1967)

1.   Culture and Collection of Cells

Paramecium aurelia, stock 51, syngen 4 (kappa-free) is grown with *Aerobacter aerogenes* as food source according to the method of Sonneborn (1950, 1970) using dried lettuce (Difco) or Cerophyll (Cerophyll Corp., Kansas City, Mo.) for the infusion medium (Wolfe, 1967). Cells are grown at room temperature in 100 ml media in a 500-ml Erlenmeyer flask or in 2.5 liters in a 4-liter Erlenmeyer flask. $CaCO_3$ is kept in a dialysis tube suspended in the medium. The tube is withdrawn at the time of inoculation with paramecia. Large cultures are stirred with a 2-inch Teflon magnetic stirring bar turning at 180 rpm.

Small volumes of cells are collected in 40-ml tubes at no more than 1000 g for 10 minutes in a refrigerated centrifuge (International Equipment Co., rotor No. 269). Large volumes of cells are collected rapidly by pouring the culture through a medium-coarse polyvinyl filter (Chemical Rubber Co.) fitted to a 4-liter vacuum flask. A motor-driven stirrer is inserted into the filter to prevent settling of the cells. Many cells of all sizes will pass through the filter; however, yields of 200 ml containing 4×10^6 cells are obtained in 10 minutes starting with 2.5 liters. The 200 ml is then centrifuged as above.

2. ISOLATION OF MACRONUCLEI

These methods have been used for structural analysis of macronuclei.

a. Method 1

1. Wash cells in 1 mM EDTA (pH 5.0).
2. Resuspend cells in at least 10 vol. of a solution containing 5 mg lysozyme per milliliter 1 mM EDTA (pH 5.0). Homogenize cells by means of a syringe with a no. 22 needle, a Teflon pestle in a tissue homogenizer, or a Waring Blendor.
3. Centrifuge the homogenate at 200 g in 1 mg per milliliter lysozyme–EDTA. Discard small debris.
4. Layer the suspension of nuclei, whole cells, and large debris on 2.5 M sucrose. Centrifuge at 10,000 rpm for 40–60 minutes in a Sorvall SS-34 rotor. The nuclei will pellet.

b. Method 2

1. Forcefully suspend a pellet of cells by rapidly adding 1% digitonin in 1 mM CaCl$_2$ with a 10-ml pipette. Repeat by squirting the suspension against the side of the tube until the majority of macronuclei are freed from the cells.
2. Centrifuge three times at 200 g in 1 mM CaCl$_2$. Discard small debris each time.
3. Pellet the nuclei in 2.5 M sucrose (in 1 mM CaCl$_2$) as in method 1, step 4.

c. Method 3.
(For some structural studies on macronuclei, sucrose is omitted in this method. See Wolfe, 1967).

1. Suspend the cell pellet in 0.2% digitonin in 2 mM MgCl$_2$ and 0.25 M sucrose.
2. Homogenize in a Waring Blendor.
3. Wash by centrifuging at 200 g in MgCl$_2$–sucrose three times. Discard small debris each time.
4. Pellet the nuclei in 2.5 M sucrose (in 2 mM MgCl$_2$) as in method 1, step 4.

Macronuclei contain two types of bodies: one measures 0.1–0.2 μm in diameter and contains DNA, and the other is the larger nucleoli. The DNA bodies join together by means of 10-nm fibrils and contain 25-nm "coils" characteristic of eukaryotic chromatin.

B. *Paramecium aurelia* (Method of Stevenson, 1967)

1. CULTURE AND COLLECTION OF CELLS

Thirty- to 100-liter cultures of *P. aurelia* (syngen 1, stock 540) are grown at 30°–31°C on the grass-extract medium of Jones (1965) with *A. aerogenes* as food-organism. After 3 days, cultures are filtered through muslin. The filtered fluid is then passed through an Alfa-Laval cream separator at a flow rate of 700 ml per minute. The concentrate is removed from the rotor at 30-minute intervals to avoid killing the paramecia, and is concentrated further by centrifugation in pear-shaped bottles at 1000 g for 2 minutes in an oil-testing centrifuge. Collected cells are washed twice in the salt solution (pH 7.1) of Dryl (1959) containing 1.0 mM NaH_2PO_4, 1.0 mM Na_2HPO_4, and 1.5 mM $CaCl_2$.

2. ISOLATION OF MACRONUCLEI

All steps are done at 0°–4°C.

1. Suspend washed cells (4–20 gm wet weight) in 50 vol. of solution containing 0.1% Tween 80 and 1.0 mM $CaCl_2$ and stir for 4–7 minutes in a homogenizer (MSE Ltd., London) running at half-speed. Seventy to 90% of the macronuclei are released.

2. Centrifuge the homogenate at 600 g for 6 minutes.

3. Wash the resultant pellet twice with 0.25 M sucrose–1.0 mM $CaCl_2$ with centrifugation at 600 g for 6 minutes each time.

4. Resuspend the resultant crude nuclear pellet in 2.4 M sucrose–1.0 mM $CaCl_2$ and centrifuge the mix at 40,000 g average in the 3 × 40 ml swing-out rotor of an MSE superspeed 50 ultracentrifuge.

5. Wash the final pellet twice with 0.25 M sucrose–1.0 mM $CaCl_2$.

The yield of macronuclei is 30–50%. Isolated nuclei are intact, free of significant cytoplasmic contamination, and capable of DNA-dependent RNA synthesis.

C. *Paramecium aurelia* (Method of Cummings, 1972)

1. CULTURE AND COLLECTION OF CELLS

Paramecium aurelia (syngen 1, 4, or 13) is grown in 28-liter quantities on a bacterized *Klebsiella*–grass infusion at 24°C according to Sonneborn (1950, 1970). The generation time for syngen 1 on this medium is 5 hours, and the culture clears in about 2 days. For syngens 4 and 13, the culture clears in about 3 days. Cultures are filtered through several layers of absorbent cotton wool and centrifuged at 2300 rpm at a flow rate of 700–800 ml per minute. The rotor is flushed twice with 350 ml of Dryl's solution (Dryl, 1959). Cell pellets are resuspended in Dryl's solution and centrifuged at 400 g for 5 minutes in pear-shaped bottles in an MSE medium oil centrifuge. Ten to 14 ml of packed cells are obtained from a 28-liter culture.

2. ISOLATION OF MACRONUCLEI

The method is the same for all three syngens. All steps are done at $0°$–$4°$C.

1. Add Dryl's solution (Dryl, 1959) to the packed cells to make a total volume of 17.6 ml. Next add 4.0 ml of solution containing 0.033 M CaCl$_2$, 1.2 M sucrose, 11 mM Tris, and 4.5 mg spermidine trihydrochloride. Mix by swirling. Add 3.0 ml of 3% Nonidet P40, and then 3.0 ml of 2% sodium deoxycholate. The final pH is 7.1–7.3.

2. Homogenize the suspension with a loose-fitting Teflon pestle by three strokes of a Tri-R stir homogenizer at a setting of 3. This minimal homogenization releases intact nuclei.

3. Mix the homogenate 1:1 with a solution containing 2.35 M sucrose, 4.8 mM CaCl$_2$, 0.02 M NaCl, and 100 μg spermidine trihydrochloride per milliliter.

4. Layer the suspension on an equal volume of a solution containing 2.1 M sucrose, 3 mM CaCl$_2$, and 100 μg spermidine trihydro chloride per milliliter, and 0.02 M NaCl. Centrifuge in the SW-21 rotor of an MSE Superspeed centrifuge to 10,000 g, hold for 10 seconds at that force, and decelerate without braking. Macronuclei are sedimented free of cellular debris, except for some unidentified crystals in some preparations.

Macronuclei consist of 22% DNA, 10% RNA, and 68% protein and are somewhat variable in size and shape, with maximum dimensions between 15 and 35 μm. The yield of macronuclei is 40–60 mg from a 28-liter culture. Under electron microscopy, isolated macronuclei appear normal with well-defined membranes. The G + C content of macronuclear DNA is about 23%. Two species of RNA having molecular weights of 1.3 $\times$ 10^6 and 2.8 $\times$ 10^6 daltons have been identified in isolated macronuclei. Isolation techniques for and properties of macronuclei are the same for all three syngens mentioned above. Of several detergents, Nonidet P40 requires the lowest amount of calcium to protect freed nuclei (step 1). Calcium protects nuclei from disruption better than Mg^{2+} does. The use of both Ca^{2+} and spermidine allows the use of low concentrations of Nonidet P40 and deoxycholate (step 1).

3. ISOLATION OF MICRONUCLEI

1. Collect the supernatant from step 4 (Section IV,C,2).

2. Layer the supernatant onto fresh sucrose solution (as in step 2, Section IV,C,2). Centrifuge at 40,000 g for 1 hour.

3. Resuspend the crude micronuclear pellet in buffer containing 9 mM NaCl, 3 mM Na$_2$HPO$_4$, 9 mM NaH$_2$PO$_4$, and 3 mM CaCl$_2$ (pH 6.2). Centrifuge for 20–30 seconds at 700 g in an MSE Super microcentrifuge. Contaminating debris is pelleted compactly.

4. Withdraw the micronuclear suspension with a Pasteur pipette. Pellet the micronuclei by centrifugation at 700 g for 10–15 minutes.

The yield of micronuclei is 1–2 mg from a 28-liter culture. Micronuclei consist of 9% DNA, 80% protein, and 11% RNA, the latter possibly being a contaminant. Isolated micronuclei vary between 1.5 and 2.5 μm in diameter, possess a membrane, and contain DNA with a G + C content of about 23%. Micronuclear preparations are contaminated by some membranous fragments and trichocysts. The details of preparation and properties of micronuclei are the same for all three syngens (Section IV,C,1).

D.   *Paramecium caudatum* (Method of Skoczylas *et al.*, 1963)

1.   CULTURE AND COLLECTION OF CELLS

Paramecium caudatum is grown in mass culture according to the method of Sonneborn (1950). As a substitute for lettuce, a Brussels sprouts infusion inoculated with *A. aerogenes* can be used (Skoczylas *et al.*, 1963). Cultures are grown at room temperature and reach densities of 300 to 400 cells/ml. They are fed three times a week by dilution with an equal volume of fresh medium.

No later than 3 days after the last feeding, 50 liters of culture are centrifuged at room temperature in a continuous-flow centrifuge at less than 200 g. The concentrated cells are diluted to 1.2 liters with fresh medium. Before and after centrifugation, cell suspensions are filtered through nylon gauze (mesh diameter, 120–160 μm) to remove detritus. About 90% of the cells are recovered. Condensed cell suspensions (1.5–4.0 × 10^4 cells/ml) are stored at 10°–14°C in 1.5-liter Fernbach flasks up to 2 days. Longer storage results in decreased final yields of nuclei.

Condensed cultures are cooled to 2°–4°C in a cold room, and sucrose and CaCl$_2$ are added to final concentrations of 0.3 M and 0.6 mM, respectively. After 20–30 minutes, cells are sedimented at 800 g for 10 minutes and then resuspended in 0.3 M sucrose–0.6 mM CaCl$_2$ solution to a density of 10^5 nuclei/ml. Some of the cells are broken during this sedimentation step. Nuclei are identified and counted as follows. Each 50-μl cell suspension used for counting is mixed with 5 μl 1% aqueous methyl green and counted with a microscope.

2.   ISOLATION OF MACRONUCLEI

All steps are done in a cold room at 0°–3°C.

1. Homogenize the cells in sucrose–CaCl$_2$ six times by passage through a

syringe needle (Luer No. 22M). Up to 10 ml of cell suspension can be homogenized with a 20-ml Luer syringe. Large volumes are sucked through the needle into a separation funnel connected to a vacuum pump.

2. The sucrose concentration of the homogenate is increased to 1.6 M, and $CaCl_2$ to 3 mM. Put samples of 35 ml each on top of a sucrose density gradient in a 100-ml centrifuge tube and centrifuge at 1300 g for 15 minutes. The density gradient (1.65–1.9 M sucrose containing 3 mM $CaCl_2$) is prepared at least 12 hours in advance (and kept in a cold room until used) by layering sucrose concentrations differing by about 0.05 M. The rotor containing the gradients is accelerated slowly.

3. Combine the 1.75–1.9 M sucrose layers, dilute with 1 M sucrose to 1.65 M sucrose, and put on top of a gradient prepared as in step 2 and containing 1.7–2.2 M sucrose in 3 mM $CaCl_2$. Centrifuge at 1400 g for 20 minutes.

4. Collect the 1.9–2.15 M sucrose layers, dilute to 1.85 M sucrose, and put on top of a gradient prepared as in step 2 and containing 1.9–2.1 M sucrose in 3 mM $CaCl_2$. Centrifuge at 1400 g for 20 minutes.

5. Collect the lower layers (2.05–2.1 M sucrose) containing the nuclei, combine, and centrifuge at 4000 g for 20 minutes. The resultant pellet contains purified nuclei.

The yield of macronuclei is about 30%. Nuclei remain intact during isolation if a sucrose concentration of 0.3 M or above is maintained. During collection and homogenization of cells, a concentration of $CaCl_2$ at 0.6 mM is critical. Lower and higher amounts of calcium result in reduced yields of nuclei.

V. *Blepharisma*

Prescott *et al.* (1966) previously gave the details of methods for isolating macronuclei from *Blepharisma japonicum* and *B. undulans* in Volume II of this series. A method for *B. intermedium* is given below.

Blepharisma intermedium (Method of Saxena, 1966)

1. CULTURE AND COLLECTION OF CELLS

The freshwater spirotrichous ciliate *B. intermedium* is cultured at 25°–27°C in a hay infusion fortified with Horlick's malted milk.

2. ISOLATION OF MACRONUCLEI

1. Starve the cells for 24 hours in Chalkley's medium (Chalkley, 1930),

wash in distilled water, and lightly centrifuge (about 5000 cells) in a 2-ml microcentrifuge tube.

2. Add ice-cold 0.02 M digitonin in a 0.3 M sucrose–0.6 mM CaCl$_2$ solution (pH 6.7). Centrifuge in a microcentrifuge for 4 minutes at 15,000 rpm.

3. Suspend the pellet in ice-cold sucrose–CaCl$_2$, transfer to a 5-ml cavity block, and keep at 4°C for 10 minutes. Wash the centrifugate repeatedly in ice-cold sucrose–CaCl$_2$ by slow rotation of the cavity block. During rotation, macronuclei remain at the bottom, and whole and broken cells and cell debris are removed by a fine pipette.

4. Collect nuclei in a microcentrifuge tube in the sucrose–CaCl$_2$ medium.

Macronuclei appear clean and morphologically intact (Saxena, 1966). Higher concentrations of CaCl$_2$ (1.0–1.5 mM) cause cells and cell fragments to clump and agglutinate. Nuclei tend to break if the digitonin concentration is raised to 0.04–0.05 M. Isolated macronuclei are capable of DNA-dependent RNA synthesis (Seshachar and Saxena, 1968).

VI.   *Didinium*

The method of Stevenson (1967) for the isolation of macronuclei from *P. aurelia* 540 (Section IV,B,2) also can be used to isolate the macronuclei of the predacious ciliate *Didinium nasutum*. The didinia are grown as in Section IV,B,1, using *P. aurelia* 540 as food organism, and harvested when no *P. aurelia* are present in 1-ml samples of the mass culture. During the procedure for isolation of macronuclei, homogenization (step 1, Section IV,B,2) is done for only 1–2 minutes. Isolated macronuclei are free of significant cytoplasmic contamination, but they become rounded, losing their characteristic horseshoe shape, and about 20% are fragmented.

VII.   *Spirostomum*

This freshwater spirotrichous ciliate is cultured (Saxena, 1966) like *B. intermedium* (Section V,1), and its macronucleus is isolated as in the procedure for *B. intermedium* (Section V,2). Another method for *Spirostomum ambiguum* was given by Prescott *et al.* (1966) in Volume II of this series.

VIII. *Amoeba*

Techniques for isolating and transplanting single nuclei from amebas have been given by Goldstein (1964), Prescott *et al.* (1966), and Jeon (1970). A method for the mass isolation of nuclei from amebas is given below.

Amoeba proteus (Method of Tautvydas, 1971)

1. CULTURE AND COLLECTION OF CELLS

Amebas are cultured essentially as described by Prescott and Carrier (1964). Cultures are maintained at 19°C in rectangular Pyrex glass dishes (13.5 × 8.75 × 1.75 inches) fitted with a sheet of Lucite to increase the surface area for cell attachment. Fifty and 100 ml of a *Tetrahymena* suspension (Klett turbidity reading = 50 units) are added to each dish containing 1.8×10^6 cells daily at 9:00 A.M. and 5:00 P.M., respectively. Amebas are transferred to clean dishes, and fresh ameba medium is added every 3–4 days.

Cells ($4–5 \times 10^6$) are collected 18 hours after the last feeding in 50-ml conical graduated centrifuge tubes by centrifugation at 500 *g* at 24°C for 6 minutes. Resultant packed cell concentration is 5×10^5 cells/ml.

2. ISOLATION OF NUCLEI

All steps are done at 3°C.

1. Eight milliliters of packed cells are transferred to a 40-ml Dounce tissue grinder (Kontes Glass Co.) with 2 ml of cold ameba homogenization medium (AHS) per milliliter of packed cells. A pestle (narrow clearance, type B) is put in the grinder before adding the cells, so that homogenization can be done with one gentle upstroke of the pestle. AHS contains 10 m*M* morpholinoethane sulfonate (MES) buffer, 0.03% 2-ethyl-1-hexanol (reduces foaming and strengthens nuclear membrane), 0.2 m*M* dithiothreitol (maintains monothiols is a reduced state), 1 m*M* KCl, 0.4 m*M* magnesium acetate, 0.05 m*M* CaCl$_2$, glass-distilled water, and KOH to adjust the pH to 5.9 at 25°C. AHS is filtered through Millipore filters (type HA) and stored frozen at −20°C until used.

2. Pour the homogenate onto the top screen of a stack of Nitex nylon screens (Tobler, Ernst and Traber, Inc., Elmsford, New York) and let it filter through the stack of screens by gravity. During filtration, shake the stack to aid movement of the homogenate through the screens. Homogenization (step 1) and filtration take 35–40 minutes. Each screen in the stack has a pore diameter smaller than the screen above it: top screen (48-μm pore diameter, Nitex 48), next screen (35 μm, Nitex 35), next screen (25 μm, Nitex

28), and bottom screen (15 μm, Nitex 20). Each screen is glued with epoxy resin to the narrow end of a polyethylene ring made by cutting away the bottom of a 250-ml Nalgene polyethylene beaker.

3. Rinse the residue on the top screen 20 times with 2 ml AHS each time and transfer the washed residue to the tissue grinder for a second homogenization. Pour this homogenate onto the top screen and rinse the residue 24 times with 2 ml nuclear purification solution (NPS) each time. Remove the top screen. Rinse the residues on the Nitex-35, -28, and -20 screens in turn 6 times, 24 times, and 10 times, respectively, with 2 ml NPS each time. Most of the nuclei are retained on the Nitex-20 screen. NPS contains 5 mM MES buffer, 0.2 mM dithiothreitol, 0.1 M sucrose (RNase-free, Mann Research Laboratories), 1% purified gum arabic, 1 mM KCl, 0.4 mM magnesium acetate, glass-distilled water, and KOH to adjust the pH to 5.9 at 25°C. Filter the NPS through Millipore filters (type HA) and keep frozen at -20°C until used. The gum arabic (Fisher Scientific Co.) is purified by centrifuging a 10% crude solution at 20,000 g for 10 minutes and then filtering the supernatant through a glass-fiber filter (Millipore, type APO) with suction. Keep the purified 10% gum arabic solution frozen at -20°C until used.

4. Collect the rinsed residue on the Nitex-20 screen in 6 ml of NPS. Layer 2 ml on each of three continuous sucrose gradients (20–60%). Centrifuge for 70 seconds at 2000–22,000 rpm in an HB-4 rotor in a Sorvall RC-2B centrifuge. Most of the nuclei collect in a distinct band about two-thirds of the way down the gradient. The gradients are prepared in flat-bottom glass tubes (2.5 × 8 cm) from NPS solutions containing 20 and 60% sucrose (RNase-free). (Note: The concentration of sucrose in the solutions is greater than 20 or 60%, since these amounts of sucrose are added to NPS which already contains 0.1 M sucrose.)

5. Add a 37% sucrose solution (in NPS) just above the band of nuclei. This helps in removing the upper part of the gradient without disturbing the band of nuclei.

6. Layer a 47% solution of sucrose (in NPS) just below the nuclear band. Transfer the nuclei from all three gradients with a Pasteur pipette onto a nylon screen (10 μm pore diameter, Nitex 10). The screen is glued to a polyethylene ring as in step 2 above. The purified nuclei are collected from the screen in NPS.

Isolated nuclei are intact and highly purified, the main cytoplasmic contaminant being a few food-waste pellets. If the amebas are homogenized earlier than 17 hours after the last feeding, the number of food-waste pellets contaminating the recovered nuclei increases. In step 3, AHS and then NPS are used since, if sucrose and gum arabic are used in AHS, cells are not broken during the second homogenization and any free nuclei are destroyed.

If more than 8.5 ml of packed cells are used in step 1, the efficiency of the isolation procedure is decreased.

The average yield of nuclei is 40.5% (range 34–53%). During isolation, nuclei lose about 3 and 15% of their RNA and protein, respectively. The average isolated nucleus has a diameter of 36.1 μm, has a dry weight of 311 pg, contains 3.4 pg DNA, and shows a mass ratio of DNA/RNA/protein of 1:7.7:60.7. Isolated nuclei are capable of DNA-dependent RNA polymerase activity with the kinetics of RNA synthesis similar to that of isolated nuclei from mammalian cells. Polymerase activity is unusual, however, in that it is inhibited by rifampicin. Also, the rate of GMP incorporation into RNA is at least 26 times greater than in isolated liver nuclei.

IX. *Euglena*

A major problem in isolating intact, functional subcellular particles from *Euglena* lies in breaking the tough cell pellicle without damaging the particles. Leedale (1958) isolated nuclei from individual *Euglena* following softening of the pellicle by overnight incubation in a saturated solution of pepsin. Parenti *et al.* (1969) reported a mass isolation of *Euglena* nuclei achieved by initially weakening cell pellicles by freezing and thawing the cells and treating them with Triton, or by exposing the cells to pancreatic protease, but gave no estimates of final yields of nuclei and did not present any microscopic evidence for purity or intactness of the final nuclear preparations. Aprille and Buetow (1973) isolated *Euglena* nuclei in a final average yield of 29% by first weakening the cell pellicle with Triton, followed by long incubation of the cells in saturated pepsin at a low pH. This method yields nuclei free of significant cytoplasmic contamination and capable of incorporating amino acids (Aprille and Buetow, 1974); however, these nuclei are not well-preserved ultrastructurally (Aprille and Buetow, 1973). A method combining freezing and thawing, citric acid (pH 2.6), and sonication yields a pellet enriched in nuclei, but not pure (Lynch, 1973; Lynch and Buetow, 1975). Also, as is the case in higher cell nuclei isolated with citric acid procedures (Busch and Smetana, 1970), these *Euglena* nuclei show an altered ultrastructure.

The following method yields ultrastructurally intact, enzymatically active nuclei from *Euglena*.

Euglena gracilis (Method of Lynch and Buetow, 1975)

1. CULTURE AND COLLECTION OF CELLS

Cultures of *Euglena gracilis* (var. bacillaris, streptomycin-bleached, strain SM-L 1) are grown in the dark at 27°C in 12-liter carboys containing 6 liters of defined medium (Buetow, 1965) with 0.2 M ethanol as carbon source (Buetow and Padilla, 1963). When cultures reach densities of $3-8 \times 10^5$ cells/ml (midlog phase of growth), they are harvested at 2°C with a Sharples Super centrifuge. The concentrated cells are washed once by resuspension in 0.9% NaCl with centrifugation at 1000 g for 10 minutes.

2. ISOLATION OF NUCLEI

Steps 2 to 7 are done at 0°–2°C. MES is 2-[N-morpholino] ethane sulfonic acid (Sigma Chemical Co.),

1. Freeze the cell pellet in a Dry Ice–acetone bath and then thaw it under running tap water at 20°C.

2. Wash the cells once by resuspension in buffer containing 50 mM MES (pH 5.55), 7 mM MgCl$_2$, and 1.25% Triton X-100. Centrifuge at 1000 g for 10 minutes.

3. Resuspend the pellet in 10 vol of buffer as in step 2 and incubate for 30 minutes at 0°–2°C.

4. Sonicate the incubate for 20–40 seconds with 10-second bursts at 238 W/in^2 until 75–80% of the cells are broken. Sonication is done with a Bronwill Biosonic III sonicator equipped with a 1.9-cm stainless-steel probe. The sample for sonication is divided into 10-ml aliquots and sonicated for 10-second bursts in a 40-ml plastic beaker cooled in an ice bath. Following each 10-second burst, the probe and solution are allowed to cool for 20 seconds.

5. Layer the sonicated mix over 10% dextran (MW 40,000, Pharmacia) in buffer containing 50 mM MES (pH 5.55), 7 mM MgCl$_2$, and 0.5% Triton X-100. Centrifuge at 1000 g for 15 minutes. The resultant crude nuclear pellet contains nuclei, paramylon, and occasional unbroken cells.

6. Resuspend the pellet by homogenization in 1 vol buffer containing 50 mM MES (pH 5.55), 7 mM MgCl$_2$, 1.25% Triton X-100, and 0.1 M freshly made sodium metabisulfite (Na$_2$S$_2$O$_5$). Layer over 4 vol of 2.4 M sucrose in the same buffer. Mix the top half of the gradient with a glass rod. Centrifuge at 21,000 g for 1 hour.

7. Collect the interface material and add to it an equal volume of buffer containing 50 mM MES (pH 5.55), 7 mM MgCl$_2$, and 1.25% Triton X-100. Mix gently with a glass rod. Layer over 2.4 M sucrose in the same buffer. Stir the top half of the gradient with a glass rod. Centrifuge at 21,000 g for 1 hour. The pellet contains nuclei.

Nuclei are obtained in a yield of 25–37%, appear intact ultrastructurally, and contain acid-soluble proteins in an amount relatively the same as that found in higher cell nuclei (Lynch and Buetow, 1975). The only contaminant in the final nuclear pellet consists of occasional paramylon granules. Pure intact nuclei are not isolated at pH 2.6 or 7.4–8.5. Over the pH range 5.5–7.0 with MES buffer, good cell breakage is combined with little apparent nuclear damage only at pH 5.55. Sodium citrate (100 mM) as buffer at pH 5.5 causes more aggregation of cellular debris than does MES buffer. Sonication (step 4) is continued until 75–80% of the cells are broken as monitored by light microscopy. Further sonication produces nuclear damage. The densities of nuclei and of undamaged whole cells are similar at step 5. The addition of 0.1 M Na$_2$ S$_2$O$_5$ (but not NaHSO$_3$) to the buffer (step 6), however, results in a decreased density of the nuclei, allowing them to be separated from the whole cells. The change in nuclear density is readily reversible, since when nuclei are subsequently suspended in buffer lacking metabisulfite, they then sediment through 2.4 M sucrose (step 7).

Whole chromatin can be obtained from isolated nuclei (Lynch *et al.*, 1975b). This chromatin, which appears under electron microscopy as uniformly condensed fibers even during the interphase period of the cell cycle, can nevertheless be subfractionated into distinct heterochromatic and euchromatic fractions. The euchromatic fraction, comprising 14% of the total nuclear DNA, contains over 80% of the endogenous RNA polymerase activity (Lynch *et al.*, 1975b). Isolated nuclei yield acid-soluble proteins which resolve into nine bands during electrophoresis on polyacrylamide gels (Lynch, 1973; Lynch *et al.*, 1975a). Five bands correspond to the five histones associated with higher eukaryotic chromatin. The other four bands may represent proteins involved in the maintenance of the continually condensed appearance of *Euglena* chromatin.

X. Algae

A. *Griffithsia globulifera* (Method of Nasatir and Brooks, 1966)

1. COLLECTION OF CELLS

Nasatir and Brooks (1966) collected the marine red alga *Griffithsia globulifera* from its natural habitat in waters surrounding the Marine Biological Laboratory at Woods Hole, Massachusetts. Algae were carefully washed and cleaned in running sea water.

2. ISOLATION OF NUCLEI

1. Homogenize clean, fresh algae for 5 minutes at about 5000 rpm in a Sorvall Highspeed Omni-Mixer. The homogenization solution contains 3 mM CaCl$_2$, 3 mM Tris (pH 7.4), and sucrose anywhere from 0.75 M (iso-

tonic) to 1.5 M. The yield of nuclei is the same over this concentration range of sucrose.

2. Filter the homogenate through flannelette to remove intact cells, cell wall fragments, and gelatinous material.

3. Pellet the nuclei by centrifuging the filtrate at 3800 g.

The yield of crude nuclei is about 1 gm from a fresh weight of 100 gm algae. Nuclear DNA has a $G+C$ content of 41.7%. This method is also useful for isolating nuclei from several species of *Polysiphonia* (Nasatir and Brooks, 1966).

B. *Peridinium trochoideum* (Method of Rizzo and Noodén, 1973)

1. Culture and Collection of Cells

Axenic stock cultures of the dinoflagellate *P. trochoideum* are maintained in 10 ml of autoclaved half-strength medium f (Guillard and Ryther, 1962) in 2 × 15 cm tubes at 20°–22°C under a daily regime of 8 hours darkness and 16 hours of light from fluorescent Cool-White bulbs (Sylvania). Sea water (Marine Biological Laboratory, Woods Hole, Massachusetts) is sterilized by filtration through Whatman GF/C glass filters. Cultures are propagated axenically by transfers to successively larger flasks up to 2800 ml wide-mouthed Erlenmeyer flasks containing 2 liters of pasteurized Erdschreiber solution (Starr, 1964) or pasteurized half-strength medium f (Guillard and Ryther, 1962). After the last transfer, cultures are grown for 5–7 days, reaching densities of about 1.5 × 10^4 cells/ml on Erdschreiber solution and 6 × 10^3 on half-strength medium f.

Fourteen to 15 liters of cells are concentrated in a DeLaval cream separator at 0°–4°C, and the concentrate is centrifuged at 2500 g for 10 minutes in the cold. The recovered pellet is suspended to a density of 1.5–2 × 10^6 cells/ml in pH 7.2 isolation medium (about 120 ml needed) containing 0.25 M sucrose, 5.0 mM CaCl$_2$, and 10% (w/v) dextran 40 (Pharmacia).

2. Isolation of Nuclei

1. To disrupt the cells, sonicate 30-ml portions of the cell suspension for a total of 2–3 minutes each with a Biosonic II ultrasonicator (Bronwill Scientific, Rochester, New York) at 90% output. To keep beakers cool, place them in an ice bath, sonicate for 1 minute, and swirl for 30 seconds between sonications. Cell breakage is monitored with a light microscope to determine the time of maximum cell disruption with minimal nuclear disruption.

2. Centrifuge sonicate for 10 minutes at 3000 g in a Sorvall SS-34 rotor. Suspend pellet in 30 ml isolation medium (Section X,B,1).

3. Prepare a three-layer discontinuous gradient in a centrifuge tube: 7 ml 2.4 M sucrose plus 10% (w/v) dextran 10 (Pharmacia); 21 ml 2.4 M sucrose; and 16 ml 2.2 M sucrose plus 0.1% (v/v) Triton X-100. Each layer also contains 10 mM Tris–HCl (pH 7.2) and 5 mM CaCl$_2$.

4. Layer each 10 ml of suspension (step 2) over a gradient (step 3) and gently stir the upper half of the 2.2 M sucrose layer. Centrifuge for 20 minutes at 27,000 g in a Spinco SW-25.2 rotor.

5. Decant the supernatant. Wipe the insides of the tubes carefully with a cotton swab to remove adhering material. The crude nuclear pellet is purified by gentle suspension in isolation medium (Section X,B,1) with the aid of a loose-fitting Potter-Elvehjem homogenizer and centrifuged again as in step 4.

Sometimes the crude nuclear pellet from step 4 is heavily contaminated as seen with light microscopy. If so, the pellet is suspended in isolation medium with a loose-fitting Potter-Elvehjem homogenizer and diluted with 1–2 vol of pH 7.6 buffer containing 0.14 M NaCl, 5 mM MgCl$_2$, and 10 mM Tris–HCl. To remove cell fragments, filter the suspension through a 25-μm Nitex nylon screen (Tolber, Ernst and Traber, Inc.) fitted to a stainless-steel filtering pan (Baruch Instruments Corp., Ossining, New York). The average yield of nuclei is 16% of the original DNA. The isolated nuclei are free of cytoplasmic tabs, and most have intact membranes. In picograms per nucleus, the amount of DNA is 34.0, RNA is 7.5, acid-insoluble protein is 41.0, and acid-soluble protein is 2.7.

C. *Peridinium cinctum* (Method of Rizzo and Noodén, 1973)

1. CULTURE AND COLLECTION OF CELLS

Axenic stock cultures of the freshwater dinoflagellate *P. cinctum* are maintained in 10 ml autoclaved 1336 medium (Carefoot, 1968) under the conditions of growth used for *P. trochoideum* (Section X,B,1), and propagated similarly by transfers to successively larger flasks up to 2800-ml, wide-mouthed Erlenmeyer flasks containing 2 liters of 1336 medium. An increased growth rate and final cell density is attained if 50 ml autoclaved soil extract (Starr, 1964) is added to the 2 liters of 1336 medium. Only low densities of cells are obtained if the cells are cultured in 5-liter carboys under the same conditions.

Cells are harvested in the cold as described for *P. trochoideum* (Section X,B,1).

2. ISOLATION OF NUCLEI

All steps are done at 0°–4°C.

1. Disrupt cells as described for *P. trochoideum* (Section X,B,2, step 1), except that the suspension is sonicated for a total of 6–8 minutes, and at maximum output of the sonicator.

2. Collect sonicate as in Section X,B,2, step 2.

3. Prepare a two-layer discontinuous gradient in a centrifuge tube: 7 ml 2.4 *M* sucrose plus 10% (w/v) dextran 10 (Pharmacia); 37 ml 2.4 *M* sucrose. Each layer also contains 10 m*M* Tris–HCl (pH 7.2) and 5 m*M* $CaCl_2$.

4. Layer each 10 ml of the sonicate suspension (step 2) over a gradient (step 3) and gently stir the upper third of the 2.4 *M* sucrose layer. Centrifuge at 9780 *g* for 20 minutes in a Spinco SW-25.2 rotor.

5. Remove the 2.4 *M* sucrose layer with a large-bore pipette. The nuclei band at the interface between 2.4 *M* sucrose and the sucrose–dextran layers. Collect nuclei with a large-bore pipette.

Alternatively, nuclei can be pelleted if the tubes are centrifuged for 60 minutes at 9780 *g*. Higher yields of nuclei result, but preparations are heavily contaminated with cell wall fragments. The nuclei can be freed of most contaminants, however, as follows:

1. Suspend the pellet in 6 ml isolation medium (Section X,B,1). Gently layer 2-ml portions over an equal volume of 2.4 *M* sucrose plus 10% (w/v) dextran 10 in each of three tubes. The interface between the two layers must not be disturbed.

2. Centrifuge at 18,000 *g* for 10 minutes in a Spinco SW-39 rotor. The nuclei band on top of the sucrose–dextran layer; most cell wall fragments are pelleted; light debris is discarded with the supernatant.

3. Suspend the nuclei in isolation medium (Section X,B,1) and pellet by centrifugation at 12,000 *g* for 10 minutes in a Sorvall SS-34 rotor. The pellet contains purified nuclei.

Most isolated nuclei are intact. The only significant contaminants in the nuclear preparations are some cell wall fragments.

D. *Gymnodinium nelsoni* (Method of Mendiola *et al.*, 1966)

1. Culture and Collection of Cells

Cultures of the marine dinoflagellate *G. nelsoni* Martin (strain GSBL) are grown in 10–15 liters of half-strength medium f (Guillard and Ryther, 1962) at 20°C with aeration and with 11,000 lumens/m² of fluorescent light. Cultures are harvested when growth ceases (at a density of about 4×10^3 cells/ml) by semicontinuous centrifugation in a 4-liter basket centrifuge at 500 rpm. Spent medium is removed by suction from the spinning rotor. About 80% of the cells in a carboy can be recovered in about 200 ml in the basket. Cells are further concentrated in 50-ml centrifuge tubes. Harvesting is done at room temperature, since the cells disintegrate during centrifugation at low temperatures. Harvested cells are resuspended in isolation medium containing 0.25 M sucrose, 5 mM CaCl$_2$, and 10% (w/v) dextran (Pharmacia).

2. ISOLATION OF NUCLEI

All steps are done at 0°–5°C.

1. Pass the cell suspension through a Logeman mill. Centrifuge the resulting brei at 1500 rpm for 10 minutes.

2. Wash the pellet twice with isolation medium (Section X,D,1) containing 0.5% Triton X-100, once with isolation medium containing 0.25% Triton X-100, and twice with isolation medium alone. At each wash, resuspend the pellet in the wash medium with a Potter-Elvehjem homogenizer and sediment again by centrifugation as in step 1. The final pellet contains purified nuclei.

The presence of dextran in the isolation medium is essential for preservation of nuclei during cell breakage at low temperature. Dextran sulfate cannot be substituted for dextran. The yield of nuclei is at least 70%. The final nuclear pellet is contaminated to a small extent with cell walls and chromatophores. In isolated nuclei, membranes appear perforated, but the internal structure is intact. In nanograms per nucleus, the amount of DNA is 1.1, RNA is 0.16, and protein is 1.0. These values are about 10 times those reported for nuclei isolated from *G. nelsoni* by the method of Rizzo and Noodén (Section X,E,2).

E. *Gymnodinium nelsoni* (Method of Rizzo and Noodén, 1973)

1. CULTURE AND COLLECTION OF CELLS

Axenic stock cultures of the naked (no cell wall) dinoflagellate *G. nelsoni* are maintained in 10 ml autoclaved half-strength medium f of Guillard and Ryther (1962) in 2 × 15 cm tubes at 23°C under constant illumination provided by fluorescent Cool-White bulbs (Sylvania). Natural sea water is sterilized as in Section X,B,1. Cultures are transferred successively as for *P. trochoideum* (Section X,B,1). Final cultures are grown for about 1 week, reaching a density of 1–2 × 10^3 cells/ml.

Concentrate cells by centrifugation at 650 g for 10 minutes at 0°–4°C. Suspend the cell pellet in 30–40 ml isolation medium (Section X,B,1) at about 2 × 10^5 cells/ml.

2. ISOLATION OF NUCLEI

All steps are done at 0°–4°C.

1. Shear cells by hand with a Potter-Elvehjem glass-Teflon homogenizer until most of the cells are broken.

2. Sonicate the broken cells for 7–10 seconds at maximum output with a Branson S-75 sonifier to free nuclei from adherent cytoplasm.

3. Layer over 44 ml of 2.2 M sucrose plus 0.1% (v/v) Triton X-100 in each of three tubes. Form a partial gradient by gently stirring the upper third of the tube.

4. Centrifuge at 9780 g for 20 minutes in a Spinco SW-25.2 rotor to pellet the nuclei.

5. Pour off the sucrose. Wipe the inside of the tube with a cotton swab to remove contaminants.

6. Suspend the pellet in 5–8 ml isolation medium (Section X,B,1) containing 0.5% (v/v) Nonidet P40 (Shell Oil Co.). Homogenize with four to five strokes in a Potter-Elvehjem glass tube with a motor-driven Teflon pestle to remove cytoplasm still adhering to the nuclei.

7. Centrifuge the homogenate at 12,000 g for 10 minutes in a Sorvall SS-34 rotor.

8. By hand, wash the pellet by suspending it in 5–8 ml isolation medium using a Potter-Elvehjem glass-Teflon homogenizer. Centrifuge the homogenate again as in step 7. The pellet contains purified nuclei.

The average yield of nuclei is 70% of the original DNA. No cytoplasmic contaminants are found in the final nuclear pellet. In picograms per nucleus, the amount of DNA is 143, RNA is 30, acid-insoluble protein is 156, and acid-soluble protein is 14. These values are about one-tenth of the values reported for nuclei isolated from the same strain of *G. nelsoni* by the method of Mendiola *et al.* (1966) (Section X,D,2).

F. *Gyrodinium cohnii* (*Cryptothecodinium cohnii*) (Method of Rizzo and Noodén, 1973)

1. CULTURE AND COLLECTION OF CELLS

Axenic stock cultures of the nonphotosynthetic heterotrophic dinoflagellate *G. cohnii* are maintained in 5 ml AXM medium (Provasoli and Gold, 1962) under the conditions described for *G. nelsoni* (Section X,E,1), except that illumination is not needed. Six milliliters of stock culture (2–4 × 10⁶ cells/ml) are used to inoculate 1 liter of modified AXM medium in a 5-liter Povitsky bottle (A. H. Thomas Co., Philadelphia, Pennsylvania). Modified AXM contains only 10% of the amount of glucose and acetate found in normal AXM. Growth on modified AXM reduces the accumulation of polysaccharides and lipids by *G. cohnii*. Cells are grown for about 3.5 days to a density of 10⁶ cells/ml.

Concentrate cells by centrifugation at 2500 g for 5 minutes at 0°–4°C. Suspend the pellet in isolation medium containing 0.25 M sucrose, 5% (w/v) dextran 40 (Pharmacia), 2.5% (w/v) Ficoll (Pharmacia), 5 mM CaCl$_2$, and 10 mM Tris–HCl (pH 7.2).

2. ISOLATION OF NUCLEI

All steps are done at 0°–4°C.

1. Sonicate the cell suspension as described for *P. trochoideum* (Section X,B,2), except sonicate for 1–1.5 minutes at maximum output of the sonifier. Generally, cells from 2 liters of culture (about 10^6 cells/ml) are sonicated in 180 ml isolation medium (1–1.5×10^7 cells/ml).

2. Centrifuge the sonicated suspension at 480 g for 5 minutes. The resultant pellet contains two layers, an upper brown one and a lower white one. Most of the nuclei are in the brown layer.

3. Add several milliliters of isolation medium to the pellet. Dislodge the top layer carefully with a stirring rod. Mix gently by touching the tube to a vortex mixer.

4. Collect the nuclear suspension and make it up to a total volume of 30 ml with isolation medium.

5. Layer the 30 ml over a two-layer discontinuous gradient containing 7 ml 2.4 M sucrose with 10% (w/v) dextran 10, overlayed with 37 ml 0.1% (v/v) Triton X-100 in 2.2 M sucrose. Form a partial gradient by gently stirring the upper third of the 2.2 M sucrose layer.

6. Centrifuge at 48,000 g for 50 minutes to pellet the nuclei.

7. Pour off the sucrose solutions. Wipe the insides of the tubes with a cotton swab to remove contaminants.

8. Resuspend the pellet gently in 30 ml isolation medium with a Potter-Elvehjem homogenizer. Repeat centrifugation as in step 6. The pellet contains purified nuclei.

The average yield of nuclei is 14% of the original DNA. Isolated nuclei are free of cytoplasmic tabs, and membranes are intact. In picograms per nucleus, the amount of DNA is 6.9, RNA is 2.2, acid-insoluble protein is 6.8, and acid-soluble protein is 0.9 (Rizzo and Noodén, 1973). A single band of histone from these nuclei has been resolved by electrophoresis on poly-acrylamide gels (Rizzo and Noodén, 1972).

G. Diatoms (Method of Mendiola et al., 1966)

The 1966 method of Mendiola *et al.* (Section X,D,2) for the isolation of nuclei from the dinoflagellate *G. nelsoni* is reported to be successful for isolating nuclei from the diatoms *Ditylum brightwelli* (West) Grun (strain

DB, Woods Hole Oceanographic Institution) and *Rhizosolenia setigera* (Brightwell) (strain Rhizo).

NOTE ADDED IN PROOF

Since this chapter was completed, the methods for the isolation of macronuclei and micronuclei from *P. aurelia* (Section IV,C) and from *T. pyriformis* (Section III,E) have been considered further, respectively, by D. J. Cummings and A. Tait [*Methods in Cell Biology* (D. M. Prescott, ed.), Vol. 9, p. 281, Academic Press, New York (1975)] and by M. A. Gorovsky, M. -C. Yao, J. B. Keevert, and G. L. Pleger [*Methods in Cell Biology* (D. M. Prescott, ed.), Vol. 9, p. 311, Academic Press, New York (1975).] Also, a new method for the isolation and purification of macronuclei from *P. aurelia* has been published [Skoczylas, B. and Soldo, A. T. *Exp. Cell Res.* **90**, 143 (1975)].

REFERENCES

Aprille, J. R., and Buetow, D. E. (1973). *Arch. Mikrobiol.* **89**, 355.
Aprille, J. R., and Buetow, D. E. (1974). *Arch. Mikrobiol.* **97**, 195.
Buetow, D. E. (1965). *J. Cell. Comp. Physiol.* **66**, 235.
Buetow, D. E., and Padilla, G. M. (1963). *J. Protozool.* **10**, 121.
Busch, H., and Smetana, K. (1970). "The Nucleolus," p. 527. Academic Press, New York.
Byfield, J. E., and Lee, Y. C. (1970). *J. Protozool.* **17**, 445.
Carefoot, J. R. (1968). *J. Phycol.* **4**, 129.
Chalkley, H. W. (1930). *Science* **71**, 442.
Cummings, D. J. (1972). *J. Cell Biol.* **53**, 105.
Dryl, S. (1959). *J. Protozool.* **6**, Suppl. 25.
Eadie, J. M., and Oxford, A. E. (1954). *Nature (London)* **174**, 973.
Godward, M. B. E. (1948). *Nature (London)* **161**, 203.
Goldstein, L. (1964). *In* "Methods in Cell Physiology" (D. M. Prescott, ed.), Vol. 1, p. 97. Academic Press, New York.
Gorovsky, M. A. (1970a). *J. Cell Biol.* **47**, 619.
Gorovsky, M. A. (1970b). *J. Cell Biol.* **47**, 631.
Gorovsky, M. A. (1973). *J. Protozool.* **20**, 19.
Gorovsky, M. A., Keevert, J. B., and Pleger, G. L. (1974). *J. Cell Biol.* **61**, 134.
Guillard, R. R. L., and Ryther, J. S. (1962). *Can. J. Microbiol.* **8**, 229.
Iwai, K., Shiomi, H., Ando, T., and Mita, T. (1965). *J. Biochem. (Tokyo)* **58**, 312.
Jensen, W. A. (1962). "Botanical Histochemistry," Freeman, San Francisco, California.
Jeon, K. W. (1970). *In* "Methods in Cell Physiology" (D. M. Prescott, ed.), Vol. 4, p. 179. Academic Press, New York.
Jones, I. G. (1965). *Biochem. J.* **96**, 17.
Kidder, G. W., and Dewey, V. C. (1951). *In* "Biochemistry and Physiology of Protozoa" (A. Lwoff, ed.), Vol. 1, pp. 323–400. Academic Press, New York.
Kimball, R. F., Vogt-Köhne, L., and Caspersson, T. O. (1960). *Exp. Cell Res.* **20**, 368.
Kumar, A. (1970). *J. Cell Biol.* **45**, 623.
Lee, Y. C., and Byfield, J. E. (1970). *Biochemistry* **9**, 3947.
Lee, Y. C., and Scherbaum, O. H. (1965). *Nature (London)* **208**, 1350.
Lee, Y. C., and Scherbaum, O. H. (1966). *Biochemistry* **5**, 2067.
Leedale, G. F. (1958). *Arch. Mikrobiol.* **32**, 32.
Leick, V. (1969). *Eur. J. Biochem.* **8**, 221.
Leick, V., and Engberg, J. (1970). *Eur. J. Biochem.* **13**, 238.

Lynch, M. J. (1973). Ph.D. Thesis, University of Illinois, Urbana.
Lynch, M. J., and Buetow, D. E. (1975). *Exp. Cell Res.* **91**, 344.
Lynch, M. J., Leake, R. E., and Buetow, D. E. (1975a). Unpublished data.
Lynch, M. J., Leake, R. E., O'Connell, K. M., and Buetow, D. E. (1975b). *Exp. Cell Res.* **91**, 349.
Mendiola, L. R., Price, C. A., and Guillard, R. R. L. (1966). *Science* **153**, 1661.
Mita, T., Shiomi, H., and Iwai, K. (1966). *Exp. Cell. Res.* **43**, 696.
Moses, M. J. (1950). *J. Morphol.* **87**, 493.
Muramatsu, M. (1970). *In* "Methods in Cell Physiology" (D. M. Prescott, ed.), Vol. 4, p. 195. Academic Press, New York.
Nasatir, M., and Brooks, A. E. (1966). *J. Phycol.* **2**, 144.
Nilsson, J. R., and Leick, V. (1970). *Exp. Cell Res.* **60**, 361.
Parenti, J., Brawerman, G., Preston, J. F., and Eisenstadt, J. M. (1969). *Biochim. Biophys. Acta* **195**, 234.
Plesner, P., Rasmussen, L., and Zeuthen, E. (1964). *In* "Synchrony in Cell Division and Growth" (E. Zeuthen, ed.), p. 543. Wiley (Interscience), New York.
Prescott, D. M., and Carrier, R. F. (1964). *In* "Methods in Cell Physiology" (D. M. Prescott, ed.), Vol. 1, p. 85. Academic Press, New York.
Prescott, D. M., Rao, M. V. N., Evenson, D. P., Stone, G. E., and Thrasher, J. D. (1966). *In* "Methods in Cell Physiology" (D. M. Prescott, ed.), Vol. 2, p. 131. Academic Press, New York.
Provasoli, L., and Gold, K. (1962). *Arch. Mikrobiol.* **42**, 196.
Ringertz, N. R., Bolund, L., and DeBault, L. E. (1967). *Exp. Cell Res.* **45**, 519.
Rizzo, P. J., and Noodén, L. D. (1972). *Science* **176**, 796.
Rizzo, P. J., and Noodén, L. D. (1973). *J. Protozool.* **20**, 666.
Roodyn, D. B. (1969). *In* "Subcellular Components, Preparation and Fractionation" (G. D. Birnie and S. M. Fox, eds.), p. 15. Plenum, New York.
Rosenbaum, J. L., and Holz, G. G., Jr. (1966). *J. Protozool.* **13**, 115.
Saxena, D. M. (1966). *Indian J. Exp. Biol.* **4**, 182.
Scherbaum, O. H., and Jahn, T. L. (1964). *Exp. Cell Res.* **33**, 99.
Scherbaum, O. H., and Zeuthen, E. (1954). *Exp. Cell Res.* **6**, 221.
Scherbaum, O. H., James, T. W., and Jahn, T. L. (1959). *J. Cell Comp. Physiol.* **53**, 119.
Schwartz, V. (1956). *Biol. Zentralbl.* **75**, 1.
Seshachar, B. R., and Saxena, D. M. (1968). *J. Protozool.* **15**, 697.
Skoczylas, B., Panusz, H., and Gross, M. (1963). *Acta Protozool.* **1**, 411.
Sonneborn, T. M. (1950). *J. Exp. Zool.* **113**, 87.
Sonneborn, T. M. (1970). *In* "Methods in Cell Physiology" (D. M. Prescott, ed.), Vol. 4, p. 241. Academic Press, New York.
Starr, R. C. (1964). *Amer. J. Bot.* **51**, 1013.
Stevenson, I. (1967). *J. Protozool.* **14**, 412.
Stone, G. E., and Cameron, I. L. (1964). *In* "Methods in Cell Physiology" (D. M. Prescott, ed.), Vol. 1, p. 127. Academic Press, New York.
Tautvydas, K. J. (1971). *Exp. Cell Res.* **68**, 299.
Whitson, G. L., Padilla, G. M., and Fisher, W. D. (1966). *Exp. Cell Res.* **42**, 438.
Wolfe, J; (1967). *Chromosoma* **23**, 59.

Subject Index

A

Acid deoxyribonuclease, 76
Acrosomal contents, 96
Acrosomal fraction, from bull sperm, 96
Acrosomal membranes, 97
Acrosome, 86–87, 95
Adipose tissue culture, 223
Albumin production, 222
Algae
    collection of cells, 303
    isolation of nuclei, 303
Ambystoma mexicanum, 215
Ameba
    harvesting of, 243
    mass culturing of, 239
    medium, 240
Aminolevulinate synthetase, 70
Ammonia-^{15}N, 157
Amoeba proteus, 239
    mass isolation of nuclei from, 299
Amphiuma means, 214
    liver culture, 225
Amylase production, 222
Angiotensin, radioimmunoassay of, 231
Aquasol, 107, 119
AR-10, silver grain formation in, 127
Autoradiogram, dry-mount, 189
Autoradiography
    of diffusible compounds, 171
    dry-mount, 174
    smear-mount, 186
    thaw-mount, 185, 190
    touch-mount, 187

B

Benzpyrene hydroxylase, 70
Biosolv, 109
Blepharisma
    collection of, 297
    culture of, 297
    isolation of macronuclei, 297
Bray's solution, 112
Bromcresyl green, 142
Buoyant densities of enzymes, 166
Bürker chamber, 32

C

Carcinogenesis *in vitro*, chemical, 233
Catecholamines, simulation of, 191
Cathepsin D, 76
Cell disruption, 199
Cell fractionation, 199
Cell fusion, 7
Cell reconstruction
    preparation of components for, 8
    protocol for, 11
Cellosolve, 107, 112
Cerenkov counting, 127
Charcoal
    acid washed, 203
    preparation of, 203
Chauveau, method of, 199
Chelation of K$^+$, 34
Chelators, use of, 33
Chemography
    negative, 188
    positive, 187
Clonal growth, 19
    of mammalian cells, 15
Clostridium histolyticum, extracts of, 36
Collagenase, 30, 36
    activation by Ca^{2+}, 40
    crude, 38
    inhibition of by Mg^{2+}, 41
Collagenase buffer, composition of, 53
Collagenase perfusion, 30, 35, 50, 56
Commerford's reaction, 136
Cream separator centrifuge, 243–244
Cryopump, 181
Cryostat, 180
    Wide Range, 185
Cryostat sectioning, 178
Cultured tissues, analysis of function in, 222
Cytochalasin B, 1
Cytoplasts, 2, 10, 12
    characterization of, 9
    fusion of, 11
Cytosol
    fractions, 200
    preparation of, 199

D

Density labeling, 156
Density markers, 164
Diazoxide, 233
Didinium, isolation of macronuclei from, 298
DNA
  counting efficiencies for ^{3}H-labeled, 110
  counting of tritium-labeled, 106
  measurement of tritium in, 105
DNA collection, effects of BSA (bovine
  serum albumin) on, 108
DNA polymerase, reaction mixtures, 115
DNA polymerase assay(s), 105–106
  effects of precipitant solutions on, 116
  optimal conditions for, 114
DNA precipitation, 117
Dryl's salt solution, 257

E

Eagle's minimal essential medium,
  composition of, 16
Endometrial cell lines, 197
Enucleated cells, 2
  percentage of, 12
Enucleation, 12
  of mammalian cells, 1
Enzyme synthesis, 154
Equilibrium density gradient sedimentation,
  158
Equilibrium gradients, 160
Estradiol
  adsorption by charcoal, 203
  adsorption of bound by DEAE, 205
  adsorption of bound by hydroxylapatite,
  204
  binding of ^{3}H, 200
  recovery of bound by protamine sulfate
  precipitation, 206
  separation of bound, 208
  separation of free, 208
Estradiol (continued)
  17β, 196
  simulation of ^{3}H, 191
Estradiol receptor, properties of, 207
Estriol, 196
Estrogen-binding proteins, characterization
  of, 195
Estrogen receptors, 196
Estrogen target cells, 198

Estrone, 196
Euglena
  collection of cells, 301
  culture of cells, 301
  isolation of nuclei, 302

F

Fatty acid synthetase, 69
Filter discs
  collection of tritiated macromolecules
  on, 119
  counting of tritiated macromolecules on,
  119
Frusemide, 233

G

Gas mask, 139
Generation times, 23
  frequency distribution of, 21
Glass beads, 96
Glass-fiber filters
  characteristics of, 117
  properties of, 118
Gloves, leaded neoprene, 141
Glucuronidase, 76
Griffithsia globulifera
  collection of cells, 303
  isolation of nuclei, 303
Gymnodinium nelsoni
  collection of cells, 306–307
  culture of cells, 306–307
  isolation of nuclei, 306–307
Gyrodinium cohnii
  Collection of cells, 308
  culture of cells, 308
  isolation of nuclei, 308

H

Heart culture, 223
Heavy water, 156
Helminthosporium dematoideum, 1
Hepatocarcinogenesis, 71
Hepatocytes, incubation of, 64
HEPES, buffering capacity of, 46
Hot lab, equipping, 138
Hyaluronidase, 35
  for liver dispersion, 38
Hydroxylapatite, preparation of, 204

I

Ilford L-4, silver grain formation in, 127
Immunoautoradiogram, 191
Insulin production, 222, 231
Intestine culture, 223
Iodination, of 5 S rRNA, 148
Iodine-125, 123
  detection efficiency of, 128
  energy and frequency distribution of
    decay of, 125
  physical properties of, 124
Iodine-125 decay, detection of, 126
Iodine-125 labeled RNA, 127
Iodine-125 radiation spectra, 127
Iodine-131
  chemical properties of, 132
  use of, 132
Isoenzymes, separation of, 167
Isoprenaline, 233

K

Kallikrein inhibitor, 232
Karyoplasts, 2, 10, 12
  fusion of, 11
  preparation of, 9
Kidney culture, 222, 227
Kodak NTB-2, silver grain formation in, 127
Kupffer cells, 75
  isolation of, 76

L

Lactic dehydrogenase assays, 200
Latex spheres, 10–12
Lineweaver-Burk method, 202
Liver, isolation and perfusion of, 51
Liver cell(s)
  biochemical characteristics of, 63
  composition of, 76
  culture of, 71
  density distribution of, 60
  isolated, 69
  preparation of nonparenchymal, 73
  primary culture of, 73
Liver cell preparation, 52
  of isolated, 29
  nonenzymatic methods for, 32
Liver cell suspensions, 56
  characteristics of, 47
Liver culture, 222, 227

effects of hormones on, 230
Liver dispersion, assay of, 35
Liver fraction L, 242
Liver slices, enzymatic treatment of, 34
Lung culture, 223
Lysosomal enzyme activities, of liver
  cells, 76

M

Macronuclear gene copies, random
  distribution of, 251
Methotrexate, 25
Methyl green-pyronine, 184
Metrizamide, 60
Metrizamide-D$_2$O, 162
Metrizamide-D$_2$O gradients, 166
Metrizamide gradients, 77
Microbiuret method, 201
Microflex vials, 139
Mitochondrial DNA, in *Tetrahymena*, 250
Mitochondrial ribosomes, in *Tetrahymena*, 250
Mucus production, 223
Multiple-filter apparatus, 205
Muscle culture, 222
Myeloma cells, 198
Myxobacter, 168

N

NCS, 112
Necturus maculosus, 215
  liver culture, 225
Nitex nylon screens, 299
Nitrate-^{15}N, 157
Nitrate reductase, 167
Nitrocellulose, solubilization of, 106
Nuclear extract, preparation of, 200
Nuclear strains, 284
Nucleated fragments, preparation of, 9
Nuclei
  isolation of from Algae, 283
  isolation of from Protozoa, 283
  purification of, 199
NUNC vials, 143
Nylon net, 51
Nytal, 51

O

Organ culture
  amphibian, 213

enzyme production in, 231
hormone production in, 231
retention of normal function in, 227
Organ culture methods, 218
Organ fragments, mitotic incidence in
    cultured, 229
Oxygen-18, 157

P

Pancreas, preparation of isolated cells
    from, 43
Pancreas culture, 222, 227
Paramecium aurelia
    collection of, 292
    culture of, 292
    isolation of macronuclei, 293
    isolation of micronuclei, 295
Paramecium caudatum
    collection of, 296
    culture of, 296
    isolation of macronuclei, 296
Parenchymal cells
    properties of isolated, 64
    purification of, 57, 61
Perforatorium, of sperm, 98
Peridinium cinctum
    collection of cells, 305
    culture of cells, 305
    isolation of nuclei, 305
Peridinium trochoideum
    collection of cells, 304
    culture of cells, 304
    isolation of nuclei, 304
Phenformin, 233
Phenotypic assortment, in *Tetrahymena*, 253
Phentolamine, 233
Pindolol, 233
Pituitary cell lines, 197
Pituitary tumors, 197
Plasma membrane, of sperm, 95
Progesterone, 196
Pronase, 34
Propranolol, 233
Protein assays, 201
Proteins
    density labeling of, 153
    isopycnic banding of, 161
Protosol, 112

R

Radiation laboratory, 138
Radioiodination
    of DNA, 121
    of macromolecules, 138
    of RNA, 121
Radioiodination reaction, 141
Radioiodine, 122
Rana temporaria, 215
Reconstruction, of mammalian cells, 7
Reeve-Angel filters, 117
Renin production, 222
Rhynchosciara, X chromosome of, 133
Rollacell apparatus, 198

S

Sendai virus
    fusion procedure, 12
    source of, 10
Separation of cell types, by zonal
    sedimentation, 59
Sequestrene, 261
Skin culture, 219, 223, 227
Solubilizing agents, 113
Sonifer
    Bronwill, 89
    Model W185D, 200
Sperm chromatin, isolated, 99
Sperm heads
    isolated, 91
    isolation of, 85
    nuclear chromatin of, 97
    separation of, 93
    subfractionation of, 85, 95
Sperm lysis, 88
Sperm tails
    isolated, 91
    isolation of, 85
    separation of, 93
    subfractionation of, 85, 100
Spermatazoa
    dense fibers of, 102
    isolation of mitochondria from, 100
    midpieces, 100
    preparation of, 89
    structural components of, 85
Spirostomum, isolation of macronuclei from,
    298

T

TES, 64
Testosterone, 196
Tetrahymena
  cell counts, 265
  cloning of, 260, 262
  complementation tests, 278
  conditions for mutant isolation, 278
  conjugation, 250
  cytoplasmic mutations in, 274
  dominant micronuclear mutants, 271
  doses of mutagens, 269
  as food organism, 241
  genetic analysis of mutants, 275
  genetics of, 249
  genomic exclusion, 270
  growth media for, 256
  harvesting of, 242
  inbred strains available from frozen
    stores, 255
  individual cell isolation, 262
  induction of mutants in, 247
  isolation of macronuclei, 285
  isolation of micronuclei, 291
  isolation of mutants in, 247
  macronuclear mutations of, 273
  mass subculturing of, 260
  mating media for, 257
  mating-type tests, 268
  meiosis in, 250
  mitochondrial mutations, 275
  recessive micronuclear mutants, 272
  storage by freezing, 265
  storage in liquid nitrogen, 254
  strains of, 253
Tetraphenylboron, 34
Tissue, freezing of, 175
Tissue mounts, preparation of, 174
Trasylol, 232
Tricine, 64
Triton-toluene scintillation mixture, 112
Triturus cristatus carnifex, 215
Trypan blue
  exclusion test, 3, 31
  solution, 32
Tryptophan oxygenase, synthesis of, 70
Tyrosine aminotransferase, 70

V

Viokase, 198

W

Whatman filters, 117

X

Xenopus laevis laevis, 215

Y

Yeast extract-lactalbumin hydrolyzate,
  composition of, 18

Z

Zea mays, pachytene chromosomes of, 134

CONTENTS OF PREVIOUS VOLUMES

Volume I

1. SURVEY OF CYTOCHEMISTRY
   R. C. von Borstel

2. METHODS OF CULTURE FOR PLASMODIAL
   MYXOMYCETES
   John W. Daniel and Helen H.
   Baldwin

3. MITOTIC SYNCHRONY IN THE PLASMODIA
   OF *Physarum polycephalum* AND
   MITOTIC SYNCHRONIZATION BY
   COALESCENCE OF MICROPLASMODIA
   Edmund Guttes and Sophie Guttes

4. INTRODUCTION OF SYNCHRONOUS
   ENCYSTMENT (DIFFERENTIATION)
   IN *Acanthamoeba* sp.
   R. J. Neff, S. A. Ray, W. F. Benton,
   and M. Wilborn

5. EXPERIMENTAL PROCEDURES AND
   CULTURAL METHODS FOR *Euplotes
   eurystomus* AND *Amoeba proteus*
   D. M. Prescott and R. F. Carrier

6. NUCLEAR TRANSPLANTATION IN AMEBA
   Lester Goldstein

7. EXPERIMENTAL TECHNIQUES WITH
   CILIATES
   Vance Tartar

8. METHODS FOR USING *Tetrahymena*
   IN STUDIES OF THE NORMAL CELL
   CYCLE
   G. E. Stone and I. L. Cameron

9. CONTINUOUS SYNCHRONOUS CULTURES
   OF PROTOZOA
   G. M. Padilla and T. W. James

10. HANDLING AND CULTURING OF
    Chlorella
    Adolf Kuhl and Harald Lorenzen

11. CULTURING AND EXPERIMENTAL
    MANIPULATION OF *Acetabularia*
    Konrad Keck

12. HANDLING OF ROOT TIPS
    Sheldon Wolff

13. GRASSHOPPER NEUROBLAST TECHNIQUES
    J. Gordon Carlson and Mary Esther
    Gaulden

14. MEASUREMENT OF MATERIAL UPTAKE
    BY CELLS: PINOCYTOSIS
    Cicily Chapman-Andresen

15. QUANTITATIVE AUTORADIOGRAPHY
    Robert P. Perry

16. HIGH-RESOLUTION AUTORADIOGRAPHY
    Lucien G. Caro

17. AUTORADIOGRAPHY WITH LIQUID
    EMULSION
    D. M. Prescott

18. AUTORADIOGRAPHY OF WATER-
    SOLUBLE MATERIALS
    O. L. Miller, Jr., G. E. Stone, and
    D. M. Prescott

19. PREPARATION OF MAMMALIAN META-
    PHASE CHROMOSOMES FOR AUTO-
    RADIOGRAPHY
    D. M. Prescott and M. A. Bender

20. METHODS FOR MEASURING THE
    LENGTH OF THE MITOTIC CYCLE
    AND THE TIMING OF DNA
    SYNTHESIS FOR MAMMALIAN
    CELLS IN CULTURE
    Jesse E. Sisken

21. MICRURGY OF TISSUE CULTURE CELLS
Lester Goldstein and Julie Micou
Eastwood

22. MICROEXTRACTION AND MICRO-
ELECTROPHORESIS FOR DETERMINA-
TION AND ANALYSIS OF NUCLEIC
ACIDS IN ISOLATED CELLULAR
UNITS
J.-E. Edström

AUTHOR INDEX—SUBJECT INDEX

Volume II

1. NUCLEAR TRANSPLANTATION IN
AMPHIBIA
Thomas J. King

2. TECHNIQUES FOR THE STUDY OF
LAMPBRUSH CHROMOSOMES
Joseph G. Gall

3. MICRURGY ON CELLS WITH POLYTENE
CHROMOSOMES
H. Kroeger

4. A NOVEL METHOD FOR CUTTING
GIANT CELLS TO STUDY VIRAL
SYNTHESIS IN ANUCLEATE CYTO-
PLASM
Philip I. Marcus and Morton E.
Freiman

5. A METHOD FOR THE ISOLATION OF
MAMMALIAN METAPHASE CHROMO-
SOMES
Joseph J. Maio and Carl L. Schild-
kraut

6. ISOLATION OF SINGLE NUCLEI AND
MASS PREPARATIONS OF NUCLEI
FROM SEVERAL CELL TYPES
D. M. Prescott, M. V. N. Rao, D. P.
Evenson, G. E. Stone, and J. D.
Thrasher

7. EVALUATION OF TURGIDITY, PLASMO-
LYSIS, AND DEPLASMOLYSIS OF
PLANT CELLS
E. J. Stadelmann

8. CULTURE MEDIA FOR *Euglena gracilis*
S. H. Hutner, A. C. Zahalsky, S.
Aaronson, Herman Baker, and Oscar
Frank

9. GENERAL AREA OF AUTORADIOGRAPHY
AT THE ELECTRON MICROSCOPE
LEVEL
Miriam M. Salpeter

10. HIGH RESOLUTION AUTORADIOGRAPHY
A. R. Stevens

11. METHODS FOR HANDLING SMALL
NUMBERS OF CELLS FOR ELECTRON
MICROSCOPY
Charles J. Flickinger

12. ANALYSIS OF RENEWING EPITHELIAL
CELL POPULATIONS
J. D. Thrasher

13. PATTERNS OF CELL DIVISION: THE
DEMONSTRATION OF DISCRETE
CELL POPULATIONS
Seymour Gelfant

14. BIOCHEMICAL AND GENETIC METHODS
IN THE STUDY OF CELLULAR SLIME
MOLD DEVELOPMENT
Maurice Sussman

AUTHOR INDEX—SUBJECT INDEX

Volume III

1. MEASUREMENT OF CELL VOLUMES BY
ELECTRIC SENSING ZONE INSTRU-
MENTS
R. J. Harvey

2. SYNCHRONIZATION METHODS FOR
MAMMALIAN CELL CULTURES
Elton Stubblefield

3. EXPERIMENTAL TECHNIQUES FOR
INVESTIGATION OF THE AMPHIBIAN
LENS EPITHELIUM
Howard Rothstein

320   CONTENTS OF PREVIOUS VOLUMES

4. CULTIVATION OF TISSUES AND LEUKO-
   CYTES FROM AMPHIBIANS
   Takeshi Seto and Donald E. Rounds

5. EXPERIMENTAL PROCEDURES FOR
   MEASURING CELL POPULATION
   KINETIC PARAMETERS IN PLANT
   ROOT MERISTEMS
   Jack Van't Hof

6. INDUCTION OF SYNCHRONY IN *Chlamy-
   domonas moewusii* AS A TOOL
   FOR THE STUDY OF CELL DIVISION
   Emil Bernstein

7. STAGING OF THE CELL CYCLE WITH
   TIME-LAPSE PHOTOGRAPHY
   Jane L. Showacre

8. METHOD FOR REVERSIBLE INHIBITION
   OF CELL DIVISION IN *Tetrahymena
   pyriformis* USING VINBLASTINE
   SULFATE
   Gordon E. Stone

9. PHYSIOLOGICAL STUDIES OF CELLS
   OF ROOT MERISTEMS
   D. Davidson

10. CELL CYCLE ANALYSIS
    D. S. Nachtwey and I. L. Cameron

11. A METHOD FOR THE STUDY OF CELL
    PROLIFERATION AND RENEWAL IN
    THE TISSUES OF MAMMALS
    Ivan L. Cameron

12. ISOLATION AND FRACTIONATION OF
    METAPHASE CHROMOSOMES
    Norman P. Salzman and John
    Mendelsohn

13. AUTORADIOGRAPHY WITH THE ELEC-
    TRON MICROSCOPE: PROPERTIES
    OF PHOTOGRAPHIC EMULSIONS
    D. F. Hülser and M. F. Rajewsky

14. CYTOLOGICAL AND CYTOCHEMICAL
    METHODOLOGY OF HISTONES
    James L. Pipkin, Jr.

15. MITOTIC CELLS AS A SOURCE OF
    SYNCHRONIZED CULTURES
    D. F. Petersen, E. C. Anderson,
    and R. A. Tobey.

AUTHOR INDEX—SUBJECT INDEX

Volume IV

1. ISOLATION OF THE PACHYTENE STAGE
   NUCLEI FROM THE SYRIAN HAM-
   STER TESTIS
   Tadashi Utakoji

2. CULTURE METHODS FOR ANURAN
   CELLS
   Jerome J. Freed and Liselotte
   Mezger-Freed

3. AXENIC CULTURE OF *Actabularia*
   IN A SYNTHETIC MEDIUM
   David C. Shephard

4. PROCEDURES FOR THE ISOLATION
   OF THE MITOTIC APPARATUS
   FROM CULTURED MAMMALIAN
   CELLS
   Jesse E. Sisken

5. PREPARATION OF MITOCHONDRIA FROM
   PROTOZOA AND ALGAE
   D. E. Buetow

6. METHODS USED IN THE AXENIC CULTI-
   VATION OF *Paramecium aurelia*
   W. J. van Wagtendonk and A. T.
   Soldo

7. PHYSIOLOGICAL AND CYTOLOGICAL
   METHODS FOR *Schizosaccharomyces
   pombe*
   J. M. Mitchison

   APPENDIX (CHAPTER 7): STAINING THE
   S. pombe NUCLEUS
   C. F. Robinow

8. GENETICAL METHODS FOR *Schizosaccharomyces pombe*
   U. Leupold

9. MICROMANIPULATION OF AMEBA NUCLEI
   K. W. Jeon

10. ISOLATION OF NUCLEI AND NUCLEOLI
    Masami Muramatsu

11. THE EFFICIENCY OF TRITIUM COUNTING
    WITH SEVEN RADIOAUTOGRAPHIC
    EMULSIONS
    Arie Ron and David M. Prescott

12. METHODS IN *Paramecium* RESEARCH
    T. M. Sonneborn

13. AMEBO-FLAGELLATES AS RESEARCH PARTNERS: THE LABORATORY BIOLOGY OF *Naegleria* AND *Tetramitus*
Chandler Fulton

14. A STANDARDIZED METHOD OF PERIPHERAL BLOOD CULTURE FOR CYTOGENETICAL STUDIES AND ITS MODIFICATION BY COLD TEMPERATURE TREATMENT
Marsha Heuser and Lawrence Razavi

15. CULTURE OF MEIOTIC CELLS FOR BIOCHEMICAL STUDIES
Herbert Stern and Yasuo Hotta

AUTHOR INDEX—SUBJECT INDEX

9. CONTINUOUS AUTOMATIC CULTIVATION OF HOMOCONTINUOUS AND SYNCHRONIZED MICROALGAE
Horst Senger, Jürgen Pfau, and Klaus Werthmüller

10. VITAL STAINING OF PLANT CELLS
Eduard J. Stadelmann and Helmut Kinzel

11. SYNCHRONY IN BLUE-GREEN ALGAE
Harald Lorenzen and G. S. Venkataraman

AUTHOR INDEX—SUBJECT INDEX

Volume V

1. PROCEDURES FOR MAMMALIAN CHROMOSOME PREPARATIONS
T. C. Hsu

2. CLONING OF MAMMALIAN CELLS
Richard G. Ham

3. CELL FUSION AND ITS APPLICATION TO STUDIES ON THE REGULATION OF THE CELL CYCLE
Potu N. Rao and Robert T. Johnson

4. MARSUPIAL CELLS *in Vivo* AND *in Vitro*
Jack D. Thrasher

5. NUCLEAR ENVELOPE ISOLATION
I. B. Zbarsky

6. MACRO- AND MICRO-OXYGEN ELECTRODE TECHNIQUES FOR CELL MEASUREMENT
Milton A. Lessler

7. METHODS WITH *Tetrahymena*
L. P. Everhart, Jr.

8. COMPARISON OF A NEW METHOD WITH USUAL METHODS FOR PREPARING MONOLAYERS IN ELECTRON MICROSCOPY AUTORADIOGRAPHY
N. M. Maraldi, G. Biagini, P. Simoni, and R. Laschi

Volume VI

1. CULTIVATION OF CELLS IN PROTEIN- AND LIPID-FREE SYNTHETIC MEDIA
Hajim Katsuta and Toshiko Takaoka

2. PREPARATION OF SYNCHRONOUS CELL CULTURES FROM EARLY INTERPHASE CELLS OBTAINED BY SUCROSE GRADIENT CENTRIFUGATION
Richard Schindler and Jean Claude Schaer

3. PRODUCTION AND CHARACTERIZATION OF MAMMALIAN CELLS REVERSIBLY ARRESTED IN G_1 BY GROWTH IN ISOLEUCINE-DEFICIENT MEDIUM
Robert A. Tobey

4. A METHOD FOR MEASURING CELL CYCLE PHASES IN SUSPENSION CULTURES
P. Volpe and T. Eremenko

5. A REPLICA PLATING METHOD OF CULTURED MAMMALIAN CELLS
Fumio Suzuki and Masakatsu Horikawa

6. CELL CULTURE CONTAMINANTS
Peter P. Ludovici and Nelda B. Holmgren

7. ISOLATION OF MUTANTS OF CULTURED MAMMALIAN CELLS
Larry H. Thompson and Raymond M. Baker

8. ISOLATION OF METAPHASE CHROMOSOMES, MITOTIC APPARATUS, AND NUCLEI
Wayne Wray

9. ISOLATION OF METAPHASE CHROMOSOMES WITH HIGH MOLECULAR WEIGHT DNA AT pH 10.5
Wayne Wray

10. BASIC PRINCIPLES OF A METHOD OF NUCLEOLI ISOLATION
J. Zalta and J-P. Zalta

11. A TECHNIQUE FOR STUDYING CHEMOTAXIS OF LEUKOCYTES IN WELL-DEFINED CHEMOTACTIC FIELDS
Gary J. Grimes and Frank S. Barnes

12. NEW STAINING METHODS FOR CHROMOSOMES
H. A. Lubs, W. H. McKenzie, S. R. Patil, and S. Merrick

AUTHOR INDEX—SUBJECT INDEX

Volume VII

1. THE ISOLATION OF RNA FROM MAMMALIAN CELLS
George Brawerman

2. PREPARATION OF RNA FROM ANIMAL CELLS
Masami Muramatsu

3. DETECTION AND UTILIZATION OF POLY(A) SEQUENCES IN MESSENGER RNA
Joseph Kates

4. RECENT ADVANCES IN THE PREPARATION OF MAMMALIAN RIBOSOMES AND ANALYSIS OF THEIR PROTEIN COMPOSITION
Jolinda A. Traugh and Robert R. Traut

5. METHODS FOR THE EXTRACTION AND PURIFICATION OF DEOXYRIBONUCLEIC ACIDS FROM EUKARYOTE CELLS
Elizabeth C. Travaglini

6. ELECTRON MICROSCOPIC VISUALIZATION OF DNA IN ASSOCIATION WITH CELLULAR COMPONENTS
Jack D. Griffith

7. AUTORADIOGRAPHY OF INDIVIDUAL DNA MOLECULES
D. M. Prescott and P. L. Kuempel

8. HELA CELL PLASMA MEMBRANES
Paul H. Atkinson

9. MASS ENUCLEATION OF CULTURED ANIMAL CELLS
D. M. Prescott and J. B. Kirkpatrick

10. THE PRODUCTION OF MASS POPULATIONS OF ANUCLEATE CYTOPLASMS
Woodring E. Wright

11. ANUCLEATE MAMMALIAN CELLS: APPLICATIONS IN CELL BIOLOGY AND VIROLOGY
George Poste

12. FUSION OF SOMATIC AND GAMETIC CELLS WITH LYSOLECITHIN
Hilary Koprowski and Carlo M. Croce

13. THE ISOLATION AND REPLICA PLATING OF CELL CLONES
Richard A. Goldsby and Nathan Mandell

14. SELECTION SYNCHRONIZATION BY VELOCITY SEDIMENTATION SEPARATION OF MOUSE FIBROBLAST CELLS GROWN IN SUSPENSION CULTURE
Sydney Shall

15. METHODS FOR MICROMANIPULATION OF HUMAN SOMATIC CELLS IN CULTURE
Elaine G. Diacumakos

16. TISSUE CULTURE OF AVIAN HEMATOPOIETIC CELLS
C. Moscovici and M. G. Moscovici

17. MEASUREMENT OF GROWTH AND RATES OF INCORPORATION OF RADIOACTIVE PRECURSORS INTO MACROMOLECULES OF CULTURED CELLS
L. P. Everhart, P. V. Hauschka, and D. M. Prescott

18. THE MEASUREMENT OF RADIOACTIVE PRECURSOR INCORPORATION INTO SMALL MONOLAYER CULTURES
C. R. Ball, H. W. van den Berg, and R. W. Poynter

19. ANALYSIS OF NUCLEOTIDE POOLS IN ANIMAL CELLS
Peter V. Hauschka

AUTHOR INDEX—SUBJECT INDEX

Volume VIII

1. METHODS FOR SELECTING AND STUDYING TEMPERATURE-SENSITIVE MUTANTS OF BHK-21 CELLS
Claudio Basilico and Harriet K. Meiss

2. INDUCTION AND ISOLATION OF AUXOTROPHIC MUTANTS IN MAMMALIAN CELLS
Fa-Ten Kao and Theodore T. Puck

3. ISOLATION OF TEMPERATURE-SENSITIVE MUTANTS OF MAMMALIAN CELLS
P. M. Naha

4. PREPARATION AND USE OF REPLICATE MAMMALIAN CELL CULTURES
William G. Taylor and Virginia J. Evans

5. METHODS FOR OBTAINING REVERTANTS OF TRANSFORMED CELLS
A. Vogel and Robert Pollack

6. MONOLAYER MASS CULTURE ON DISPOSABLE PLASTIC SPIRALS
N. G. Maroudas

7. A METHOD FOR CLONING ANCHORAGE-DEPENDENT CELLS IN AGAROSE
C. M. Schmitt and N. G. Maroudas

8. REPETITIVE SYNCHRONIZATION OF HUMAN LYMPHOBLAST CULTURES WITH EXCESS THYMIDINE
H. Ronald Zielke and John W. Littlefield

9. USES OF ENUCLEATED CELLS
Robert D. Goldman and Robert Pollack

10. ENUCLEATION OF SOMATIC CELLS WITH CYTOCHALASIN B
Carlo M. Croce, Natale Tomassini, and Hilary Koprowski

11. ISOLATION OF MAMMALIAN HETEROCHROMATIN AND EUCHROMATIN
Walid G. Yasmineh and Jorge J. Yunis

12. MEASUREMENTS OF MAMMALIAN CELLULAR DNA AND ITS LOCALIZATION IN CHROMOSOMES
L. L. Deaven and D. F. Petersen

13. LARGE-SCALE ISOLATION OF NUCLEAR MEMBRANES FROM BOVINE LIVER
Ronald Berezney

14. A SIMPLIFIED METHOD FOR THE DETECTION OF MYCOPLASMA
Elliot M. Levine

15. THE ULTRA-LOW TEMPERATURE AUTORADIOGRAPHY OF WATER AND ITS SOLUTES
Samuel B. Horowitz

16. QUANTITATIVE LIGHT MICROSCOPIC AUTORADIOGRAPHY
Hollis G. Boren, Edith C. Wright, and Curtis C. Harris

17. IONIC COUPLING BETWEEN NONEXCITABLE CELLS IN CULTURE
Dieter F. Hülser

18. METHODS IN THE CELLULAR AND MOLECULAR BIOLOGY OF *Paramecium*
Earl D. Hanson

19. METHODS OF CELL TRANSFORMATION BY TUMOR VIRUSES
Thomas L. Benjamin

AUTHOR INDEX—SUBJECT INDEX

Volume IX

1. PREPARATION OF LARGE QUANTITIES OF PURE BOVINE LYMPHOCYTES AND A MONOLAYER TECHNIQUE FOR LYMPHOCYTE CULTIVATION
   J. Hinrich Peters

2. METHODS TO CULTURE DIPLOID FIBROBLASTS ON A LARGE SCALE
   H. W. Rüdiger

3. PARTITION OF CELLS IN TWO-POLYMER AQUEOUS PHASES: A METHOD FOR SEPARATING CELLS AND FOR OBTAINING INFORMATION ON THEIR SURFACE PROPERTIES
   Harry Walter

4. SYNCHRONIZATION OF CELL DIVISION *in Vivo* THROUGH THE COMBINED USE OF CYTOSINE ARABINOSIDE AND COLCEMID
   Robert S. Verbin and Emmanuel Farber

5. THE ACCUMULATION AND SELECTIVE DETACHMENT OF MITOTIC CELLS
   Edwin V. Gaffney

6. USE OF THE MITOTIC SELECTION PROCEDURE FOR CELL CYCLE ANALYSIS: EMPHASIS ON RADIATION-INDUCED MITOTIC DELAY
   D. P. Highfield and W. C. Dewey

7. EVALUATION OF S PHASE SYNCHRONIZATION BY ANALYSIS OF DNA REPLICATION IN 5-BROMODEOXYURIDINE
   Raymond E. Meyn, Roger R. Hewitt, and Ronald M. Humphrey

8. APPLICATION OF PRECURSORS ADSORBED ON ACTIVATED CHARCOAL FOR LABELING OF MAMMALIAN DNA *in Vivo*
   George Russev and Roumen Tsanev

9. GROWTH OF FUNCTIONAL GLIAL CELLS IN A SERUMLESS MEDIUM
   Sam T. Donta

10. MINIATURE TISSUE CULTURE TECHNIQUE WITH A MODIFIED (BOTTOMLESS) PLASTIC MICROPLATE
    Eliseo Manuel Hernández-Baumgarten

11. AGAR PLATE CULTURE AND LEDERBERG-STYLE REPLICA PLATING OF MAMMALIAN CELLS
    Toshio Kuroki

12. METHODS AND APPLICATIONS OF FLOW SYSTEMS FOR ANALYSIS AND SORTING OF MAMMALIAN CELLS
    H. A. Crissman, P. F. Mullaney, and J. A. Steinkamp

13. PURIFICATION OF SURFACE MEMBRANES FROM RAT BRAIN CELLS
    Kari Hemminki

14. THE PLASMA MEMBRANE OF KB CELLS; ISOLATION AND PROPERTIES
    F. C. Charalampous and N. K. Gonatas

15. THE ISOLATION OF NUCLEI FROM *Paramecium aurelia*
    Donald J. Cummings and Andrew Tait

16. ISOLATION OF MICRO- AND MACRONUCLEI OF *Tetrahymena pyriformis*
    Martin A. Gorovsky, Meng-Chao Yao, Josephine Bowen Keevert, and Gloria Lorick Pleger

17. MANIPULATIONS WITH *Tetrahymena pyriformis* ON SOLID MEDIUM
    Enore Gardonio, Michael Crerar, and Ronald E. Pearlman

18. THE ISOLATION OF NUCLEI WITH CITRIC ACID AND THE ANALYSIS OF PROTEINS BY TWO-DIMENSIONAL POLYACRYLAMIDE GEL ELECTROPHORESIS
    Charles W. Taylor, Lynn C. Yeoman, and Harris Busch

19. ISOLATION AND MANIPULATION OF SALIVARY GLAND NUCLEI AND CHROMOSOMES
    M. Robert

SUBJECT INDEX

Volume X

1. NUCLEIC ACID HYBRIDIZATION TO THE DNA OF CYTOLOGICAL PREPARATIONS
Mary Lou Pardue and Joseph G. Gall

2. MICROANALYSIS OF RNA FROM DEFINED CELLULAR COMPONENTS
Bo Lambert and Bertil Daneholt

3. STAINING OF RNA AFTER POLYACRYLAMIDE GEL ELECTROPHORESIS
K. Marcinka

4. ISOLATION AND FRACTIONATION OF MOUSE LIVER NUCLEI
Ph. Chevaillier and M. Philippe

5. THE SEPARATION OF CELLS AND SUBCELLULAR PARTICLES BY COLLOIDAL SILICA DENSITY GRADIENT CENTRIFUGATION
David A. Wolff

6. ISOLATION OF SUBCELLULAR MEMBRANE COMPONENTS FROM *Tetrahymena*
Y. Nozawa

7. ISOLATION AND EXPERIMENTAL MANIPULATION OF POLYTENE NUCLEI IN *Drosophila*
James B. Boyd

8. METHODS FOR MICROSURGICAL PRODUCTION OF MAMMALIAN SOMATIC CELL HYBRIDS AND THEIR ANALYSIS AND CLONING
Elaine G. Diacumakos

9. AUTOMATED CELL CYCLE ANALYSIS
Robert R. Klevecz

10. SELECTION OF SYNCHRONOUS CELL POPULATIONS FROM EHRLICH ASCITES TUMOR CELLS BY ZONAL CENTRIFUGATION
Hans Probst and Jürgen Maisenbacher

11. METHODS WITH INSECT CELLS IN SUSPENSION CULTURE I. *Aedes albopictus*
Allan Spradling, Robert H. Singer, Judith Lengyel, and Sheldon Penman

12. METHODS WITH INSECT CELLS IN SUSPENSION CULTURE II. *Drosophila melanogaster*
Judith Lengyel, Allan Spradling, and Sheldon Penman

13. MUTAGENESIS IN CULTURED MAMMALIAN CELLS
N. I. Shapiro and N. B. Varshaver

14. MEASUREMENT OF PROTEIN TURNOVER IN ANIMAL CELLS
Darrell Doyle and John Tweto

15. DETECTION OF MYCOPLASMA CONTAMINATION IN CULTURED CELLS: COMPARISON OF BIOCHEMICAL, MORPHOLOGICAL, AND MICROBIOLOGICAL TECHNIQUES
Edward L. Schneider

16. A SIMPLE BIOCHEMICAL TECHNIQUE FOR THE DETECTION OF MYCOPLASMA CONTAMINATION OF CULTURED CELLS
Edward L. Schneider and E. J. Standbridge

17. PURITY AND STABILITY OF RADIOCHEMICAL TRACERS IN AUTORADIOGRAPHY
E. Anthony Evans

18. ^{125}I IN MOLECULAR HYBRIDIZATION EXPERIMENTS
Lewis C. Altenburg, Michael J. Getz, and Grady F. Saunders

19. RADIOIODINE LABELING OF RIBOPOLYMERS FOR SPECIAL APPLICATIONS IN BIOLOGY
Neal H. Scherberg and Samuel Refetoff

20. AUTORADIOGRAPHIC ANALYSIS OF TRITIUM ON POLYACRYLAMIDE GEL
Paola Pierandrei Amaldi

21. A Radioautographic Method for Cell Affinity Labeling with Estrogens and Catecholamines
José Uriel

Subject Index—Cumulative Subject Index

Volume XI

1. The Preparation of Yeasts for Light Microscopy
C. F. Robinow

2. Electron Microscopy of Yeasts
Friedrich Kopp

3. Methods in Sporulation and Germination of Yeasts
James E. Haber and Harlyn O. Halvorson

4. Synchronous Mating in Yeasts
Elissa P. Sena, David N. Radin, Juliet Welch, and Seymour Fogel

5. Synchronous Zygote Formation in Yeasts
T. Biliński J. Litwińska, J. Żuk, and W. Gajewski

6. Continuous Cultivation of Yeasts
A. Fiechter

7. Methods for Monitoring the Growth of Yeast Cultures and for Dealing with the Clumping Problem
John R. Pringle and Juan-R. Mor

8. Preparation and Growth of Yeast Protoplasts
S.-C. Kuo and S. Yamamoto

9. Dissecting Yeast Asci without a Micromanipulator
P. Munz

10. Use of Micromanipulators in Yeast Studies
Fred Sherman

11. Cell Cycle Analysis
J. M. Mitchison and B. L. A. Carter

12. Genetic Mapping in Yeast
R. K. Mortimer and D. C. Hawthorne

13. The Use of Mutants in Metabolic Studies
F. Lacroute

14. Isolation of Regulatory Mutants in *Saccharomyces cerevisiae*
Helen Greer and G. R. Fink

15. Methods for Selecting Auxotrophic and Temperature-Sensitive Mutants in Yeasts
Barbara Shaffer Littlewood

16. Isolation and Characterization of Mutants of *Saccharomyces cerevisiae* Able to Grow after Inhibition of dTMP Synthesis
M. Brendel, W. W. Fäth, and W. Laskowski

17. Mutants of *Saccharomyces cerevisiae* That Incorporate Deoxythymidine 5'-Monophosphate into DNA *in Vivo*
Reed B. Wickner

18. Mutants of Meiosis and Ascospore Formation
Michael S. Esposito and Rochelle E. Esposito

Subject Index

Volume XII

1. Growth and Handling of Yeasts
Anthony H. Rose

2. Inhibitors of Macromolecular Synthesis in Yeast
Daniel Schindler and Julian Davies

3. Isolation of Yeast DNA
D. R. Cryer, R. Eccleshall, and J. Marmur

4. Preparation of RNA and Ribosomes from Yeast
Gerald M. Rubin

5. DNA-Dependent RNA Polymerases from Yeasts
   H. Ponta, U. Ponta, and E. Wintersberger

6. The Isolation of Yeast Nuclei and Methods to Study Their Properties
   John H. Duffus

7. Isolation of Vacuoles from Yeasts
   Andres Wiemken

8. Analytical Methods for Yeasts
   P. R. Stewart

9. Methods for Avoiding Proteolytic Artefacts in Studies of Enzymes and Other Proteins from Yeasts
   John R. Pringle

10. Induction of Haploid Glycoprotein Mating Factors in Diploid Yeasts
    Marjorie Crandall and Joan H. Caulton

11. Mutagenesis in Yeast
    B. J. Kilbey

12. Induction, Selection, and Experimental Uses of Temperature-Sensitive and Other Conditional Mutants of Yeast
    John R. Pringle

13. In Vivo and in Vitro Synthesis of Yeast Mitochondrial DNA
    L. J. Zeman and C. V. Lusena

14. Isolation of Mitochondria and Techniques for Studying Mitochondrial Biogenesis in Yeasts
    Anthony W. Linnane and H. B. Lukins

15. Separation and Some Properties of the Inner and Outer Membranes of Yeast Mitochondria
    W. Bandlow and P. Bauer

16. The Use of Fluorescent DNA-Binding Agent for Detecting and Separating Yeast Mitochondria DNA
    D. H. Williamson and D. J. Fennell

17. Cytoplasmic Inheritance and Mitochondrial Genetics in Yeasts
    D. Wilkie

18. Synchronization of the Fission Schizosaccharomyces pombe Using Heat Shocks
    Birte Kramhøft and Erik Zeuthen

Subject Index